Microwave Engineering

Microwave Engineering

G. S. N. Raju

M.E., Ph.D (IIT-KGP)
Department of Electronics and Communication Engineering
Andhra University College of Engineering, (Autonomous)
Visakhapatnam

ISBN: 978-93-89698-77-0

Edition: 2020

Printed at: Rekha Printers

Dedicated to My Parents

Shri G. Venkatrama Raju
&
Smt. G. Sitamma

*Who bestowed on me the best possible life with
innocence, health and intellect*

Preface

The use of microwaves has penetrated into all walks of life. They have wide applications starting from wireless toy car to radar systems. In fact, there is no wireless communication, no wireless networks, no radar and no satellite without microwaves.

Microwaves are the carriers of information for short and long distance communications. Such microwaves are nothing but electromagnetic waves with short wavelengths.

Microwaves are invisible to naked eye and they can neither be sensed, nor heard and not even smelt.

Microwave Engineering is a core subject for all communication engineers. It is offered at both U.G. and P.G. levels in all universities throughout the world.

There are a good number of books available on this subject. But there is no single book which can fit into a semester course. Moreover, some books are either too fundamental or too advenced.

In view of the above facts, this book on "Microwave Engineering" is written specifically to cover both fundamentals and advanced concepts. This covers the course content of B.E./B.Tech program in Electronics and Communication Engineering.

It also meets the requirements of M.Sc (Electronics), M.Sc (Tech), AMIETE, AMIE (Telecommunication) and other such courses.

The book contains 9 chapters starting from introduction, applications of microwaves and ending with the chapter on measurements.

The Chapter 1 provides introduction and detailed applications. Chapters 2-3 provide the details of active devices which generate microwave signals. These include both microwave tubes and solid state devices.

Chapter 4 deals with scattering, impedance and admittance matrix parameters.

Chapter 5 provides the details of all the microwave passive components. Chapter 6 covers contemporary microwave transmission lines. Chapter 7 covers the details of microwave integrated circuits. Chapter 8 contains analysis and design aspects of microwave antennas.

Chapter 9 covers details of measurements of characteristics of active and passive devices and also different parameters.

A key feature of this textbook is many illustrative examples, problems and about 1000 objective and multiple choice questions and answers. A number of exercise problems are included and the solution manual is also available.

The book is written in simple English and the concepts are presented in a logical and orderly manner.

It is also reader friendly and is most suitable for self-learning of Microwave Engineering.

It is hoped that the book is useful for students, teachers, professionals, engineers, technicians, designers, paper setters, examiners and for short-term course organizers.

Author will be grateful if for any suggestions to improve the book.

G. S. N. Raju

Acknowledgements

I take this opportunity to express my respects to my brother Prof. G. Krishnam Raju who made me what I am today. He, a lover of social work, is one of the best mathematicians.

I am grateful to my gurus Prof. B.N. Das and Prof. Ajoy Chakraborty of IIT Kharagpur who taught me the art of continuous learning and research.

I express my regards to Prof. L. Venugopala Reddy, our Hon'ble Vice-Chancellor for his dynamic leadership in bringing Andhra University to greater heights.

I am extremely grateful to Prof. Allam Appa Rao our beloved Principal for his encouragement and full support in all my academic and research pursuits. He has been on a continuous work and is striving hard from day one to make the Andhra University College of Engineering (Autonomous) the best institute by global standards.

I thank all my colleagues, Prof. Raja Rajeswari, Prof. Satyanarayana Reddy, Prof. Mallikarjuna Rao, Prof. Gopala Rao, Prof. G. Sasibhushana Rao, Mrs. Santa Kumari, Dr. Srinivasa Baba, Dr. Sridevi, Dr. Rajesh Kumar, Dr. Panduranga Reddy, Ms. Anuradha and Ms. Aruna for their cooperation throughout.

I cherish the association of all my research scholars Mallikarjuna Rao, Sridevi, Gopala Rao, Misra, Sudhakar, Srinivasa Baba, Padma Raju, Chandra Bhushana Rao, G.M.V. Prasad, Srinivasa Rao, Subrahmanyam, Habibullah Khan, Narayana, Sadasiva Rao, Gutti, Mallikarjuna Prasad, Ramana Reddy, Jayalakshmi, Dora Babu, Karunakar, K.V.S.N. Raju, Babu, Surendra, Ramesh, Srinivasa Raju, Chakravarty, Sunil Prakash, Vali, Ms. Ujvala Prabha, Satyanarayana, who are sincere and hardworking to excel in teaching and research.

I thank my friends, Prof. Appa Rao, Prof. K.V.L.P. Raju, K. Subba Raju, Raju Indukuri, Dr. G. Prasad, D.M, Dr. Shivaji, Er. Rama Mohan, Prof. D.G.K. Raju, Er. Gopala Krishnam Raju, Bhaskar Raju, Kishore Babu, Indukuri Venkata Rama Raju, Indukuri Ganesh and Prof. Appa Reddy for the moral support given to me throughout.

My special thanks Mr. Ramana Reddy who has done a great job during the preparation of the manuscript of the book. I thank Ms. Jayalakshmi, Sunil Prakash and Vali for assisting in verifying the manuscript with interest and devotion.

I love all my students who are always fond of my teaching, guidance and discipline.

I have great regards to all senior authors and experts in Microwave Engineering for their invaluable contributions.

I am extremely grateful to the Department of Electronics and Communication Engineering, Andhra University College of Engineering (Autonomous), Andhra University for the encouragement given throughout my teaching and Research career.

I extend my warm regards to Sri S.V.S.S. Ramachandra Raju, former P.F. commissioner, Sri K. Raghu, Chairman, Raghu Engineering College, Sri Tapovarthan, Secretary and Correspondent of ANITS Engineering College, Sri Kanakayya, Chairman, Odalarevu Engineering College, Sri Rajesh, Chairman and Sri Upandra Reddy, Secretary, Sri Raja Rajeswari College of Engineering for their motivation and great appreciation of my teaching and research efforts.

I express my happiness for the support and respect extended by our technical and office staff, Koteswara Rao, Sahu, Vali, Prasada Reddy, Appala Raju, Ramesh, Adilakshmi, Ayodhya Ramayya, Ramana, Pallam Raju, Someswara Rao, Tavudu and Appa Rao. The assistance given by Sankar Panda is acknowledged with thanks for an excellent job done in preparing the manuscript.

I take this opportunity to convey my respects to Saripella Satyanarayana Raju, Kucharlapati Suryanarayana Raju for their continuous encouragement given to me throughout.

I thank my wife Kanaka Durga, daughter Narmada Devi, M.S., son Venkata Krishna Varma, B.Tech (IIT-Delhi), M.S. and a highly matured and principled person Rajesh, M.S.-my son-in-law for their love, affection, help and patience throughout the preparation of the book.

Finally, I thank all those who helped me directly and indirectly in bringing out this book excellently.

G. S. N. Raju

Contents

Preface   *vii*
Acknowledgements   *ix*

1. Introduction to Microwaves and Their Applications   **1**

   1.1 Definition of Microwave   1
   1.2 The Characteristics of Microwaves   1
   1.3 Advantages of Microwaves   2
   1.4 Parameters of Microwave   3
   1.5 Microwave Regions and Band Designations   3
   1.6 Microwave Electromagnetic Spectrum Domain   4
   1.7 Electric and Magnetic Fields   4
   1.8 Static Electric Field   4
   1.9 Static Magnetic Field (H)   5
  1.10 Time Varying Electric Field   6
  1.11 Time Varying Magnetic Field   10
  1.12 Electromagnetic Field Equations   10
  1.13 Maxwell's Equations for Time-varying Fields   10
  1.14 Meaning of Maxwell's Equations   11
  1.15 Characteristics of Free Space   11
  1.16 Maxwell's Equations for Free Space   12
  1.17 Sinusoidal Time Varying Fields   12
  1.18 Maxwell's Equations In Phasor Form   12
  1.19 Wave Equations   12
  1.20 Power Flow by Microwaves   15
  1.21 Frequency   16
  1.22 Velocity of Microwave   16
  1.23 Wavelength (λ)   17
  1.24 Propagation Constant (γ)   18
  1.25 Expression for Propagation Constant of
       a Microwave in Conductive Medium   19

1.26 Attenuation Constant (α) 19
1.27 Phase Shift Constant (β) 19
1.28 Polarization 20
1.29 Applications of Microwaves 21
1.30 Representation of Small and Large Numbers 22
1.31 Relation between Power, dB, dB_m, dB_μ 22
1.32 Operation of Microwave Devices 23
1.33 Microwave Ovens 23
1.34 Radars 25
1.35 Satellite 26
1.36 Points to Remember 26
1.37 Multiple Choice Questions 27
1.38 Answers 33
1.39 Exercise Problems 33

2. Microwave Tubes for Microwave Signal Generation 34

2.1 Introduction to Tubes 34
2.2 Limitations of Conventional Tubes 34
2.3 Microwave Tubes 37
2.4 Velocity Modulation 38
2.5 Klystron 39
2.6 Reflex Klystron 41
2.7 Traveling Wave Magnetrons 52
2.8 Slow Wave Devices 55
2.9 Traveling Wave Tube 56
2.10 Backward-wave Oscillator 60
2.11 Backward Wave Amplifier (BWA) 62
2.12 Points to Remember 65
2.13 Multiple Choice Questions 67
2.14 Answers 71
2.15 Exercise Problems 72

3. Microwave Semiconductor Devices 74

3.1 Introduction 74
3.2 Microwave Bipolar Transistor 75
3.3 Field Effect Transistors 75
3.4 Tunnel Diode 76
3.5 Transferred Electron Devices (TEDs) 78
3.6 Gunn Diodes 79
3.7 LSA Diode 80
3.8 IMPATT Diode 81
3.9 TRAPATT Diode 85

3.10 BARITT Diode   88
3.11 PIN Diode   90
3.12 Points to Remember   93
3.13 Multiple Choice Questions   94
3.14 Answers   96
3.15 Exercise Problems   96

4. Scattering Matrix Parameters   **97**

4.1   Introduction   97
4.2   Properties of Scattering Matrix   97
4.3   Proof of Symmetric Property   98
4.4   Proof of Unitary Property   98
4.5   Definition of Scattering Matrix   99
4.6   Characteristics of S-matrix   99
4.7   Scattering Matrix of a Two-Port Network   99
4.8   Salient Features of S-matrix   101
4.9   Scattering Matrix of Multi-Port Network   102
4.10 Losses In Microwave Circuits   102
4.11 Return Loss (RL)   103
4.12 Insertion Loss (IL)   103
4.13 Transmission Loss (TL)   103
4.14 Reflection Loss (Γ)   103
4.15 Impedance Matrix   104
4.16 Admittance Matrix   105
4.17 Summary of S, Z and Y Matrics   106
4.18 Shunt Element in a Transmission Line   107
4.19 S-matrix of Series Element in the Transmission Line   108
4.20 Points to Remember   110
4.21 Multiple Choice Questions   111
4.22 Answers   112
4.23 Exercise Problems   112

5. Microwave Passive Components   **113**

5.1   Common Passive Components   113
5.2   Two-wire Lines   113
5.3   Coaxial Lines   117
5.4   Rectangular Waveguides   124
5.5   Cavity Resonators   127
5.6   Circular Waveguides   130
5.7   Ridge Waveguides   131
5.8   Attenuators   132
5.9   Corners   134

5.10 Bends   135
5.11 Twists   136
5.12 Circulators   137
5.13 Terminations   140
5.14 Matched Load   140
5.15 Isolators   142
5.16 Directional Couplers   145
5.17 Parameters of Directional Couplers   146
5.18 Scattering Matrix of Directional Coupler   147
5.19 Scattering Matrix of 3 dB Directional Coupler   148
5.20 Tee Junctions   149
5.21 H-plane Tee Junction   149
5.22 E-plane Tee Junction   152
5.23 Hybrid Tee (Magic Tee)   155
5.24 Ferrite Devices   158
5.25 Phase Shifters   159
5.26 Hybrid Rings (Rat Race Coupler)   159
5.27 Choke Joints   160
5.28 Tuning Screws or Posts   161
5.29 Flanges   162
5.30 Transitions   163
5.31 Microwave Filters   163
5.32 Points to Remember   164
5.33 Multiple Choice Questions   167
5.34 Answers   172
5.35 Exercise Problems   172

6. Microwave Transmission Lines   **173**

6.1   Definition   173
6.2   Functions of Transmission Lines   173
6.3   Examples of Transmission Lines/Media   173
6.4   Application of Transmission Lines   174
6.5   Salient Features of Transmission Lines   174
6.6   Transmission Medium – Free Space   176
6.7   Two-wire Parallel Open Lines   177
6.8   Definitions of the Parameters of
      Two-wire Transmission Lines   178
6.9   Definitions of Secondary Constants   179
6.10 Propagation Constant   180
6.11 Characteristic Impedance   180
6.12 Lossless Lines and Parameters   183
6.13 Distortionless Lines and Parameters   184

6.14 Input Impedance of a Transmission Line 186
6.15 Loading of Transmission Lines 188
6.16 Reflection Coefficient in Terms of Z_L and Z_0 189
6.17 Standing Waves on the Transmission Lines 190
6.18 Standing Wave Ratio 191
6.19 Relation between Standing Wave Ratio
and Reflection Coefficient 191
6.20 Matched Transmission Line 192
6.21 Mismatched Transmission Line 192
6.22 Losses in Transmission Lines 192
6.23 Summary of Transmission Line Parameters 193
6.24 Summary of Definitions of Transmission Line Parameters 194
6.25 Smith Chart 195
6.26 Applications of Smith Chart 196
6.27 Stubs 196
6.28 Design of Single Stub Matching 197
6.29 Shorted and Open Stubs 197
6.30 Applications of Two-wire Open Lines 203
6.31 Twisted Pair Lines 204
6.32 Shielded Pair Line 204
6.33 Coaxial Lines 205
6.34 Fields in Coaxial Cable 207
6.35 Parallel Plates 209
6.36 Rectangular Waveguide 210
6.37 Transverse Electric Waves (*TE* Waves) 212
6.38 Dominant Mode 214
6.39 Degenerate Modes 214
6.40 Transverse Magnetic Waves (*TM* Waves) 214
6.41 Transverse Electromagnetic Waves (*TEM* Waves) 215
6.42 Circular Waveguides 217
6.43 Waveguide Resonators 218
6.44 Striplines 225
6.45 Microstripline 227
6.46 Optical Fibres 230
6.47 Points to Remember 232
6.48 Multiple Choice Questions 234
6.49 Answers 239
6.50 Exercise Problems 240

7. Microwave Integrated Circuits **241**

7.1 Introduction 241
7.2 Salient Features of MICs 241
7.3 Types of Electonic Circuits 242

7.4	Discrete Circuit (DC)	242
7.5	Integrated Circuit (IC)	242
7.6	Monolithic Microwave Integrated Circuit (MMIC)	243
7.7	Hybrid Integrated Circuit (HIC)	243
7.8	Film Integrated Circuit (FIC)	243
7.9	Quasi-Monolithic Integrated Circuit (QMIC)	243
7.10	Merits of MIC	243
7.11	Applications of MMICs	244
7.12	MMIC Materials	244
7.13	Examples of Substrate Materials	244
7.14	Characteristics of Ideal Substrates	244
7.15	Applications and Properties of Substrate Materials At 10 Ghz	245
7.16	Examples of Metals (Conductor Materials)	246
7.17	Ideal Characteristics of Metals	246
7.18	Applications of Metals	246
7.19	Properties of Common Metals	246
7.20	Examples of Dielectric Materials used in MICs	248
7.21	Applications	248
7.22	Examples of Resistive Materials used in MICs	248
7.23	Properties	248
7.24	Methods of MMIC Fabrication	249
7.25	Steps Involved in Fabrication	250
7.26	Transmission Lines in MICs	250
7.27	Fabrication of MOSFET in MMICs	251
7.28	Advantages of MOSFET in MMICs	252
7.29	Disadvantages of BJT in MMICs	252
7.30	Fabrication of CMOS	252
7.31	Steps Involved in the Fabrication of CMOS	252
7.32	NMOS Fabrication	253
7.33	Applications of MMICs	253
7.34	Fabrication of Passive Components	253
7.35	Design of Planar Resistors	253
7.36	Application of Planar Resistors	254
7.37	Resistance of Planar Resistors	254
7.38	Design of Planar Inductor	254
7.39	Expressions for Inductance of Different Geometries	255
7.40	Planar Capacitors	259
7.41	Hybrid Integrated Circuits (HICs)	260
7.42	Points to Remember	264
7.43	Multiple Choice Questions	265
7.44	Answers	266
7.45	Exercise Problems	267

8. Microwave Antennas

8.1 Introduction — 268
8.2 Antenna Parameters — 269
8.3 Directional Characteristics — 269
8.4 Antenna Impedance, Z_a — 270
8.5 Radiation Resistance, R_r — 270
8.6 Effective Length of Antenna (L_{eff}) — 270
8.7 Effective Length of Transmitting Antenna — 270
8.8 Effective Length of Receiving Antenna — 270
8.9 Radiation Intensity (ψ) — 270
8.10 Directive Gain g_d — 271
8.11 Directivity, D — 271
8.12 Power Gain, g_p — 271
8.13 Antenna Efficiency (η) — 271
8.14 Effective Area — 272
8.15 Admittance Characteristics — 272
8.16 Antenna Bandwidth — 273
8.17 Front-to-Back Ratio (FBR) — 273
8.18 Polarization — 273
8.19 Main Lobe of Radiation Pattern — 273
8.20 Side Lobes of the Radiation Pattern — 273
8.21 Beam Width of Main Lobe — 274
8.22 Classification of Antennas — 274
8.23 Primary Antennas — 274
8.24 Half-wave Dipole — 274
8.25 Quarter-wave Monopole — 275
8.26 Uniform Linear Arrays — 276
8.27 Non-uniform Arrays — 277
8.28 Slot Antennas — 278
8.29 Wave Guide Slots — 280
8.30 Horn Antenna — 282
8.31 Corrugated Horns — 287
8.32 Microstrip or Patch Antennas — 288
8.33 Secondary Antennas — 292
8.34 Lens Antennas — 292
8.35 Reflector Antennas — 297
8.36 Plane Reflector — 297
8.37 Corner Reflector — 297
8.38 Paraboloid — 299
8.39 Shaped Beam Antennas — 309
8.40 Points to Remember — 324
8.41 Multiple Choice Questions — 326

268

   8.42 Answers   330
   8.43 Exercise Problems   331

9. Microwave Measurements   333

   9.1   Introduction   333
   9.2   Microwave Sources   333
   9.3   Oscilloscopes and Sampling Oscilloscopes   334
   9.4   Frequency Meters / Wavemeters   334
   9.5   Absorption Type Wavemeter   334
   9.6   Reaction Type Wavemeter   334
   9.7   Transmission Type Wavemeter   334
   9.8   Coaxial Wavemeters   334
   9.9   Spectrum Analyzer   335
  9.10 Salient Features of Spectrum Analyzer   335
  9.11 Applications of Spectrum Analyzer   336
  9.12 Power Meters   336
  9.13 Network Analyzer   336
  9.14 Precaution in Microwave Measurements   336
  9.15 Reflex Klystron Characteristics   337
  9.16 Characteristics of Crystal Detector   342
  9.17 Measurement of Guide Wavelength and
       Source Frequency   346
  9.18 V-I Characteristics of Gunn Diode   351
  9.19 Frequency Measurements   354
  9.20 VSWR Measurement   356
  9.21 Measurement of Attenuation   359
  9.22 Measurement of Parameters of Directional Coupler   361
  9.23 Study of Magic Tee   366
  9.24 Study the Isolator and Circulator   369
  9.25 Calibration of Attenuators   373
  9.26 Measurement of Unknown Impedance   377
  9.27 Microwave Power Measurement   380
  9.28 Measurement of Dielectric Constant   383
  9.29 Points to Remember   387
  9.30 Multiple Choice Questions   389
  9.31 Answers   394
  9.32 Exercise Problems   394

Objective Questions and Answers   395

References   424

Index   427

Introduction to Microwaves and Their Applications

Microwaves are not visible, not heard and also not sensed by human beings.

1.1 DEFINITION OF MICROWAVE

Microwave wave is an electromagnetic wave which has a wavelength of 3 mm to 1 meter and a frequency of 0.3 GHz to 100 GHz.

That is, microwave domain exists between 0.3 GHz and 100 GHz.

The frequency and wavelength of a microwave are inversely proportional to each other. That is,

$$f = \frac{v_0}{\lambda}$$

Here, f = frequency, Hz
v_0 = velocity of microwave in free space, m/sec
 = 3×10^8 m/sec
λ = wavelength, m

1.2 THE CHARACTERISTICS OF MICROWAVES

The characteristics of microwaves are:

1. Their wavelength is small.
2. Frequency is high.
3. They propagate in free space freely.
4. The attenuation in free space is small.
5. Their depth of penetration is infinite in dielectrics.
6. They do not penetrate into good conductors.
7. They are reflected from good conductors.
8. They are transmitted through good dielectrics.
9. They are reflected and transmitted if incident on a material.
10. They consist of electric and magnetic fields which are perpendicular to each other.

11. The electric and magnetic fields of microwaves in free space are related by

$$\left|\frac{E}{H}\right| = \eta_0 = 120\,\pi\ \Omega$$

    Here, E = electric field, V/m

    H = magnetic field, A/m

    η_0 = Intrinsic impedance or characteristic impedance of free space, ohm

12. The power transfer takes place through microwaves.

13. The power flow by microwaves is

$$P = E \times H$$

    Here, P = instantaneous power flow, W/m^2

    P is called Poynting vector

14. Microwaves carry power in free space like current carries power through transmission lines. But the microwaves and current have opposite characteristics. The microwave does not propagate in conductors but current propagates. On the other hand, microwave propagates in insulators but current does not propagate.

15. Microwaves lead to atomic and molecular resonant characteristics in several substances.

16. Microwaves propagate along a straight line.

17. Microwaves are produced by antennas.

18. They cannot propagate in water for long distances.

19. They vary sinusoidally in free space.

20. They are received/detected by antennas.

1.3 ADVANTAGES OF MICROWAVES

The microwaves exhibit several advantages over low frequency electromagnetic waves. They are:

- ➢ Large bandwidth
- ➢ High directivity of microwave radiation pattern
- ➢ Antenna size becomes small
- ➢ Low fading effect
- ➢ Propagate through ionosphere providing effective satellite communication
- ➢ Effective for radar communication
- ➢ Effective for radiation therapy application
- ➢ Effective for TV transmission and reception

1.4 PARAMETERS OF MICROWAVES

The microwaves are described by the following parameters.

> ➤ Electric and magnetic fields
> ➤ Power
> ➤ Frequency
> ➤ Wavelength
> ➤ Polarization
> ➤ Velocity
> ➤ Attenuation
> ➤ Phase constant

1.5 MICROWAVE REGIONS AND BAND DESIGNATIONS

Microwave domain exists between 0.3 GHz and 100 GHz.

The conventional frequency range of microwaves is 0.3 GHz to 100 GHz. However, the frequencies from 0.1 GHz to 10^6 GHz is also called microwave frequency range.

Different types of band designations are available. For example, the U.S. military microwave bands consists of P, L, S, C, X, K, Q, V and W bands. This region is from 0.225 GHz to 100 GHz. The band P represents the frequency range of 0.225 GHz – 0.390 GHz and W band represents the frequency of 56.00 GHz – 100 GHz. The U.S. new military microwave bands consists of A, B, C, D, E, F, G, H, I, J, K, L and M bands. This region is from 0.1 GHz to 100 GHz. The band A represents the frequency range of 0.1 GHz to 0.250 GHz. The band M represents the frequency of 60 GHz – 100 GHz.

The third designation of bands is given by IEEE (USA). These are popular and are given in table 1.1.

Table 1.1   IEEE microwave frequency bands

Band name	Frequency Range (GHz)	Wavelength Range (m)
High frequency (HF)	0.003 – 0.030	100 – 10
Very high frequency (VHF)	0.03 – 0.300	10 – 1
Ultra high frequency (UHF)	0.300 – 1.00	1 – 0.3
L	1.00 – 2.00	0.3 – 0.15
S	2.00 – 4.00	0.15 – 0.075
C	4.00 – 8.00	0.075 – 0.0375
X	8.00 – 12.00	0.0375 – 0.025
Ku	12.00 – 18.00	0.025 – 0.0166

K	18.00	–	27.00	0.0166 – 0.011
Ka	27.00	–	40.00	0.011 – 0.0075
Millimeter	40.00	–	300.00	0.0075 – 0.0001
Sub-Millimeter	Greater than 300			less than 0.0001

I.6 MICROWAVE ELECTROMAGNETIC SPECTRUM DOMAIN

The microwave spectrum in the overall electromagnetic frequency range is shown in fig. 1.1.

Cosmic rays
Gamma rays
X-rays
Ultra light
Visible light
Infrared
Visible light
Visible light

Fig. 1.1 Electromagnetic spectrum

I.7 ELECTRIC AND MAGNETIC FIELDS

A microwave is characterized by electric and magnetic fields which are perpendicular to each other.

I.8 STATIC ELECTRIC FIELD

Definition 1: The electric field, E is defined as the coulombs force due to a fixed charge, Q on a unit test charge.

i.e.
$$E \equiv \frac{F}{Q_t} = \frac{Q}{4\pi \in r^2} \ \text{N/C}$$

Here, F is the force, Newtons

Q = charge, coulombs

Definition 2: It is also defined as

$$E = -\nabla V \ \text{V/m}$$

Here, ∇ = vector differential operator

$$\equiv a_x \frac{\partial}{\partial x} + a_y \frac{\partial}{\partial y} + a_z \frac{\partial}{\partial z}$$

a_x, a_y and a_z are the unit vectors along x, y and z axis respectively.

Definition 3:

$$E \equiv \frac{D}{\in} = \frac{c}{F-m}$$

Here, D = electric flux density, c/m^2

$\in$ = Permittivity of the medium, F/m

It may be noted that electric field, electric field intensity and electric field strength mean the same.

The static electric field is produced by a charge at rest.

It is evident from the above definitions, the unit of E is Newton/coulomb or volt/meter or coulomb/farad-m.

Volt/m is most popular as the unit of E and is commonly used.

1.9 STATIC MAGNETIC FIELD (H)

The static magnetic field is produced by the charges in motion.

Definition 1: H is defined as the force due to a magnetic pole on a unit magnetic pole.

That is,

$$H \equiv \frac{F}{m_t} \equiv \frac{m}{4\pi\mu r^2} \ \text{N/Wb}$$

m_t is test magnetic strength, Wb

Definition 2:

$$H \equiv \oint \frac{IdL \times a_r}{4\pi r^2} \ \text{A/m}$$

Here, $I\,dL$ = current element

r = the distance of the point from the current element.

I = current through the current element

dL = length of the current element

Definition 3: H is defined as

$$H \equiv \frac{B}{\mu} \ \frac{\text{Wb}}{\text{Henry}-m}$$

Here, B = magnetic flux density, Wb/m^2

μ = Permeability of the medium, Henry/m

It may be noted that magnetic field magnetic field intensity and magnetic field strength mean the same.

It is evident from the above definitions, the unit of H is Newton/Wb or ampere/meter or wb/Henry-m.

But ampere/meter is the most popular unit of H and is commonly used.

1.10 TIME VARYING ELECTRIC FIELD

Definition 1:

The time varying electric field is defined as

$$E = -\nabla V - \frac{\partial A}{\partial t}$$

Here, A = vector magnetic potential, Wb/m

Definition 2:

It is also defined as

$$E \equiv -\mu(V \times H)$$

V = Velocity of the wave

PROBLEM 1.1 When the force, $F = 2a_x + a_y + a_z$ Newtons is acting on a charge of 1.0 C, calculate the electric field, its magnitude and direction.

Solution

$$\text{Force, } F = 2a_x + a_y + a_z, N$$

$$Q = 1.0 \text{ C}$$

$$\therefore \quad E = \frac{F}{Q}$$

$$= \frac{2a_x + a_y + a_x}{1.0}$$

$$\therefore \quad \boxed{E = 2\,a_x + a_y + a_z, \, N/C}$$

The magnitude of E is

$$E = |E|$$

$$= \sqrt{(2)^2 + (1)^2 + (1)^2}$$

$$= 2.449 \text{ V/m}$$

The direction of E is

$$a_E = \frac{E}{E}$$

$$= \frac{2a_x + a_y + a_z}{2.449}$$

$$\therefore \quad \boxed{a_E = 0.816\,a_x + 0.408\,a_y + 0.408\,a_z}$$

PROBLEM 1.2   When a charge of 2 μC is located at P_1 and a second charge of 1 μC is located at P_2 in free space, determine the electric field at P_2. P_1 is given by a_x and P_2 is given by a_y.

Solution

$$Q_f = 2 \ \mu C = 2 \times 10^{-6} \ C$$
$$Q_t = 1 \ \mu C = 1 \times 10^{-6} \ C$$
$$r_f = (1, 0, 0) = a_x$$
$$r_t = (0, 1, 0) = a_y$$
$$r_{tf} = r_t - r_f = a_y - a_x$$
$$r_{tf} = \sqrt{1^2 + 1^2} = \sqrt{2}$$

$$\therefore \quad a_{tf} = \frac{r_{tf}}{r_{tf}} = \frac{(a_y - a_x)}{\sqrt{2}}$$

$\therefore$ Electric field at P_2 is

$$E = \frac{Q_f \, Q_t}{4\pi \in_0 r_{tf}^2} a_{tf}$$

$$= \frac{2 \times 10^{-6} \times 1 \times 10^{-6} \, (a_{tf})}{4\pi \in_0 (\sqrt{2})^2}$$

$$= \frac{2 \times 10^{-6} \times 1 \times 10^{-6}}{4\pi \in_0 (\sqrt{2})^2} \frac{(a_y - a_x)}{\sqrt{2}}$$

$$\frac{1}{4\pi \in_0} = 9 \times 10^9$$

$$E = 9 \times 10^9 \times 10^{-12} \times \frac{2}{2} \left(\frac{a_y - a_x}{\sqrt{2}} \right)$$

$$\therefore \quad \boxed{E = (-6.364 \ a_x + 6.364 \ a_y) \ \text{mN/C}}$$

PROBLEM 1.3   Considering three charges $Q_1 = 1 \ \mu C$, $Q_2 = 2 \ \mu C$ and $Q_3 = 3 \ \mu C$, the field due to each charge at a point, P in free space is $a_x + 2a_y - a_z$, $a_y + 3a_z$ and $2a_x - a_y$ N/C. Determine the total field at the point, P due to all the three charges.

Solution

$$\text{at } P \text{ due to } (1 \ \mu C) = a_x + 2a_y - a_z, \ \text{N/C}$$
$$E_2 \text{ due to } (2 \ \mu C) = a_y + 3a_z, \ \text{N/C}$$
$$E_3 \text{ due to } (3 \ \mu C) = 2a_x - a_y, \ \text{N/C}$$

The total field at P,

$$E = E_1 + E_2 + E_3$$

$$= a_x + 2a_y - a_z + a_y + 3a_z + 2a_x - a_y$$

$$\therefore \quad \boxed{E = 3a_x + 2a_y + 2a_z, \quad \text{N/C}}$$

PROBLEM 1.4   A charge, $Q_1 = -10$ nC s at the origin in free space. If the x-component of E is to be zero at the point (3, 1, 1), what charge, Q_t should be kept at the point (2, 0, 0)?

Solution

$$Q_1 = -10 \text{ nC at the origin}$$
$$r_1 = (3, 1, 1) - (0, 0, 0)$$
$$r_1 = (3 - 0)a_x + (1 - 0)a_y + (1 - 0)a_z$$
$$= 3a_x + a_y + a_z$$
$$r_1 = \sqrt{9 + 1 + 1} = \sqrt{11}$$
$$r_1 = (3, 1, 1) - (2, 0, 0)$$
$$r_2 = (3 - 2)a_x + (1 - 0)a_y + (1 - 0)a_z$$
$$= a_x + a_y + a_z$$
$$r_2 = \sqrt{1 + 1 + 1} = \sqrt{3}$$

$$\therefore \quad E = E_1 + E_2$$

$$= \frac{Q_1}{4\pi \epsilon_0 r_1^2} a_{r1} + \frac{Q_1}{4\pi \epsilon_0 r_2^2} a_{r2}$$

$$a_{r1} = \frac{r_1}{r_1} = \frac{3a_x + a_y + a_x}{\sqrt{11}}$$

$$a_{r2} = \frac{r_2}{|r_2|} = \frac{a_x + a_y + a_z}{\sqrt{3}}$$

$$\therefore E \text{ at } (3, 1, 1)$$

$$= \frac{-10 \times 10^{-9} \times 9 \times 10^9}{(11)^{3/2}} (3a_x + a_y + a_z) + \frac{Q_t \times 9 \times 10^9}{(3)^{3/2}} (a_x + a_y + a_z)$$

$$\therefore \quad E = \frac{-90}{(11)^{3/2}} (3a_x + a_y + a_z) + \frac{Q_t \times 9 \times 10^9}{(3)^{3/2}} (a_x + a_y + a_z)$$

For $E_x = 0$, we have

$$\frac{-90 \times 3}{(11)^{3/2}} + \frac{Q_t \times 9 \times 10^9}{(3)^{3/2}} = 0$$

$$\therefore \quad \boxed{Q_t = 4.27 \text{ nC.}}$$

PROBLEM 1.5   The charges $Q_1 = 1.0$ nC and $Q_2 = 5.0$ nC are at $(-1, 1, -3)$ m and $(3, 1, 0)$ m respectively. Find the electric field at Q_1.

Solution
$$Q_1 = 1.0 \text{ nC is at } (-1, 1, -3) \text{ m}$$
$$Q_2 = 5.0 \text{ nC is at } (3, 1, 0) \text{ m}$$

$\therefore$ Electric field, E at Q_1 $(-1, 1, -3)$ m is

$$E = \frac{Q_2}{4\pi \in_0 r^2} a_r$$

$$r = (-1 - 3)a_x + (1 - 1)a_y \, (-3 - 0)a_z$$
$$= -4a_x - 3a_z$$

$$r = |r_1| = \sqrt{16 + 9} = 5$$

$$a_r = \frac{-4a_x - 3a_z}{5}$$

$$E = \frac{1.0 \times 10^{-9} \times 9 \times 10^9}{25 \times 5}(-4a_x - 3a_z)$$

$\therefore$
$$\boxed{E = -0.0576 \, a_x - 0.0432 \, a_z}$$

PROBLEM 1.6   A point charge exerts a force of 12 mN on a charge of 3 coulombs. Determine the electric field at the location of 3 C charge.

Solution
$$F = 12 \text{ mN}$$
$$Q_t = 3 \text{ C}$$

$$E = \frac{F}{Q_t} = \frac{12 \times 10^{-3}}{3}$$

$\therefore$
$$\boxed{E = 4 \text{ mN/C}}$$

PROBLEM 1.7   If a magnetic pole exerts f force of 3 mN on magnetic pole of 2 Wb, find the magnetic field at a distance of 2 meters in free space.

Solution
$$F = 3 \text{ mN}$$
$$m_t = 2$$
$$r = 2 \text{ meter}$$

$$H = \frac{F}{m_t}$$

$$H = \frac{3 \times 10^{-3}}{2}$$

$\therefore$
$$\boxed{H = 1.5 \text{ mN/Wb}}$$

1.11 TIME VARYING MAGNETIC FIELD

Definition 1:

> The time varying magnetic field is defined as
>
> $$H \equiv \frac{1}{\mu} \nabla \times A$$

Definition 2:

> It is also defined as
>
> $$H \equiv \in (V \times E)$$

Here, V = velocity of the wave.

1.12 ELECTROMAGNETIC FIELD EQUATIONS

The electromagnetic field equations are given by Maxwell. The electromagnetic fields of microwaves are basically time-varying. They are expressed in both differential and integral form.

1.13 MAXWELL'S EQUATIONS FOR TIME-VARYING FIELDS

Maxwell's equations in differential form for a general medium are given by:

$$\nabla \times H = \dot{D} + J$$
$$\nabla \times E = -\dot{B}$$
$$\nabla \cdot D = \rho_v$$
$$\nabla \cdot B = 0$$

Here, H = Magnetic field strength (A/m)

D = Electric flux density, (C/m^2)

$\dot{D} = \dfrac{\partial D}{\partial t}$ = Displacement electric current density (A/m^2)

J = Conduction current density (A/m^2)

E = Electric field (V/m)

B = Magnetic flux density (Wb/m^2 or Tesla)

$\dot{B} = \dfrac{\partial B}{\partial t}$ = Time-derivative of magnetic flux density (Wb/m^2-sec)

$\dot{B}$ is called magnetic current density (V/m^2) or Tesla/sec

ρ_v = Volume charge density (C/m^3)

Maxwell's equations for time-varying fields in integral form are given by:

$$\oint_L \boldsymbol{H} \cdot d\boldsymbol{L} = \oint_S (\dot{\boldsymbol{D}} + \boldsymbol{J}) \cdot d\boldsymbol{S}$$

$$\oint_L \boldsymbol{E} \cdot d\boldsymbol{L} = -\oint_S \dot{\boldsymbol{B}} \cdot d\boldsymbol{S}$$

$$\oint_S \boldsymbol{D} \cdot d\boldsymbol{S} = \oint_O \rho_v \, d_v$$

$$\oint_S \boldsymbol{B} \cdot d\boldsymbol{S} = 0$$

Here, $d\boldsymbol{L}$ is differential length and
$d\boldsymbol{S}$ is the differential area. Its direction is always outward normal to the surface.

1.14 MEANING OF MAXWELL'S EQUATIONS

It is easy to understand the meaning of Maxwell's equations from their integral form.

1. The first Maxwell's equation means that the magnetomotive force around a closed path is equal to the sum of the displacement current and conduction current through any surface enclosed by the path.

2. The 2nd Maxwell's equation means that the electromotive force around a closed path is equal to the minus of time derivative of magnetic flux flowing through any surface enclosed by the path.

3. The third Maxwell's equation means that the total electric displacement flux passing through a closed surface is equal to the total charge enclosed.

4. The fourth Maxwell's equation means that the net magnetic flux passing through any closed surface is zero.

1.15 CHARACTERISTICS OF FREE SPACE

Free space is characterized by the following parameters:

Relative permittivity,	$\epsilon_r = 1$
Relative permeability,	$\mu_r = 1$
Conductivity,	$\sigma = 0$
Conduction current density,	$\boldsymbol{J} = 0$
Volume charge density,	$\rho_v = 0$
Intrinsic impedance or characteristic impedance	$\eta = 120\,\pi$ or $377\,\Omega$

1.16   MAXWELL'S EQUATIONS FOR FREE SPACE

$$\nabla \times H = \dot{D} \leftrightarrow \oint_L H \cdot dL = \oint_S \dot{D} \cdot dS$$

$$\nabla \times E = -\dot{B} \leftrightarrow \oint_L E \cdot dL = \oint_S \dot{B} \cdot dS$$

$$\nabla \cdot D = 0 \leftrightarrow \oint_S D \cdot dS = 0$$

$$\nabla \cdot B = 0 \leftrightarrow \oint_S B \cdot dS = 0$$

1.17   SINUSOIDAL TIME VARYING FIELDS

In practice, electric and magnetic fields vary sinusoidally. It is well known that any periodic variation can be described in terms of sinusoidal variations with fundamental and harmonic frequencies.

The fields can be represented by

$$\tilde{E} = E_m \cos \omega t$$

or
$$\tilde{E} = E_m \sin \omega t$$

Here, $\omega = 2\pi f$, f = frequency variation of the field, E_m is the maximum field strength.

1.18   MAXWELL'S EQUATIONS IN PHASOR FORM

The Maxwell's equations in phasor form are given by:

$$\nabla \times H = j\omega D + J$$
$$\nabla \times E = -j\omega B$$
$$\nabla \cdot D = \rho_v$$
$$\nabla \cdot B = 0$$

1.19   WAVE EQUATIONS

Microwaves in free space are represented by wave equations and are given by:

$$\nabla^2 E = \mu_0 \in_0 \ddot{E}$$

$$\nabla^2 H = \mu_0 \in_0 \ddot{H}$$

Microwaves in a general medium are represented by the wave equations

$$\nabla^2 E = \mu\epsilon \frac{\partial^2 E}{\partial t^2} + \mu\sigma \frac{\partial E}{\partial t} \quad \text{and}$$

$$\nabla^2 H = \mu\epsilon \frac{\partial^2 H}{\partial t^2} + \mu\sigma \frac{\partial H}{\partial t^2}$$

The wave equations in terms of propagation constant are given by

$$\nabla^2 E = \gamma^2 E$$
$$\nabla^2 H = \gamma^2 H$$

Here, γ = propagation constant, $\left(\dfrac{1}{m}\right)$

$$= \sqrt{-\omega^2 \mu\epsilon + j\omega\mu\sigma}$$

PROBLEM 1.8 If $E = 2 \sin(\omega t - \beta z)a_y$ V/m in free space, find D, B, H.

Solution
$$E = 2 \sin(\omega t - \beta z)a_y \text{ V/m}$$
$$D = \epsilon_0 E, \; \epsilon_0 = 8.854 \times 10^{-12} \text{ F/m}$$

$\therefore$
$$D = 2\epsilon_0 \sin(\omega t - \beta z)a_y \text{ C/m}^2$$

2nd Maxwell's equation is

$$\nabla \times E = -\dot{B}$$

i.e.,
$$\nabla \times E = \begin{vmatrix} a_x & a_y & a_z \\ \dfrac{\partial}{\partial x} & \dfrac{\partial}{\partial y} & \dfrac{\partial}{\partial z} \\ 0 & E_y & 0 \end{vmatrix}$$

or
$$\nabla \times E = a_x\left[-\frac{\partial}{\partial z}E_y\right] + 0 + a_z\left[\frac{\partial}{\partial x}E_y\right]$$

As
$$E_y = 2 \sin(\omega t - \beta z) \text{ V/m}$$

$$\frac{\partial E_y}{\partial x} = 0$$

Now $\nabla \times E$ is given by

$$\nabla \times E = -\frac{\partial E_y}{\partial z}a_x$$

$$= 2\beta \cos(\omega t - \beta z)a_x$$

$$= -\frac{\partial \mathbf{B}}{\partial t}$$

$$\therefore \qquad \mathbf{B} = -\int 2\beta \cos (\omega t - \beta z) \, dt \; \mathbf{a}_x$$

or
$$\mathbf{B} = \frac{2\beta}{\omega} \sin (\omega t - \beta z) \mathbf{a}_x, \; \text{wb/m}^2$$

and
$$\mathbf{H} = \frac{\mathbf{B}}{\mu_0} = \frac{-2\beta}{\mu_0 \omega} \sin (\omega t - \beta z) \mathbf{a}_x, \; \text{A/m}$$

PROBLEM 1.9   When the electric field strength, $\mathbf{E}$ of an electromagnetic wave in free space is given by $\mathbf{E} = \cos \omega \left(t - \dfrac{z}{v_0} \right) \mathbf{a}_y$ V/m, determine the magnetic field, $\mathbf{H}$.

Solution   We have $\quad \dfrac{\partial \mathbf{B}}{\partial t} = -\nabla \times \mathbf{E}$

$$= - \begin{vmatrix} \mathbf{a}_x & \mathbf{a}_y & \mathbf{a}_z \\ \dfrac{\partial}{\partial x} & \dfrac{\partial}{\partial y} & \dfrac{\partial}{\partial z} \\ 0 & E_y & 0 \end{vmatrix}$$

$$= - \left[\mathbf{a}_x \left(-\frac{\partial}{\partial z} E_y \right) + \mathbf{a}_y (0) + \mathbf{a}_z \left(\frac{\partial}{\partial x} E_y \right) \right]$$

$$= \frac{\partial E_y}{\partial z} \mathbf{a}_x$$

$$= \frac{\omega}{v_0} \sin \omega \left(t - \frac{z}{v_0} \right) dt \; \mathbf{a}_x$$

$$\therefore \qquad \mathbf{B} = \frac{\omega}{v_0} \int \sin \omega \left(t - \frac{z}{v_0} \right) dt \; \mathbf{a}_x$$

$$\mathbf{B} = \frac{-\omega}{v_0 \omega} \cos \omega \left(t - \frac{z}{v_0} \right) \mathbf{a}_x$$

$$\mathbf{H} = \frac{\mathbf{B}}{\mu_0} = \frac{-1.0}{v_0 \mu_0} \cos \omega \left(t - \frac{z}{v_0} \right) \mathbf{a}_x, \; \eta = \sqrt{\frac{\mu_0}{\epsilon_0}} = 120\pi \; \Omega$$

i.e.
$$H = \frac{-1.0}{\eta_0} \cos \omega \left(t - \frac{z}{v_0} \right) a_x, \quad \left[v_0 = \frac{1}{\sqrt{\mu_0 \, \epsilon_0}} \right]$$

$$\therefore \quad H = -\frac{1}{120\pi} \cos \omega \left(t - \frac{z}{v_0} \right) a_x \text{ A/m}$$

1.20 POWER FLOW BY MICROWAVES

The flow of power by microwaves is expressed by Poynting vector.
Poynting vector is given by

$$P \equiv E \times H$$

Here, P = power, ω/m^2
E = V/m, H = A/m
The direction of power flow is perpendicular to both E and H.

The average power is given by $P_{av} = \dfrac{1}{T} \displaystyle\int_0^T p(t)\, dt$.

T = time period
If E and H are complex, the complex Poynting vector is given by

$$p_c = \frac{1}{2} E \times H^*$$

In this case, $p_{av} = \dfrac{1}{2} R_e (EH^*)$

PROBLEM 1.10   If E is along y axis, and H is along x-axis, find out the direction of power flow of microwave.

Solution E is along y-axis.
i.e. its direction is a_y.
Similarly, the direction of H is a_x.
The direction of power flow = $a_y \times a_x = -a_x$

PROBLEM 1.11   If $E = 1.0\, a_x$, $H = 2.0\, a_y$, find the Poynting vector and the direction of power flow of microwave.

Solution
$$E = 1.0\, a_x$$
$$H = 2.0\, a_y$$
$$\therefore \quad P = E \times H$$
$$\therefore \quad P = 2.0\, a_x$$

The direction of power flow is along z-axis.

1.21 FREQUENCY

Definition:

Frequency (f) of microwave is defined as the number of cycles of the wave per second. Its unit is Hertz (Hz). The frequency remains constant irrespective of medium. It is inversely proportional to the time period (T) of the wave.

$$\therefore \qquad f \equiv \frac{1}{T} \text{ Hz}$$

T is the time period in seconds corresponding one cycle.

PROBLEM 1.12 If the time period of one cycle is (a) 100 ps (b) 500 ps (c) 1.0 ns, find the frequency of the wave.

Solution (a) $\qquad T = 100$ ps

$$f = \frac{1}{T} = \frac{1}{100 \times 10^{-12}}$$

$$\therefore \qquad f = 10 \text{ GHz}$$

(b) $\qquad T = 500$ ps

$$f = \frac{1}{T} = \frac{1}{500 \times 10^{-12}}$$

$$\therefore \qquad f = 2 \text{ GHz}$$

(c) $\qquad T = 1.0$ ns

$$f = \frac{1}{T} = \frac{1}{10^{-9}}$$

$$\therefore \qquad f = 1 \text{ GHz}$$

1.22 VELOCITY OF MICROWAVE

Definition 1:

The velocity of a microwave is defined as the velocity with which the wave propagates. Mathematically, it is defined as

$$v = \frac{1}{\sqrt{\mu \epsilon}}$$

Here, μ = permeability, H/m

ϵ = permittivity F/m

Velocity of the microwave depends on permittivity and permeability of the medium. It is equal to 3×10^{8} m/s.

Definition 2:

$$v \equiv \frac{\omega}{\beta}$$

Here, $\omega = 2\pi f$

β = phase constant

PROBLEM 1.13   If a wave propagates through a medium where $\epsilon_r = 4$, $\mu_r = 1$. Find the velocity of propagation of the microwave.

Solution

$$\mu_r = 1$$
$$\epsilon_r = 4$$

$\therefore$

$$v = \frac{1}{\sqrt{\mu \epsilon}} = \frac{1}{\sqrt{\mu_0 \epsilon_0}\sqrt{\epsilon_r \mu_r}}$$

$$= \frac{3 \times 10^8}{2} = 1.5 \times 10^8 \text{ m/s}$$

$\therefore$

$$v = 1.5 \times 10^8 \text{ m/s}$$

1.23   WAVELENGTH (λ)

Definition 1:

The wavelength of microwave is defined as a variable which varies directly as the velocity of the wave and inversely as the frequency. That is,

$$\lambda = \frac{v}{f}$$

Here, v = velocity of the wave

f = frequency of the wave

The wavelength is expressed in terms of

- Physical length in linear units
- Electrical length in wavelengths
- Distance where the phase changes 2π radians

Definition 2:

$$\lambda \equiv \frac{2\pi}{\beta}$$

Here, β = phase constant

PROBLEM 1.14   Find the wavelength of microwave in free space at frequencies (a) 1 MHz (b) 10 MHz (c) 100 MHz (d) 1 GHz (e) 10 GHz.

Solution Free space velocity, $v_0 = 3 \times 10^8$ m/s

(a) $f = 1$ MHz

$$\therefore \qquad \lambda = \frac{v_0}{f} = \frac{3 \times 10^8}{1 \times 10^6}$$

i.e. $\boxed{\lambda = 300 \text{ m}}$

(b) $f = 10$ MHz

$$\lambda = \frac{v_0}{f} = \frac{3 \times 10^8}{10 \times 10^6}$$

i.e. $\lambda = 30$ m

(c) $f = 100$ MHz

$$\lambda = \frac{3 \times 10^8}{100 \times 10^6}$$

i.e. $\boxed{\lambda = 3 \text{ m}}$

(d) $f = 1$ GHz

$$\lambda = \frac{v_0}{f} = \frac{3 \times 10^8}{10^9}$$

i.e. $\boxed{\lambda = 0.3 \text{ m}}$

(e) $f = 10$ GHz

$$\lambda = \frac{v_0}{f} = \frac{3 \times 10^8}{10 \times 10^9}$$

i.e. $\boxed{\lambda = 0.03 \text{ m}}$

1.24 PROPAGATION CONSTANT (γ)

Definition 1:

The propagation constant of a microwave is defined as a complex quantity that expresses the effect of free space for the wave propagation.

Definition 2:

The propagation constant, γ is defined as

$$\boxed{\gamma \equiv \alpha + j\beta, \ \left(\frac{1}{m}\right)}$$

Here, α = attenuation constant, (dB/m)

β = phase constant, (rad/m)

1.25 EXPRESSION FOR PROPAGATION CONSTANT OF A MICROWAVE IN CONDUCTIVE MEDIUM

$$\gamma = \sqrt{-\omega^2 \mu \in + j\omega\mu\sigma}$$

Here, $\omega = 2\pi f$, f = frequency

μ = permeability of the medium, H/m

$\in$ = permittivity of the medium, F/m

σ = conductivity of the medium, mho/m

1.26 ATTENUATION CONSTANT (α)

Definition:

The attenuation constant, α is defined as the rate at which the amplitude of microwave reduces as it progresses in the medium.

It is the real part of the propagation constant. Its unit is $\left(\dfrac{1}{m}\right)$ or $\left(\dfrac{dB}{m}\right)$. It is given by

$$\alpha = \omega \sqrt{\frac{\mu \in}{2} \left[1 + \frac{\sigma^2}{\omega^2 \in^2} - 1 \right]}, \quad dB/m$$

1.27 PHASE SHIFT CONSTANT (β)

Definition 1:

The phase shift constant is defined as a measure of the phase shift of the microwave as it propagates. It is expressed in radian/m.

Definition 2:

$$\beta \equiv \frac{2\pi}{\lambda}$$

It is the imaginary part of propagation constant and is given by

$$\beta = \omega \sqrt{\frac{\mu \in}{2} \left(\sqrt{1 + \frac{\sigma^2}{\omega^2 \in^2}} + 1 \right)}, \quad rad/m$$

PROBLEM 1.15 Find the phase shift of a wave whose frequency is 1 GHz in free space.

Solution

$$f = 1 \text{ GHz}$$

$$\lambda = \frac{v_0}{f} = \frac{3 \times 10^8}{10^9} = 0.3 \text{ m}$$

$$\beta = \frac{2\pi}{\lambda} = \frac{2\pi}{0.3} = \frac{2 \times 3.14}{0.3} = \frac{6.28}{0.3} = 20.93$$

$$\therefore \qquad \boxed{\beta = 20.93 \text{ rad/m}}$$

1.28 POLARIZATION

Definition:

> The polarization of a wave is defined as the direction of the electric field at a given point as a function of time.

Types of Polarizations

These are of 3 types:

(a) Linear polarization.
(b) Circular polarization.
(c) Elliptical polarization.

(a) Linear Polarization

A wave is said to be linearly polarized if the electric field remains along a straight line as a function of time at some point in the medium. Linear polarization of a wave is again of 3 types:

(i) Horizontal polarization.
(ii) Vertical polarization.
(iii) Theta polarization.

When a wave travels in z-direction with $\tilde{E}$ and $\tilde{H}$ fields lying in xy-plane, if $\tilde{E}_y = 0$ and $\tilde{E}_x$ is present, it is said to be **x-polarized** or **horizontally polarized.**

If $\tilde{E}_y$ is only present and $\tilde{E}_x = 0$ the wave is said to be **vertically (y-polarized) polarized.** On the other hand, if $\tilde{E}_x$ and $\tilde{E}_y$ are present and are in phase then the wave is said to be θ polarized. This is given by

$$\theta = \left(\tan^{-1} \frac{E_y}{E_x} \right)$$

(b) Circular Polarization

A wave is said to be circularly polarized when the electric field traces a circle. If $\tilde{E}_x$ and $\tilde{E}_y$ have equal magnitudes and a 90 degree phase difference, the locus of the resultant $\tilde{E}$ is a circle and the wave is **circularly polarized.**

(c) Elliptical Polarization

If $\tilde{E}_x$ and $\tilde{E}_y$ are not equal in magnitude and they differ by $90°$ phase, then the tip of resultant electric vector traces an ellipse. It is said to be elliptically polarized.

1.29 APPLICATIONS OF MICROWAVES

The microwaves are used extensively in

- Satellite communications
- Tropospheric communications
- Telephone wireless networks
- Radio broadcast
- TV communications
- Police wireless
- Civil aviation and railway communications
- Cellular and mobile communications
- Weather forecast radars
- Remote sensing radars
- Speed trap radars
- Radio astronomy radars
- Airport surveillance radars
- Traffic control radars
- Missile guidance radars
- Fire control radars
- Early warning radars
- IFF radars
- Side looking radars
- Over the horizon radars
- Synthetic aperture radars
- Wireless LAN networks
- Tracking radars
- MTI radars
- Pulse Doppler radars
- Marine radars

- Altimeters
- Global positioning system (GPS)
- Navigational aids
- Spectroscopy
- Testing of materials
- Thickness detectors
- The measurement of concentration of gases
- Industrial heating
- Processing of food and chemicals
- Mining
- Drying in the printing industry
- Breaking of concrete and steel
- Cooking food by microwave ovens
- Medical diagnosis
- Radiation therapy of tumors and cancerous tissues

1.30 REPRESENTATION OF SMALL AND LARGE NUMBERS

The large and small frequency values are represented with prefixes conveniently as shown in table 1.2.

Prefix	Meaning	Value
K	Kilo	10^3
M	Mega	10^6
G	Giga	10^9
T	Tera	10^{12}
P	Peta	10^{15}
E	Exa	10^{18}
m	Milli	10^{-3}
μ	Micro	10^{-6}
n	Nano	10^{-9}
p	Pico	10^{-12}
f	femto	10^{-15}

1.31 RELATION BETWEEN POWER, dB, dB_m, dB_μ

Power	DB	dB_m	dB_μ
1 W	0	30	60
10 W	10	40	70
100 W	20	50	80
1 kW	30	60	90
1 mW	− 30	0	30
10 mW	− 20	10	40

100 mW	– 10	20	50
1 μW	– 60	– 30	0
10 μW	– 50	– 20	10
1 nW	– 90	– 60	– 30
10 nW	– 80	– 50	– 20
100 nW	– 70	– 40	– 10

1.32 OPERATION OF MICROWAVE DEVICES

Although, microwaves are used in several devices, the most common devices and their basic operations are presented below:

1.33 MICROWAVE OVENS

Microwave ovens operate at a frequency of 2.45 GHz.

Microwave oven is a domestic kitchen electronic unit which is useful for cooking food quickly.

1.33.1 Principle of Operation

It contains a magnetron which converts electric energy into electromagnetic energy. When this energy comes in contact with food, it excites the molecules within and causes rapid vibration of the molecules. As a result, friction is developed and hence heat is produced. The resultant heat cooks the food.

Microwave energy cooks the food from the surface to inwards. The inner portion of the food is cooked by radiation of heat from the outer layers.

Microwaves are dispersed by the metallic stirrer fans and reflected from the metal surfaces within the oven cavity. This phenomenon makes the cooking even. The absorption of microwaves by the food is made uniform by the use of rotating trays. These trays are called turnables or carousels.

1.33.2 Specifications of Microwave Ovens

- Power level
- Oven size in terms of litres
- Weight

The power level varies for each size. The compact size oven (up to 20 litres) has power levels ranging between 800 – 1200 watts.

The medium size oven (20 to 25 litres) has power levels ranging between 1200 – 1800 watts.

The full size oven (25 litres or more) has power levels more than 1800 watts.

The cost of microwave oven ranges between Rs. 6,000/- and Rs. 30,000/-.

1.33.3   Types of Microwave Ovens

- Basic microwave oven
- Microwave oven with grill
- Microwave oven with grill and convection

1.33.4   Functions of the Oven

The basic microwave oven is useful for cooking, heating and defrosting.

The microwave oven with grill is useful for cooking, heating, defrosting and baking.

The microwave oven with grill and convection is useful for cooking, heating, defrosting, baking and crisping.

1.33.5   Some Manufacturers

- Glima
- Godrej
- LG
- Onida
- Pigion
- Samsung
- Videocon etc.

1.33.6   Features of the Ovens

The main features available for ovens are:

Child lock

Turnable tray

Autodefrost

Clock setting

Feather-touch operation

Stainless steel cavity

Auto menu

Quick convection fan

Push button door

Turn table

Power level controls

Express cooling

Memory function

Multistage cooking facility

Baking plate

Auto roast facility etc.

Speed auto defrost

Speed auto cook

Multi rotisserie

Exhaust fan

Work light

Preset control

Quick start setting

Interactive displays

Dual-mode timer

Keep-warm or simmer functions

Language operation

Cooking sensors

Integrated browning element etc.

Bells and whistles

Depending on the model, make and price the features of oven change. Most of the features are available on high end costly models.

I.34 RADARS

RADAR represents **RA**dio **D**etection **A**nd **R**anging. It operates with the help of microwaves.

Radar is used to find the position, altitude and velocity of the object.

Microwaves are produced by Radar antennas.

Radar is considered to be an electronic eye and it can see any object whether it is hidden in water, cloud, ice, or anywhere except when the object is enclosed by a perfect conductor. It is considered to be a powerful eye and more powerful than human eye.

The radar equation is given by

$$R = \left[\frac{P_T G_T G_R \sigma \lambda^2 C_m C_p}{(4\pi)^2 (P_R)} \right]^{\frac{1}{4}}$$

Here, R = the distance between the target and the radar

P_T = transmitted power

(P_R) = received power

λ = operating wavelength

G_T = gain of transmitting antenna

G_R = gain of the receiving antenna

$$\sigma = \text{radar target cross section}$$
$$C_m = \text{factor of medium loss}$$
$$C_p = \text{factor of isolation loss}$$

Different types of radars are:

> ➢ Navigation radars
> ➢ Early warning radars
> ➢ Speed trap radars
> ➢ Meteorology radars
> ➢ Airport control radars
> ➢ Missile tracking radars
> ➢ Gunfire control radars
> ➢ Radio astronomy radars
> ➢ Remote sensing radars etc.

1.35 SATELLITE

It is basically a spacecraft placed in an orbit around a planet. It operates using microwaves to serve a variety of functions. A few functions are communication, GPS, remote sensing, weather forecast, space exploration, surveying, meteorological observations etc. Communication satellites are placed in geostationary orbits. They are used for passing telephone messages, TV programs, computer data all across the globe.

1.36 POINTS TO REMEMBER

- ➢ Microwaves propagate in free space with the velocity of light.
- ➢ The wavelength of microwaves is very small.
- ➢ Microwaves propagate through dielectrics.
- ➢ Microwaves cannot propagate through conductors.
- ➢ Power transfer takes place through microwaves.
- ➢ The instantaneous power flow by microwaves is given by $E \times H$
- ➢ X-band frequency range is 8 GHz – 12 GHz.
- ➢ KU-band frequency range is 12 GHz – 18 GHz.
- ➢ Satellite communication is by microwaves.
- ➢ Mobile communication is by microwaves.
- ➢ Propagation constant is $\alpha + j\beta$.
- ➢ The unit of propagation is m^{-1}.
- ➢ The unit of attenuation constant is dB/m
- ➢ The unit of phase constant is rad/m
- ➢ 10^{12} Hz is represented by THz.
- ➢ 10^{-15} is represented by femto.
- ➢ Electric field is F/Q_t or ∇V.
- ➢ Magnetic field is F/m_t or B/μ.

- ➤ The unit of electric field is V/m.
- ➤ The unit of magnetic field is A/m

- ➤ The time varying electric field is $E = -\nabla V - \dfrac{\partial A}{\partial t}$ or $-\mu(V \times H)$.

- ➤ The time varying magnetic field is $H = \dfrac{1}{\mu} \nabla \times A$ or $\in (V \times E)$.

- ➤ The wave equation in free space is $\nabla^2 E = \mu \in \dfrac{\partial^2 E}{\partial t^2}$.

- ➤ The frequency, $f = \dfrac{1}{T}$.

- ➤ Poynting vector, $p = E \times H$.
- ➤ The wavelength, $\lambda = v/f$ or $2\pi/\beta$.
- ➤ Different polarizations of microwaves are linear, circular and elliptical.
- ➤ Microwaves offer large bandwidth.
- ➤ Microwaves are radiated by antennas.
- ➤ Microwaves are received by antennas.

1.37 MULTIPLE CHOICE QUESTIONS

1. The relation between frequency and wavelength of a microwave is

   (a) $f = \dfrac{v_0}{\lambda}$ (b) $f = v_0 \lambda$

   (c) $f = \dfrac{\lambda}{v_0}$ (d) $f = \dfrac{v_0}{\lambda^2}$

2. The velocity of a microwave in free space is
   (a) 3×10^8 cm/s   (b) 3×10^8 m/s
   (c) 3×10^{10} cm/s   (d) 3×10^{10} mm/s

3. The microwaves propagate in good conductors
   (a) excellently   (b) moderately
   (c) reasonably   (d) with zero penetration

4. E and H of a microwave
   (a) are perpendicular to each other   (b) are parallel to each other
   (c) has no relation   (d) make 45° between them

5. $\left| \dfrac{E}{H} \right|$ of microwave in free space is

   (a) 120 Ω (b) 120π Ω
   (c) 377 kΩ (d) 120π mho

6. $E \times H$ of microwave gives
   - (a) instantaneous power
   - (b) average power
   - (c) peak power
   - (d) reactive power

7. Circularly polarized antenna produces
   - (a) elliptically polarized waves
   - (b) circularly polarized waves
   - (c) linearly polarized waves
   - (d) unpolarized wave

8. If the wavelength of microwave is 10 cm, the preferred antenna size is
   - (a) 5 cm
   - (b) 20 cm
   - (c) 2.5 cm
   - (d) 100 cm

9. The unit of $E \times H$ is
   - (a) volt-amp
   - (b) watts
   - (c) watts/m^2
   - (d) watts/m

10. X-band frequency range is
    - (a) 8 – 12 GHz
    - (b) 4 – 8 GHz
    - (c) 2 – 4 GHz
    - (d) 1 – 12 GHz

11. Ku band frequency range is
    - (a) 8 – 12 GHz
    - (b) 8 – 18 GHz
    - (c) 12 – 18 GHz
    - (d) 18 – 27 GHz

12. Unit of E is
    - (a) volt
    - (b) volt/m
    - (c) C/m
    - (d) Netwotn/m

13. The unit of H is
    - (a) amp/m
    - (b) wb/m
    - (c) wb/m^2
    - (d) tesla

14. E is given by
    - (a) $-\nabla V$
    - (b) ∇V
    - (c) D/ϵ_r
    - (d) $\epsilon_r D$

15. Del (∇) has
    - (a) no units
    - (b) unit of m^{-1}
    - (c) unit of m
    - (d) unit of m^{-2}

16. Permittivity of a medium has
    - (a) no units
    - (b) unit of F/m
    - (c) F-m
    - (d) F/m^2

17. $E = 1$ V/m, $\epsilon_r = 2$, D is
    - (a) 2 V/m
    - (b) $2 \epsilon_r$ C/m^2
    - (c) 2 C/m^2
    - (d) ϵ_0 C/m^2

18. If $E = 3 a_x + 4 a_y$, $|E|$ is
    - (a) 6 V/m
    - (b) 5 V/m
    - (c) 4 V/m
    - (d) 3 V/m

19. The relation B and H is
    (a) $B = \mu H$

    (b) $H = \mu B$

    (c) $B = \dfrac{H}{\mu}$

    (d) $B = \in H$

20. For time-varying field, E is given by

    (a) $-\nabla V - \dfrac{\partial A}{\partial t}$

    (b) $-\nabla V$

    (c) $-\dfrac{\partial A}{\partial t}$

    (d) $-\nabla V + \dfrac{\partial A}{\partial t}$

21. E is given by
    (a) $-\mu \nabla \times H$

    (b) $\mu \nabla \times H$

    (c) $\mu \nabla \times B$

    (d) $\in \nabla \times D$

22. The units of H are
    (a) Wb

    (b) amp

    (c) volt/m

    (d) amp/m

23. The relation between H and A is

    (a) $H = \dfrac{1}{\mu} \nabla \times A$

    (b) $H = \nabla \times A$

    (c) $H = \nabla \cdot A$

    (d) $H = \nabla A$

24. In a dielectric medium,

    (a) $\nabla \times H = J$

    (b) $\nabla \times H = \dot{D}$

    (c) $\nabla \times H = 0$

    (d) $\nabla \times H = \rho_v$

25. $\nabla \cdot B$ is
    (a) zero

    (b) ρ_v

    (c) $\dot{D} + J$

    (d) m_i

26. $\nabla \times E$ is
    (a) $-j\omega B$

    (b) $-j\omega H$

    (c) $j\omega B$

    (d) $\mu\omega D$

27. $\nabla \cdot D$ is
    (a) zero

    (b) ρ_v

    (c) m_t

    (d) μH

28. $\nabla^2 E$ is
    (a) γE

    (b) $\gamma^2 E$

    (c) E/γ

    (d) $\gamma \dfrac{\partial E}{\partial t}$

29. $\nabla^2 H$ is
    (a) $\gamma^2 H$

    (b) γH

    (c) $\dfrac{H}{\gamma}$

    (d) $\gamma^2 \dfrac{\partial H}{\partial t}$

30. Complex potential vector is

   (a) $E \times H^*$
   (b) $\dfrac{1}{2} E \times H^*$

   (c) $\dfrac{1}{2} EH^*$
   (d) $\dfrac{1}{2} H \times E^*$

31. If the time period of microwave is 100 ps, the frequency is
   (a) 100 ps
   (b) 10 GHz
   (c) 1 GHz
   (d) 100 GHz

32. The velocity of a microwave in a waveguide is

   (a) $\dfrac{v_0^2}{v_p}$
   (b) $v_p v_0$

   (c) $\dfrac{v_p}{v_0}$
   (d) $\dfrac{v_0}{v_p}$

33. The unit of propagation constant is

   (a) $\dfrac{1}{m}$
   (b) $\dfrac{1}{m^2}$

   (c) dB
   (d) neper

34. The unit of attenuation constant is

   (a) $\dfrac{1}{m^2}$
   (b) $\dfrac{1}{m}$

   (c) dB
   (d) neper

35. β is given by

   (a) $\dfrac{2\pi}{\lambda}$
   (b) $\dfrac{\lambda}{2\pi}$

   (c) $2\pi\lambda$
   (d) $\dfrac{2\pi f}{\lambda}$

36. Dissipation factor is

   (a) $\dfrac{\sigma}{\omega \in}$
   (b) $\dfrac{\omega \in}{\sigma}$

   (c) $\sigma \omega \in$
   (d) $\dfrac{1}{\omega \in}$

37. Horizontal polarization is nothing but
   (a) x-polarization
   (b) y-polarization
   (c) circular polarization
   (d) elliptical polarization

38. Group velocity of a microwave in a waveguide is

    (a) $\dfrac{v_0^2}{v_p}$

    (b) $v_p v_0$

    (c) $\dfrac{v_p}{v_0}$

    (d) $\dfrac{v_p^2}{v_0}$

39. Femtometer is
    (a) 10^{-15} m
    (b) 10^{15} m
    (c) 10^{-12} m
    (d) 10^{-18} m

40. 10 W in dB is
    (a) 10 dB
    (b) 2 dB
    (c) 1 dB
    (d) 100 dB

41. 1 mW in dBm is
    (a) 10 dBm
    (b) 0
    (c) 30 dBm
    (d) 20 dBm

42. Terameter is
    (a) 10^{12} m
    (b) 10^{-12} m
    (c) 10^{9} m
    (d) 10^{-15} m

43. 100 mW in dBm is
    (a) 10 dBm
    (b) 50 dBm
    (c) 100 dBm
    (d) 20 dBm

44. In domestic houses, microwaves are used in
    (a) mixies
    (b) refrigerators
    (c) microwave ovens
    (d) hair driers

45. Nano means
    (a) 10^{-9}
    (b) 10^{9}
    (c) 10^{12}
    (d) 10^{15}

46. Exa means
    (a) 10^{15}
    (b) 10^{18}
    (c) 10^{12}
    (d) 10^{-18}

47. 10 mW in dB is
    (a) -20 dB
    (b) 30 dB
    (c) 20 dB
    (d) 10 dB

48. 1 kW in dBm is
    (a) 90 dBm
    (b) 100 dBm
    (c) 10 dBm
    (d) 20 dBm

49. The unit of permittivity is
    (a) F/m^2
    (b) F/m
    (c) H/m
    (d) F-m

50. The unit of permeability is
    (a) H/m
    (b) H/m^2
    (c) H-m
    (d) F/m

51. L-band frequency range is
    (a) 1 – 2 GHz
    (b) 2 – 4 GHz
    (c) 4 – 8 GHz
    (d) 8 – 12 GHz

52. X-band frequency range is
    (a) 2 – 4 GHz
    (b) 8 – 12 GHz
    (c) 12 – 18 GHz
    (d) 4 – 8 GHz

53. The magnetic field is
    (a) $H = \in (D \times E)$
    (b) $H = \mu E$
    (c) $H = \dfrac{1}{\mu} \nabla \times A$
    (d) $H = \in (E \times V)$

54. The electrical field is
    (a) $E = -\nabla V$
    (b) $E = \mu(V \times H)$
    (c) $E = \in H$
    (d) $E = \mu H$

55. If the output of a microwave amplifier is 125 W when the input is 45 W, power gain in dB is
    (a) 10 dB
    (b) 4.44 dB
    (c) 44.4 dB
    (d) 80 dB

56. If the input to an attenuator is 100 mW and output power is 3.5 mW, attenuation in dB is
    (a) – 14.6 dB
    (b) 14.6 dB
    (c) 1.46 dB
    (d) 15 dB

57. The difference between –15 dBm and 20 dBm is
    (a) 5 dB
    (b) 35 dB
    (c) 3.5 dBm
    (d) 20 dB

58. The time for a wave to travel one unit length of line is
    (a) $T(s) = \sqrt{LC}$
    (b) $T(s) = \sqrt{L/C}$
    (c) $T(s) = \dfrac{1}{\sqrt{LC}}$
    (d) $T(s) = \dfrac{L}{C}$

59. The time required for the wave to travel 1 m of length of line, for $L = 2$ μH/m, $C = 10$ pF/m, is
    (a) 20 ns
    (b) 2.0 ns
    (c) 4.47 ns
    (d) 44.7 ns

60. The reflection coefficient due to mismatch in a transmission line is
    (a) $\rho = \dfrac{Z_L - Z_0}{Z_L + Z_0}$
    (b) $\rho = \dfrac{Z_L + Z_0}{Z_L - Z_0}$
    (c) $\rho = Z_L / Z_0$
    (d) $\rho = Z_0 / Z_L$

61. The reflection coefficient in a transmission line, when $Z_L = 130\ \Omega$, $Z_0 = 75\ \Omega$, is
    (a) 0.268
    (b) 2.68
    (c) 0.18
    (d) 0.36

1.38 ANSWERS

1. a	2. b	3. d
4. a	5. b	6. a
7. b	8. a	9. c
10. a	11. c	12. b
13. a	14. a	15. b
16. b	17. b	18. b
19. a	20. a	21. a
22. d	23. a	24. b
25. a	26. a	27. b
28. b	29. a	30. b
31. b	32. a	33. a
34. b	35. a	36. a
37. a	38. a	39. a
40. a	41. b	42. a
43. b	44. c	45. a
46. b	47. a	48. a
49. b	50. a	51. a
52. b	53. c	54. a
55. b	56. a	57. b
58. a	59. c	60. a
61. a		

1.39 EXERCISE PROBLEMS

1. What is the phase constant of a wave if the frequency is 10 GHz
2. If $D = 4a_x + 5a_y$ in free space, find E.
3. If $H = 3a_x + 5a_y$, find B in free space.
4. Find E when the potential is given by $x + 5y^2 + 6z$.
5. Determine E at 5 C. If a force due to a charge of 2 C exists on 5 C charge.
6. Find the instantaneous power flow if $E = 6a_x + 7a_y$ and $H = 2a_x + 4a_x$.
7. Find the frequency of the wave if the time period is 2 ns.
8. What is the velocity of propagation of wave in a medium of $\epsilon_r = 9$, $\mu_r = 4$.
9. Find the wavelength if the phase constant is 0.1 rad/m.
10. Find the phase constant of a wave whose wavelength is 3 cm in free space.

Chapter 2

Microwave Tubes for Microwave Signal Generation

2.1 INTRODUCTION TO TUBES

The conventional vacuum tubes are diode, triode, tetrode and pentode. The vacuum tube diode contains a filament, cathode and anode. The triode contains 3 elements i.e. a cathode, control grid and an anode. The tetrode contains a cathode, two grids and an anode. The pentode contains 5 elements i.e. a cathode, 3 grids and an anode. The vacuum tube diode is used as a rectifier. However, triode, tetrode and pentode are used as amplifiers and oscillators. But they are useful at low microwave frequencies as they have some limitations at microwave frequencies.

2.2 LIMITATIONS OF CONVENTIONAL TUBES

The conventional tubes—triode, tetrode and pentode—are useful as signal sources for frequencies less than 1.0 GHz. But they cannot be used at microwave frequencies as they exhibit a few undesirable effects and limitations.

The main limitations of the conventional tubes are due to the presence of

> ➤ lead inductance effect
> ➤ interelectrode capacitance effect
> ➤ transit time effect
> ➤ gain-band width product limitation

The Lead Inductance

The lead inductance contains 3 components.

> ➤ Grid lead inductance, L_g
> ➤ Cathode lead inductance, L_c
> ➤ Plate lead inductance, L_p

These are shown in fig. 2.1.

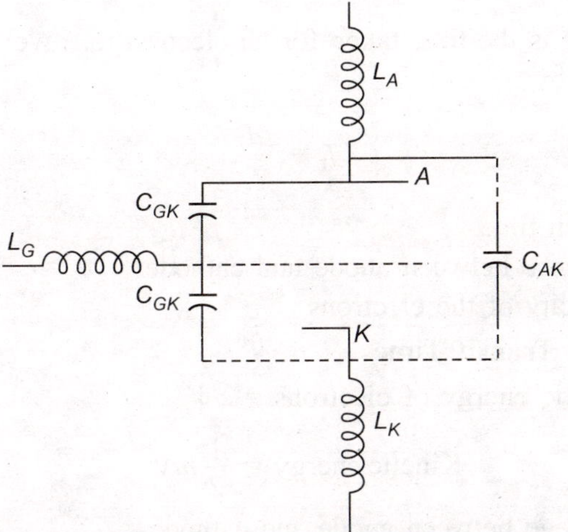

Fig. 2.1 Lead-inductances and capacitances in conventional tubes

Effect of Lead Inductance

The lead-inductance becomes large at microwave frequencies and it affects the input impedance of the device. This in turn reduces the gain of the tube amplifier.

Method of Minimization of Lead Inductance

The lead inductance can be minimized by

> ➤ reducing the lead length and
> ➤ the electrode area

The above methods, however, reduce power handling capacity.

The Lead Capacitance

The lead capacitance contains 3 components in a triode. They are

> ➤ plate-grid capacitance
> ➤ grid-cathode capacitance
> ➤ plate-cathode capacitance

These are shown in fig. 2.1.

Effect of Lead Capacitance

At high frequencies, the reactance of each lead capacitance decreases and hence output voltage decreases due to shunt effect.

Minimization of Lead Capacitance

The lead capacitance can be reduced by

> ➤ decreasing the electrode area
> ➤ increasing the distance between electrodes.

Transit–Time

Definition

The transit time is the time taken for an electron to travel from cathode to a anode. It is given by

$$t_t = \frac{d}{v_e}$$

Here, t_t = transit time

d = distance between anode and cathode

v_e = velocity of the electrons

Expression for Transit Time

Static energy of electrons = $e\,V_a$

Kinetic energy = $\frac{1}{2}mv^2$

Here, V_a = voltage between anode and cathode

e = electron charge

m = electron mass

Under equilibrium condition,

Static Energy = Kinetic Energy

i.e.

$$e\,V_a = \frac{1}{2}mv_e^2$$

$$v_e = \sqrt{\frac{2eV_a}{m}}$$

$\therefore$

$$t_t = \frac{d}{\sqrt{\dfrac{2eV_a}{m}}}$$

It may be noted that

$$v_e = 0.593 \times 10^6 \sqrt{V_a}$$

The Effect of Transit Time

The transit time effect reduces the efficiency of the tube.

At microwave frequencies, the transit time is large compared to the period of microwave signal. The signal between the cathode and grid changes several times during the transit of electron.

The positive half-cycle of the grid potential supplies the energy to the electrons. This energy is removed during the negative cycle. This results in the oscillation of electrons in the cathode-grid region. The electrons may also come back to the cathode. This process reduces the efficiency of the tube.

Minimization of Transit Time Effect

The transit time effect can be reduced by

➤ reducing the distance between the electrodes
➤ increasing the plate-cathode voltage, V_a

Gain-Bandwidth Limitation

Consider the equivalent circuit of a triode (fig. 2.2).

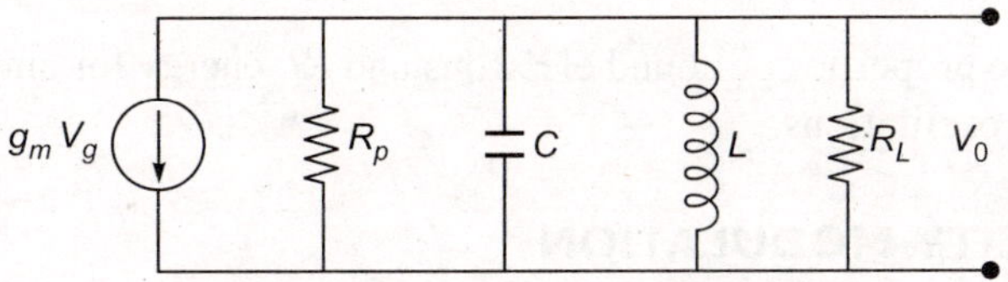

Fig. 2.2 Equivalent circuit of triode

From the analysis of this circuit, we get an expression for gain-bandwidth product as

$$G \times BW = \frac{g_m}{C}$$

From this expression, it is evident that gain-bandwidth product is independent of frequency. G can be increased only at the cost of bandwidth for fixed values of g_m and C.

Method for Large Gain with Large Bandwidth

It is possible to get large gain by

➤ using re-entrant cavities
➤ using slow wave tubes

2.3 MICROWAVE TUBES

The microwave signal generators are also called microwave sources. The signals at microwave frequencies are produced from active devices. These devices are classified into two groups.

1st Group

A few examples of this group are

Klystron
Magnetron
traveling wave tube
Backward wave oscillators

Salient Features of 1st Group of Active Devices

1. They use the properties of free electrons and d.c. energy for amplification and for producing oscillations.
2. The operation of these devices depends on a process called 'velocity modulation'.

3. The velocity modulation is obtained by the acceleration and retardation of free electrons.

4. The acceleration and retardation of free electrons are obtained by the applied alternating voltage between the electrodes or electric field.

5. Velocity modulation produces a density modulated stream of electrons.

2nd Group

They use the properties of bound electrons and *RF* energy for amplification and for producing oscillations.

2.4 VELOCITY MODULATION

Definition

> The velocity modulation is a modulation process in which the velocity of electrons is varied and controlled, keeping the number of electrons in the beam constant.

2.4.1 Method of Producing the Velocity Modulation

A typical tube in which velocity modulation is produced is shown in fig. 2.3.

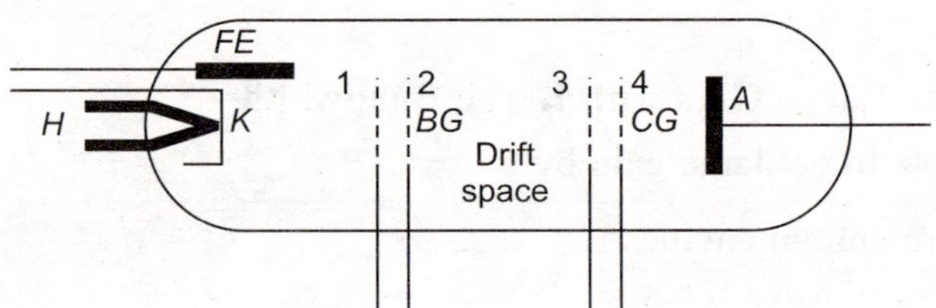

Fig. 2.3 A typical microwave tube in which velocity modulation is produced

H : Heater, FE : Focusing electrode, K : Cathode, A : Anode,
BG : Buncher grid, CG : Cathode grid

➢ When the cathode is heated, a beam of electrons is produced. It is focused by focusing electrode.

➢ The anode is maintained at positive voltage with respect to cathode.

➢ The beam of electrons move towards the anode.

➢ An *r-f* signal is given to buncher grids. The buncher grids are closely spaced and hence the transit time is very small.

➢ If the second buncher grid is positive with respect to 1st, the electrons are accelerated.

➢ If the second buncher grid is negative with respect to 1st grid, the electrons are slowed.

➢ In the drift space, the electrons travel with different velocities.

➢ It is obvious that the fast electrons overtake the slow ones and hence bunches of electrons are created.

➤ The bunching becomes maximum at some point in drift space.

➤ When the electron bunches pass between the catcher grids, a voltage appears across them. This voltage is proportional to the number of electrons.

➤ As the number of electrons varies, the voltage across catcher grids also varies at the rate of radio frequency.

➤ As the bunching at catcher grids is greater than that at buncher grids, the r-f voltage at the catcher grids is more than the applied r-f voltage. That means, amplification is produced to the extent of about 1000.

The frequency of operation is increased by minimizing lead inductances stray capacitances and transit time. The lead inductance is reduced by reducing lead length with large diameter. The stray capacitance is minimized by modifying the physical configuration of the anode, cathode and grid.

The transit time is reduced by reducing the distance between anode and cathode by increasing anode-cathode voltage.

All the above modifications made to increase the operating frequency reduce the power ratings of the tube. Hence, the triodes are not preferred for use at microwave frequencies.

2.5 KLYSTRON

The klystron is a microwave electronic device and it is basically a vacuum tube which operates on the concept of velocity modulation of electrons.

The klystron can be of two-cavity or a multicavity type.

2.5.1 Two-Cavity Klystron

A two-cavity klystron is used as an amplifier. A typical structure is shown in fig. 2.4.

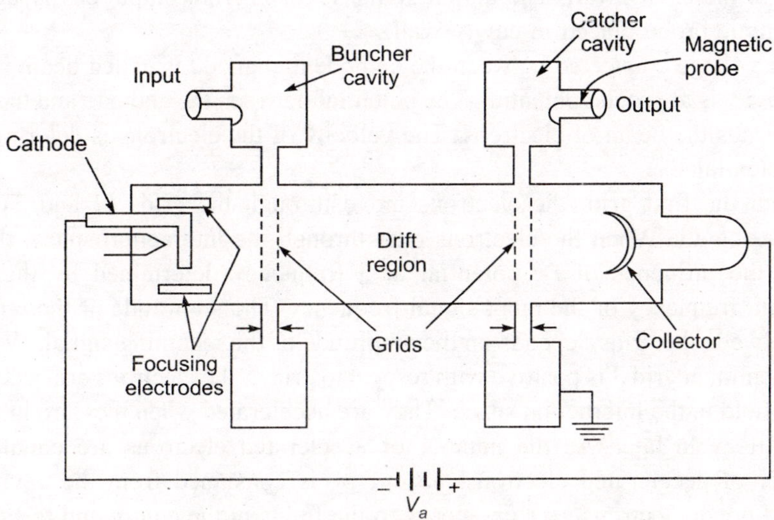

Fig. 2.4 Two-cavity klystron amplifier

2.5.2 Parts in Two-Cavity Klystron Amplifier

It consists of the following parts.

> ➢ Cathode
> ➢ Collector
> ➢ A buncher cavity
> ➢ Catcher cavity
> ➢ A drift region
> ➢ Focussing electrodes
> ➢ Cavity interaction regions
> ➢ Input port
> ➢ Output port
> ➢ Grids
> ➢ Magnetic probes

2.5.3 Principle of Operation

A d.c. voltage is connected between the anode and cathode. This voltage is called beam voltage. The cavity is maintained at ground potential and hence the voltage between the cathode and cavity accelerates the beam of electrons coming from cathode. The velocity of the electrons depends on this potential. The electrons are under the influence of r-f signal given at the input port.

The buncher cavity is an input cavity and the catcher cavity is an output cavity. The region between these cavities is known as drift region. The electron beam moves from cathode to anode. The external magnetic field sets the beam down at the center of the tube. The r-f signal is coupled to the input cavity by a magnetic probe in the cavity wall. The output signal is taken from output cavity again by a magnetic probe placed in cavity wall.

The voltage connected between the cathode and anode is called beam voltage. The cavity is at ground potential. The potential between the cathode and the cavity accelerates the beam of electrons. The velocity of the electrons is determined by this potential.

From the first grid, the electrons move through the grids (2 and 3) of the buncher cavity. When the electrons pass through the interaction space, they are under the influence of r-f potential at a frequency determined by the cavity resonant frequency or the input signal frequency. The amplitude of the r-f potential between the grids depends on the amplitude of the incoming signal. When the r-f potential at grid 3 is positive with respect to grid 2, the electrons are accelerated in the field in the interaction space. They are accelerated when they are in the gap one-half cycle later. As the number of accelerated electrons are equal to the number of decelerated electrons, no energy is consumed from the cavity. The slowed down electrons give up energy to the fields in the cavity and the speeded up electrons absorb energy from the fields in the cavity.

From the interaction space, the electrons enter the drift space (bunching space). Here, speeded up electrons overtake the slower ones and form bunchers producing density modulation.

A second cavity resonator (catcher) is located at the place of maximum bunching. This cavity is tuned to the frequency equal to that of input cavity resonator. The output is obtained by slowing down the electron bunches.

With an alternating field at the output cavity resonator, if the grid 4 is positive with respect to grid 5, the bunches of electrons are slowed down and deliver energy to the output cavity. The output power is the difference in kinetic energy of the electrons averaged before and after passing the interaction space.

The anode collects electrons. The secondary emission may exist due to the impact of the electrons on the end wall.

Microwave signal is applied to the first cavity, output of the second cavity is taken. The output is greater than the input and hence there is amplification. If the bunching is narrow, more energy is released to the second cavity. Klystrons are used as amplifiers up to 25 GHz.

2.5.4 Applications of Klystron

The klystron has the following applications.

1. It is used as a power amplifier
2. It is used as a frequency multiplier

2.6 REFLEX KLYSTRON

The reflex klystron is a vacuum tube microwave signal generator and its operation depends on the principle of velocity modulation, and transit time. In this, the feedback is re-entrant.

2.6.1 Construction

The reflex klystron has the following components:

- ➤ cavity
- ➤ filament
- ➤ cathode
- ➤ repeller
- ➤ output port
- ➤ focusing electrode
- ➤ control grids

The schematic diagram of reflex klystron is shown in fig. 2.5.

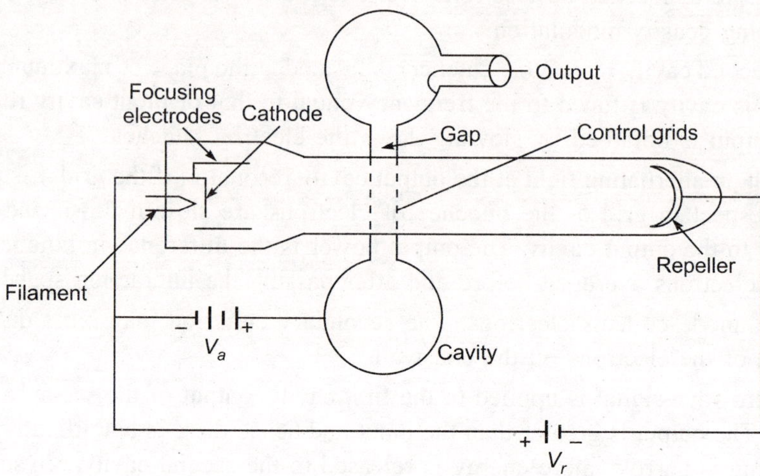

Fig. 2.5 Schematic diagram of reflex klystron

The cavity is at a positive potential and the reflector electrode is at a high negative potential.

2.6.2 Principle of Operation

The cathode emits a steam of electrons and a beam is formed by the focusing electrodes. The beam is accelerated towards cavity. The velocity of the electrons in the beam is given by

$$v_e = \sqrt{\frac{2eV_a}{m}}$$

Here, e = charge of electron

V_a = acceleration voltage or beam voltage

m = mass of electron

The electrons travel towards the reflector electrode. However, the electrons do not reach the reflector as it is at a negative potential. These electrons repel back towards the gap. The reflected electrons enter the cavity which now acts as a catcher for these reflected electrons. At the same time, the cavity acts as a buncher for the new electrons. The point at which the electrons are repelled back is controlled by reflector voltage. The repelling of electrons becomes fast if the reflector voltage is more negative.

The beam of electrons passing through the interaction region between the grids is velocity modulated. The velocity of this beam is given by

$$v = v_0 \left[1 + \frac{V_1}{V_2} \sin \omega t \right]^{\frac{1}{2}}$$

Here, V_1 is $\sin \omega t$ is an oscillating voltage caused by a small disturbance in the electron beam.

The velocity modulated beam leaving the interaction region is repelled by the negative voltage on the reflector electrode (mode) and returned through the grid pair in the reverse direction. The energy of the beam is released to the cavity as it decelerates on its return through the interaction region. The energy is added to the cavity and sustained oscillations are produced. The output is taken from the cavity using a probe.

The reflector voltage is adjusted to repel the beam into the interaction space when the electric field opposes it. When the beam is slowed down, the kinetic energy is released.

The output power is maximum when the bunching of the reflected beam at the catcher is complete. This condition occurs at more than one reflector voltage. These reflector voltages are called repeller modes. The maximum output power in the klystron occurs when the reflector voltage is at the center of the mode. For each frequency of operation of the klystron, there exist different sets of reflector modes.

The frequency of oscillations is controlled by the cavity dimensions, and reflector voltage. But the frequency and output power are controlled by reflector voltage to a small extent. The resonant frequency is varied to a small extent by moving the cavity mechanically in and out.

Frequency modulation of the reflex klystron is obtained by changing reflector voltage with the modulating signal. On the other hand, pulse modulation is obtained by switching the reflector voltage between two voltages V_1 and V_2 as in fig. 2.6.

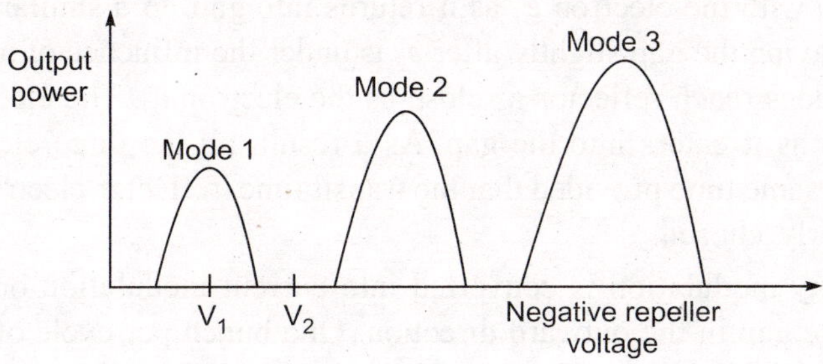

Fig. 2.6   Mode voltage selection for pulse modulation of klystron

The reflex klystron provides the followings:

➢ Output power less than 500 mW.

> ➤ Efficiency of about 25%

> ➤ The operating frequency is below 16 GHz.

2.6.3 Velocity Modulation in Reflex Klystron

Velocity modulation diagram is also called Applegate diagram. This is shown in fig. 2.7.

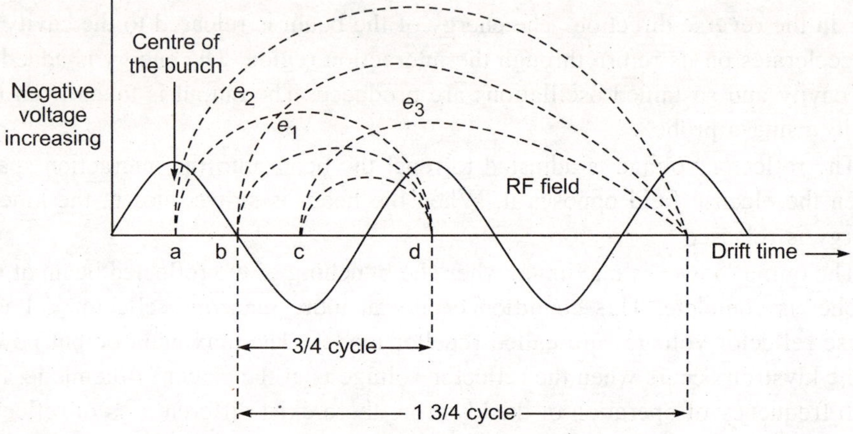

Fig. 2.7 Applegate diagram of reflex klystron

In the figure, the paths of electrons e_1, e_2 and e_3 are shown.

Let e_1 be a reference electron. It passes the gap towards the reflector and there is no effect of the gap voltage and is returned to the anode without reaching the reflector. Now consider an electron, e_2 which passes the gap slightly before e_1. If there is no gap voltage, e_2 returns back before e_1. But RF voltage velocity modulates the electron. The electron e_2 accelerates by the influence of positive voltage and it moves more close to reflector than e_1. However, it is possible for e_2 to catch up with the electron e_1 as it returns into gap. In a similar manner, the electron e_3 leaving the gap slightly after e_2 is under the influence of negative field and hence it does reach reflector as close as the electron e_2. The electron, e_3 also catches up e_2 as it enters into the gap. As a result, all the three electrons reach the gap at the same time provided that the transit time, reflector electrode potential etc are properly chosen.

The velocity modulation is converted into current modulation once the electrons leave the gap in the outward direction. One bunch per cycle of oscillations is formed around a reference electron. The energy is delivered to the gap by these bunches.

As few electrons arriving at the gap are out of phase, high noise is present and hence efficiency is less.

2.6.4   Transit Time (T_t)

> The transit depends on reflector and anode voltages.

Transit time is defined as the time taken for the electron to travel into reflector space and back to the gap. Sustained oscillations are obtained when the transit time is optimum. The optimum departure time is on the reference electron, e_2 which is at π radian point of sine wave voltage across the resonator gap.

The accelerated electrons takes some energy and decelerated electrons release some energy. So the net energy is zero.

When the gap voltage is a positive maximum, the maximum retardation takes place. This results in the electrons to fall through maximum negative voltage across the gap which in turn gives up the maximum amount of energy to the gap. The best time, as it appears, for electrons to return to the gap is at the $T/2$ radian point of the sinusoidal gap voltages. The electron returning after $1^3/4$ cycles satisfies these requirements.

The transit time in reflector space is given by

$$T_t = n + \frac{3}{4}$$

Here, n = integer

2.6.5   Voltage Characteristics of Reflex Klystron

The voltage characteristics of reflex klystron are the variation of reflector voltage and beam voltage for sustained oscillations with transit time as a variable.

Typical characteristics are shown in fig. 2.8.

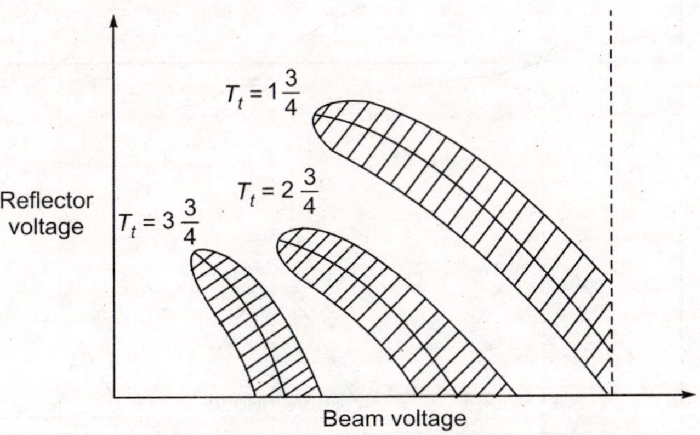

Fig. 2.8   Voltage characteristics of reflex klystron

In fig. 2.8, thick lines represent optimum combinations of beam and reflector voltages for sustained oscillations. The shaded areas represent possible combinations.

The transit time is represented by $T_t = \left(n + \dfrac{3}{4}\right)$, Here, $n = 1, 2, 3, \ldots$ represents different modes. For lower modes, the output power is high. But it requires high voltages. At high modes, the efficiency becomes low and there exist insulation problems due to high voltages. In view of this, the modes corresponding to $n = 2$ and $n = 3$ are extensively used.

2.6.6 Modes in Reflex Klystron

Transit time depends on beam and reflector voltages. The cavity is tuned to a required frequency. Both the voltages are adjusted to provide correct value of transit time from the data supplied by the manufacturer. Several combinations of beam-reflector voltages provide oscillations for the particular value of n. Each value of n corresponds to a different mode. Modes corresponding $n = 2$ and $n = 3$ are often used for optimum efficiency.

2.6.7 Power-Frequency Characteristics

The variations of frequency and output power as a function of reflector voltage with modes as variable are shown in fig. 2.9. The power characteristics are often called mode curves of reflex klystron. The variation of reflector voltage changes the frequency marginally. This corresponds to electronic tuning of the device. As a result the reflex klystron is used as voltage-tuned oscillator or frequency modulated oscillator.

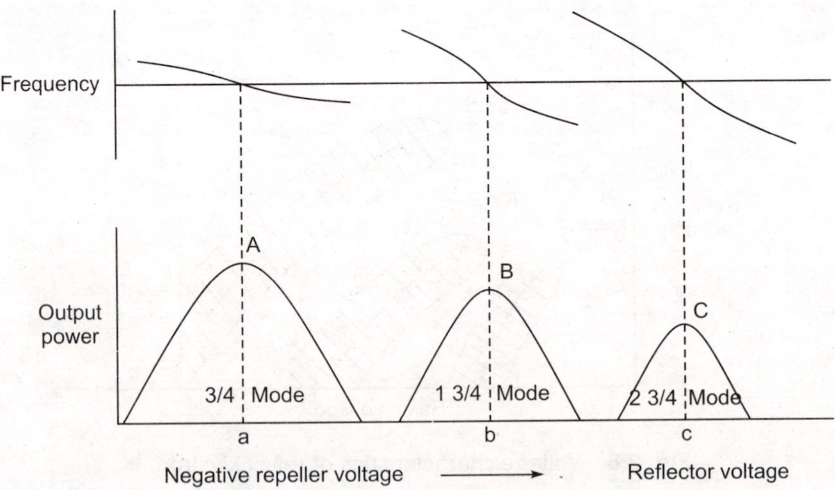

Fig. 2.9 Variation of frequency and power in the reflector voltage

2.6.8   Analysis of Reflex Klystron

The velocity of accelerated electron beam is given by

$$v_0 = 0.593 \times 10^6 \sqrt{V_a}$$

Here, V_a = anode voltage

The velocity with which the electron leaves the gap at an instant t_1, is given by

$$v = v_0(1 + m \sin \omega t_1)$$

Here, $m = \dfrac{V_1}{V_a}$ = depth of modulation

V_1 = r-f voltage
V_a = accelerating voltage

This expression is valid for negligible transit time across the gap.
For a finite transit time, it is given by

$$v = v_0 \left(1 + \frac{C_c m}{2} \sin \omega t_1 \right) \qquad (2.1)$$

Here, C_c = beam coupling coefficient

$$= \frac{\sin(\omega T_t/2)}{(\omega T_t/2)}$$

T_t = transit time across the gap
ωT_t = electron transit angle across the gap

The expression (2.1) indicates velocity modulation

If $\omega t_1 = 2\pi n$, $n = 0, 1, 2, 3, \ldots$, there will be no change in velocity. However, at different instants of time, electrons leave the gap with different velocities in repeller space.

Let ℓ be the distance between resonator and repeller and potential difference be $(V_r - V_a)$. Let the exit plane of the gap be at $x = 0$. Then the force on electron in the repeller space is given by

$$m_e \frac{\partial^2 x}{\partial t^2} = \frac{e(V_r - V_a)}{\ell}$$

or $\qquad \dfrac{d^2 x}{dt^2} = \left(\dfrac{e}{m_e} \right)\left(\dfrac{V_r - V_a}{\ell} \right) \qquad (2.2)$

Here, e = electron charge
m_e = mass of the electron

∴ Integrating both sides twice, we get

$$x = K_e \frac{(V_r - V_a)}{\ell}(t - t_1)^2 + v(t - t_1)$$

Here, $K_e = \dfrac{e}{m_e} = 1.76 \times 10^{11}$ coulomb/kg

If the electron returns back to $x = 0$ at $t = t_2$, the total transit time is

$$(t_2 - t_1) = -\frac{2\ell v}{K_e(V_r - V_a)} \tag{2.3}$$

or

$$t_2 = t_1 - \frac{2\ell v_0}{K_e(V_r - V_a)}\left(1 + \frac{C_c m \sin \omega t_1}{2}\right)$$

i.e.

$$t_2 = t_1 + T_0\left(1 + \frac{C_c m}{2}\sin \omega t_1\right)$$

Here,

$$T_0 = -\frac{2\ell v_0}{K_e(V_r - V_a)} \tag{2.4}$$

This is the total transit time of an electron which leaves the gap when RF voltage is zero. This electron does not experience any change in its velocity, v_0. Such an electron is known as center of bunch electron.

The phase angle at t_2 is

$$\omega t_2 = \omega t_1 + \omega T_0\left(1 + \frac{C_c m}{2}\sin \omega t_1\right)$$

$$= \omega t_1 + \theta_0\left(1 + \frac{C_c m}{2}\sin \omega t_1\right)$$

Here, $\theta_0 = \omega T_0$. This is the total transit angle of center of bunch angle.
The optimum transit time T_0 is given by

$$T_0 = \left(n + \frac{3}{4}\right)T_t \tag{2.5}$$

and the optimum transit angle is

$$\theta_0 = 2\pi\left(n + \frac{3}{4}\right) \tag{2.6}$$

It is evident that there exist several values for total transit time and total transit angle to satisfy the conditions for oscillations.

Now consider the following condition for oscillations

$$T_0 = -\frac{2\ell v_0}{K_e(V_r - V_a)} = \left(n + \frac{3}{4}\right)T_t$$

As $v_0 = \sqrt{2K_e V_a}$, we have

$$\frac{v_0}{(V_r - V_a)} = \frac{K_e}{2\ell}\left(n + \frac{3}{4}\right)T_t$$

or

$$\frac{\sqrt{2K_e V_a}}{(V_r - V_a)} = \frac{K_e}{2\ell}\left(n + \frac{3}{4}\right)T_t$$

$$\frac{2K_e V_a}{(V_r - V_a)^2} = \frac{K_e^2}{4\ell}\left(n + \frac{3}{4}\right)^2 T_t^2$$

or

$$\frac{V_a}{(V_r - V_a)^2} = \frac{K_e}{8\ell}\left(n + \frac{3}{4}\right)^2 T_t^2$$

$$(V_r - V_a)^2 = \frac{V_a \, 8\ell f^2}{K_e\left(n + \frac{3}{4}\right)^2}$$

or

$$V_r - V_a = \frac{6.744 \times 10^{-6} \, \ell f \sqrt{V_a}}{\left(n + \frac{3}{4}\right)}$$

$$\boxed{V_r = \frac{6.744 \times 10^{-6} \, \ell f \sqrt{V_a}}{\left(n + \frac{3}{4}\right)} + V_a}$$

It is evident from the above expression, V_r is high when n is small.

For a fixed ℓ and V_a, the repeller voltage required for oscillation can be obtained or for a fixed V_r and V_a, ℓ can be determined.

2.6.9 Power Output

Let $I_1 \, dt_1$ is the net charge leaving the resonator cavity gap at a time interval of dt_1. Let it arrive back cavity gap at a time interval of dt_2. Let it induce a current I_2. Then by the law of conservation of charge, we have

$$I_1 \, dt_1 = I_2 \, dt_2$$

or

$$I_2 = I_1 \frac{dt_1}{dt_2}$$

As

$$t_2 = t_1 + T_0 \left(1 + \frac{C_c m}{2} \sin \omega t_1 \right)$$

$$I_2 = \frac{I_1}{(1 + C_c\, m\, \theta_0 \cos \omega t_1)}$$

$$\approx I_1 \left[1 + \frac{C_c m \theta_0}{2} \cos \theta t_1 \right]$$

or

$$I_2 \approx I_1 [1 + X \cos \omega t_1]$$

Here,

$$X = -\frac{C_c\, m\, \theta_0}{2}$$

$$= -\theta_0\, C_c\, \frac{V_1}{2V_a}$$

X is known as bunching parameter.

The fundamental component of the induced current in the cavity by density modulated electron beam is given by

$$I_f = 2I_1\, J_1(X)$$

Here, $J_1(X)$ is Bessel function of the first kind.

If a beam voltage V_a supplies a d.c. power, P_{dc}, then

$$P_{dc} = I_1\, V_a$$

and $r\text{-}f$ power supplied to the load,

$$P_{RF} = \frac{V_1 I_f}{2} = V_1\, I_1\, J_1(X)$$

$\therefore$

$$\frac{V_1}{V_a} = \frac{2(X)}{(2\pi n + 3\pi/2)}$$

or

$$V_1 = \frac{2(X)V_a}{(2\pi n + 3\pi/2)}$$

$$P_{RF} = \frac{2I_1 V_a\,(X)\, J_1(X)}{2\pi(n + 3/4)}$$

$\therefore$

$$P_{RF} = \frac{I_1 V_a\,(X)\, J_1(X)}{\pi(n + 3/4)}$$

From the variation of $X\, J_1(X)$ with X, it is found that $[X\, J_1(X)]_{\max} = 1.25$ at $X = 2.408$ and $J_1(X) = 0.52$.

Hence
$$(P_{RF})_{\max} = \frac{0.3986 I_1 V_a}{\left(n + \frac{3}{4}\right)}$$

If
$$N = \left(n + \frac{3}{4}\right)$$

$$\boxed{(P_{RF})_{\max} = \frac{0.3986 I_1 V_a}{N}}$$

2.6.10 Efficiency of Reflex Klystron

Efficiency is defined as $\eta = \dfrac{P_{RF}}{P_{dc}}$

$$= \frac{X J_1(X)}{\pi \left(n + \frac{3}{4}\right)}$$

$$\boxed{(\eta)_{\max} = \frac{0.3986}{\left(n + \frac{3}{4}\right)}}$$

If $n = 1$, $n + \dfrac{3}{4} = 1\dfrac{3}{4}$

This gives maximum $(P_{r-f})_{\max}$

$$\therefore \quad (P_{r-f})_{\max} = 0.227 I_1 V_a$$

$$\eta_{\max} = \frac{(P_{RF})}{P_{dc}}$$

$$= 0.227$$

$$\boxed{\eta_{\max} = 22.7\%}$$

PROBLEM 2.1 A reflex klystron operates at a frequency of 9 GHz with a beam voltage of 300 V. The repeller space is 20 mm for $1\dfrac{3}{4}$ mode. Find out the maximum power when the beam current is 20 mA.

Solution The maximum *r-f* power is

$$(P_{r-f})_{\max} = \frac{0.3986 I_1 V_a}{\left(n + \frac{3}{4}\right)}$$

$$= \frac{0.3986 \times 300 \times 20 \times 10^{-3}}{(1.75)}$$

$$(P_{r-f})_{max} = 1.366 \text{ watts.}$$

2.7 TRAVELING WAVE MAGNETRONS

The magnetron is a high power microwave oscillator. It is a vacuum tube and contains a cathode and an anode with several resonant cavities.

2.7.1 Principle of Operation

It operates on the principle of feedback. The feedback is interactive and re-entrant. The interaction is by the cross-field.

The cross-section of a magnetron is shown in fig. 2.10. The cathode is a rod in the centre of the tube and anode is a solid block. The anode contains the several resonant cavities. The space between cathode and anode is the interaction space.

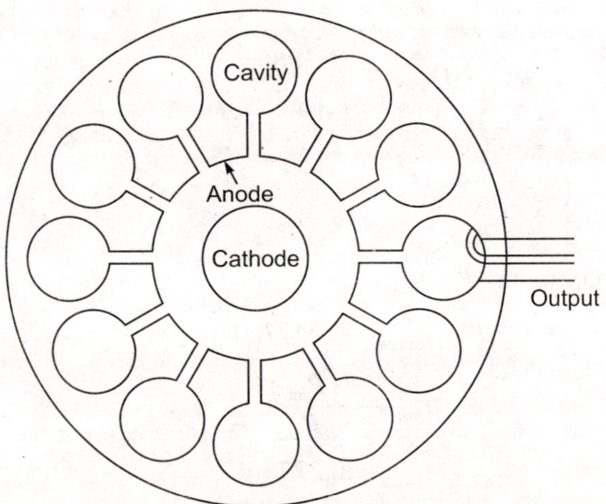

Fig. 2.10 Magnetron cross-section

A magnetic field is applied with flux lines parallel to the axis of the cathode. The poles of the magnet are at each end of the cathode and outside the vacuum. The field is provided by a horseshoe magnet or by an electromagnet.

The electrons from the hot cathode reach the anode radially. The electrons describe arcs by the applied critical magnetic field.

These electrons graze the anode and bend back towards the cathode. The electrons have spiral motion at higher values of the magnetic field.

When an electron approaches the anode in phase with the RF signal, it completes a cycle. The total phase shift around the periphery of the anode must be a multiple of 2π. The anode contains even number of cavities. The number of frequencies generated depends on the number of cavities.

An electron gives energy to one part of the anode and drop to a smaller orbit. On its return to the anode, it approaches adjacent section or a section away from it.

These are called different modes of operation and produce different frequencies. The adjacent anode poles have a phase difference of π radians and it is called the π mode.

The frequency separation is achieved by strapping. In this method, two rings are connected to the ends of the anode and each ring is connected only to alternate anode poles.

Each ring is at a constant potential at the π mode and the two rings have appropriate potential. Due to this, it results in capacitive loading to the cavities which reduces the frequency of this mode.

On the other hand, inductive loading exists for other modes as each ring is not at a uniform potential. The inductive loading raises the frequencies for other modes. Thus the π mode is separated from others.

These are most widely used devices in centimeter-to-millimeter microwave region. It has an efficiency of about 70%. Average power produced by this type of magnetron is a kW and peak powers of up to 40 kW are produced.

Common magnetrons are

- Cyclotron-frequency magnetron
- Split ring or negative resistance magnetron
- Traveling-wave magnetron

2.7.2 Cyclotron Frequency Magnetron

This contains a cylindrical anode, axial cathode with a perpendicular d.c. magnetic field. It uses transit time of the electron between the cathode and anode.

2.7.3 Operation

The emitted electrons from the cathode are accelerated towards the anode by the magnetic field. The field exerts a force on the electrons and they take curved path. When the field is increased, the electrons will come back to the cathode in a circular path.

The device will oscillate at a frequency and the frequency is determined by the time of travel of the electron.

This type of magnetron is useful at the lower end of microwave spectrum.

2.7.4 Split-ring or Negative Resistance Magnetron

This contains an anode split into two anodes of different potentials (fig. 2.11).

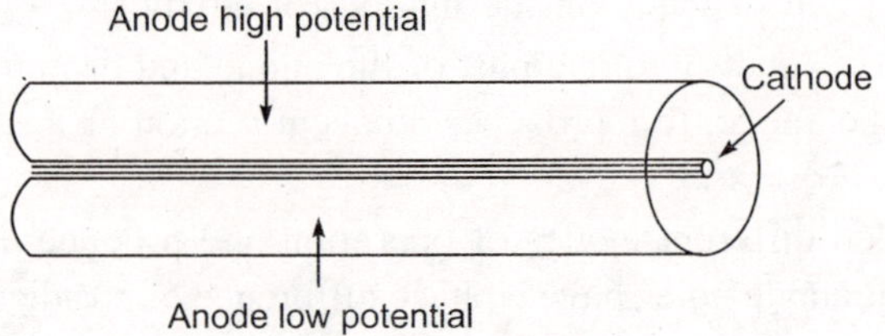

Fig. 2.11 Split-ring or negative resistance magnetron

The electron from the cathode is accelerated to the anode of high potential. It takes a curved path by the perpendicular magnetic field and the electron travel in smaller circles. The circle ends on the anode of lower voltage.

The interaction of electric and magnetic fields on the electron results in negative resistance which in turn creates oscillations in magnetron.

2.7.5 Applications

These are used in

- Radar transmitters
- Microwave ovens
- Industrial heating

2.7.6 Electron Trajectory in Magnetron

The trajectory of electron depends on Lorentz force equation given by

$$F = -e\,(E + V \times B)$$

and the cyclotron frequency is given by

$$\omega_c = eB/m$$

Here, F = force on the electron

E = electric field

$B = \mu H$, H is magnetic field

μ = permeability

m = mass of the electron

2.7.7 Formula for Electron Motion

The electron motion in magnetron is described by cyclotron angular frequency and it is given by

$$\omega_c = \frac{e}{m} B_m$$

Here, $\frac{e}{m} = 1.759 \times 10^{11}$ C/kg

B_m = Magnetic flux density

Depending on the magnetic field and anode voltage, the electron acquires a tangential and radial velocities.

The hull cut-off magnetic equation is given by

$$B_c = \frac{\left(8V_a \frac{m}{e}\right)^{1/2}}{b\left(1 - \frac{a^2}{b^2}\right)}$$

and cut-off voltage is given by

$$V_c = \frac{e}{8m} B_m^2 b^2 \left(1 - \frac{a^2}{b^2}\right)^2$$

Here, V_a = anode voltage

a = radius of cathode cylinder

b = radius of vane edge

2.8 SLOW WAVE DEVICES

A slow wave device is a device in which an electron beam is velocity modulated without a resonant cavity.

The principle involved in this device is the interaction of electron beam continuously with a weak electric field which is traveling along the beam.

A typical slow wave structure is shown in fig. 2.12.

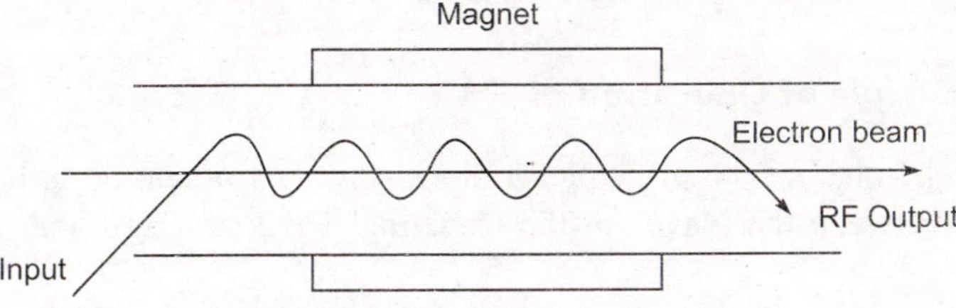

Fig. 2.12 Slow wave structure.

In this, the electrons in the electron beam have the same velocity. The external magnetic field confines the beam to the centre and along the axis of the tube. The beam travels through the center of helical conductor.

The wave interacts with the electron beam continuously if their velocities are equal. The wave velocity is equal to the speed of light. But the axial component of the wave has the velocity less than that of light and hence it is called 'slow wave'.

The interaction between the wave and the beam becomes continuous if the velocity of the electron is made slightly greater than the axial velocity of the helix wave. The interaction region is within the helix. As the fast moving electrons encounter the retarded axial helix wave, they give up energy to the helix due to their reduced speed. If the electrons encounter accelerating axial component of the wave, they take energy from the helix.

This process results in velocity modulation of the beam and hence bunching is formed.

2.8.1 Advantages of Slow Wave Devices

* Better amplification
* Wide-band characteristics due to non-resonant nature

2.8.2 Examples of Slow Wave Devices

* Traveling wave tubes (TWTs)
* Backward wave amplifiers (BWAs)
* Backward wave oscillators (BWOs)
* Coupled cavity traveling wave tubes (CCTWTs)

2.9 TRAVELING WAVE TUBE

The traveling wave tube is a broadband slow wave device. Its operation is based on the interaction between the waves in the traveling wave structure and the electron beam.

It consists of (i) an electron beam and (ii) a structure supporting a slow electromagnetic wave.

2.9.1 Principle of Operation of TWT

The TWT operates on the principle of slow wave. Its operation is based on the interaction between the waves in the traveling wave structure and the electron beam.

It consists of

* two slow-wave structures along the helix
* the RF input near the cathode
* the RF output near the collector

- magnet
- electron beam
- cathode focusing plates

The structure of TWT is shown in fig. 2.13.

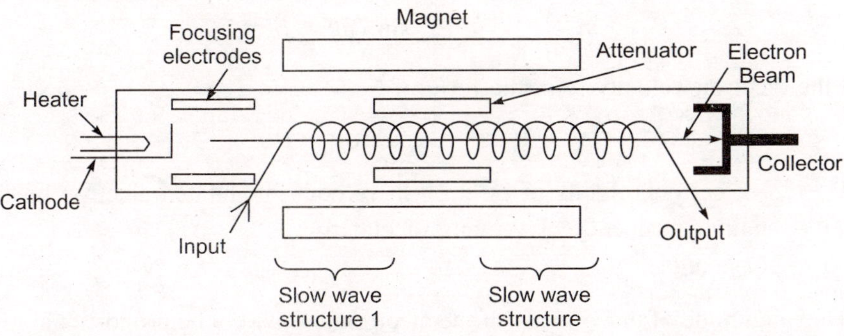

Fig. 2.13   TWT structure

The released electron beam from the cathode is focused by focusing electrodes. The collector collects the electrons. The attenuator shown isolates the input wave structure from the output wave structure. The magnet provides the magnetic field which continues the electron beam in the helix.

The beam velocity is made greater than the velocity of the axial electric field of helix wave. The interaction takes place between them. When the RF input is applied, interaction with the beam takes place until it reaches the attenuator. After the attenuator, the velocity modulation becomes current modulation in the output slow wave structure. This results in amplified output at the same frequency.

The interactions between electron beam and electric field is described in fig. 2.14.

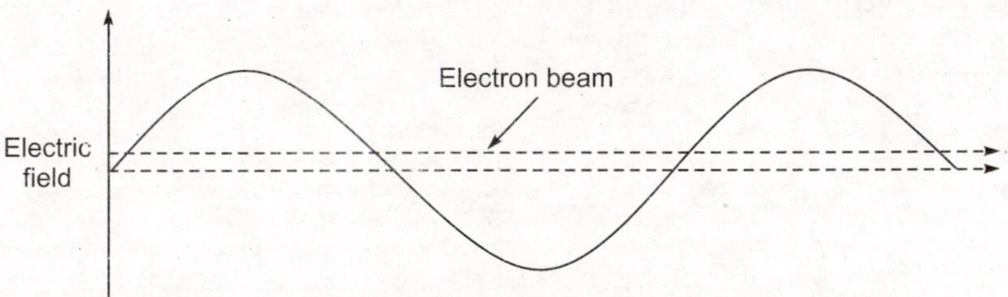

Fig. 2.14   Interaction of electron beam with electric field

If the traveling wave propagates along z-direction, the z-component of electric field is

$$E_z = E_m \sin (\omega t - \beta_a z)$$

Here, E_m = magnitude of electric field

$$\beta_a = \omega/v_a$$

v_a is the axial phase velocity of the wave

The equation of motion of the electron is found to be

$$m\frac{dv}{dt} = -e\,E_m\,\sin\,(\omega t - \beta_a z)$$

and the electron velocity is in the form of

$$v = v_{dc} + v_e\,\cos\,(\omega_e t + \theta_c)$$

v_{dc} = dc electron velocity of electron in velocity modulated electron beam

ω_c = angular frequency of velocity of electron

θ_c = phase angle

The magnitude of the velocity of the electron is found to be proportional to the magnitude of axial field. That is,

$$v_e = \frac{e\,E_m}{m\,\omega_e}$$

The conversion current in the electron beam induced by axial electric field is given by

$$i = j\frac{\beta_e\,I_{dc}}{2V_{dc}\,(j\beta_e - \gamma)^2}\,E_m$$

Here, $\beta_e = \omega/v_{dc}$

$$v_{dc} = \sqrt{(2e/m)V_{dc}}$$

The above equation for i is called **electronic equation.**

The equivalent circuit of a slow wave is given in fig. 2.15.

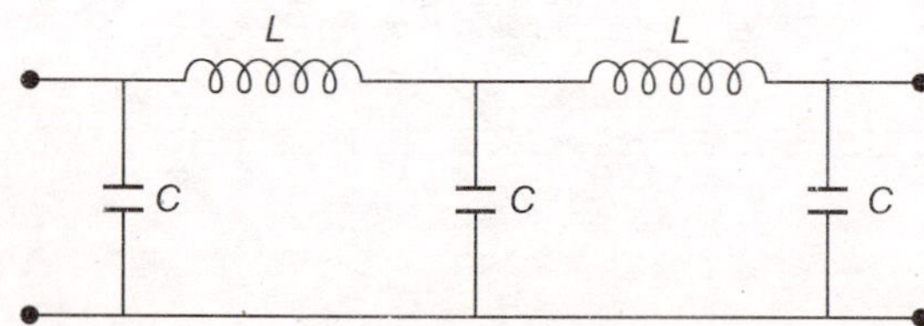

Fig. 2.15 Equivalent circuit of slow-wave helix

The axial electric field is given by

$$E_m = -\frac{\gamma^2\,\gamma_0\,Z_0}{(\gamma^2 - \gamma_0^2)}\,i$$

Here, $\gamma_0 = j\omega\sqrt{LC}$

$\qquad Z_0 = \sqrt{LC}$

The above equation for E_m is called **circuit equation.**

The equation describes the effect of electron beam current on the axial electric field.

2.9.2 Gain Parameter of TWT

The gain parameter of TWT is defined as

$$G \equiv \left(\frac{I_{dc}Z_0}{4V_{dc}}\right)^{\frac{1}{3}}$$

2.9.3 Output Power Gain, A_P

It is defined as

$$A_p \equiv 10\ \log\left|\frac{V(\ell)}{V(0)}\right|^2$$

$$= -9.54 + 47.3\ LC,\ dB$$

Here, LC is numerical number

L = circuit length

C = gain parameter

$V(\ell)$ = the amplitude of the output voltage

$V(0)$ = voltage at $Z - 0$.

PROBLEM 2.2   The parameters of a TWT are beam voltage, $V_{dc} = 2.5$ W, Beam current $I_{dc} = 25$ mA. The characteristics impedance of the helix, $Z_0 = 10\ \Omega$. It is operated at a frequency of $f = 9.5$ GHz, circuit length = 40. Find out the gain parameter, output power gain and β_e.

Solution   The gain parameter, $G = \left(\dfrac{I_{dc}Z_0}{4V_{dc}}\right)^{\frac{1}{3}}$

$$= \left(\frac{25\times10^{-3}\times10}{4\times2.5\times10^3}\right)^{\frac{1}{3}} = \left(\frac{10^{-4}}{4}\right)^{\frac{1}{3}}$$

$\therefore\qquad\qquad\boxed{G = 0.0292}$

The output power gain

$$A_p = -9.54 + 4.73 \ LC$$
$$= -9.54 + 47.3 \times 40 \times 0.0292$$

$$\therefore \quad \boxed{A_p = 45.7 \ dB}$$

$$\beta_e = \frac{\omega}{v_e}$$

Here, $\omega = 2\pi f$, $f = 9.5$ GHz

$$v_{dc} = 0.593 \times 10^6 \sqrt{V_{dc}}$$
$$V_{dc} = 2.5 \times 10^3 \ V$$

i.e.
$$\beta_e = \left(\frac{2\pi \times 9.5 \times 10^9}{0.593 \times 10^6 \sqrt{2.5 \times 10^3}} \right)$$

$$= \frac{5.969 \times 10^{10}}{29.65 \times 10^6}$$

$$= 0.201 \times 10^4$$

$$\boxed{\beta_e = 2.01 \times 10^3 \ rad/m}$$

2.9.4 Advantages

- It provides an octave bandwidth.
- It provides a gain of 40 dB and more.
- It has low noise figure and is about 6 dB.

2.10 BACKWARD WAVE OSCILLATOR

The backward wave oscillator (BWO) is a slow wave device and operates on the principle of velocity modulation. It operates with *RF* wave traveling back down the helix in the opposite direction to the electron beam.

2.10.1 Principle of Operation

BWO is a vacuum tube and contains a cathode and anode (fig. 2.16.).

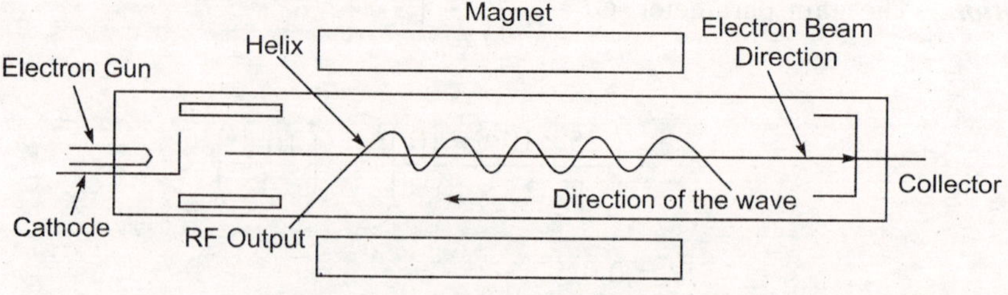

Fig. 2.16 Backward-wave oscillator

BWO has larger diameter compared to TWT. It consists of helix in the center. The output is available near the cathode. The electron beam terminates at the collector. The magnet provides magnetic field which confines the beam to the helix.

The velocity of the beam is made slightly more than the phase velocity by adjusting beam voltage. The interaction of axial component of the helix wave with the electron beam produces velocity modulation of the electrons and bunching. The interaction depends on the phase and amplitude of the feedback loop. The bunching decelerates the electrons near the collector and give up energy to the RF wave. small amounts of energy are transferred to the wave from noise signals in the tube and the wave travels back towards the cathode. This leads to velocity modulation and bunching. This creates oscillations. The frequency of oscillation depends on voltage between the cathode and helix.

BWO is a broad band oscillator and continuous wave generator. It is basically a non-resonant oscillator.

The velocity of electron in BWO is given by

$$v_0 = \sqrt{\frac{2e}{m} V_a}$$

The phase constant of electron beam is

$$\beta_e = \frac{\omega}{v_e}$$

The condition for oscillations is given by

$$DN = \frac{2k + 1}{4}$$

Here, $D = D$ factor

k = integer

$$N = \frac{\beta_e \ell}{2\pi}$$

2.10.2   Applications

1. It is used as continuous wave generator
2. It is used to generate a wide range of frequencies.

2.10.3   Salient Features of BWO

- It operates on the principle of velocity modulation of electrons.
- It is a non-resonant oscillator

- It is a broad band oscillator
- The velocity modulation is produced by the interaction between the electrons and the helix.
- The interaction produces bunching.
- The velocity of electrons is controlled by cathode-to-helix voltage.
- The frequency ranges greater than five to one are achieved.
- It is a voltage tunable oscillator.

2.11 BACKWARD WAVE AMPLIFIER (BWA)

The principle involved in backward wave amplifier is the velocity modulation of electrons.

The velocity modulation is created by continuously bunching the electrons in the beam traveling in the reverse direction from the wave. in this the input RF wave is applied to the helix and it interacts with the electron beam.

2.11.1 Principle of Operation

BWA has two terminals, electron gun, cathode, focusing electrodes, collector, magnetic and RF input and RF output (fig. 2.17). The input to the helix is located at the collector side. The output is located at the cathode side.

The operation BWA and BWO are similar except that the helix wave depends on the input RF wave in BWA. With the application of RF input, the wave travels in the reverse direction to the beam direction. The interaction between the electron beam and helix wave produces velocity modulation. This velocity modulation is created by continuous bunching of electrons. The phase difference between the input wave and bunched electrons decelerates the electrons and hence energy is given up to the wave.

The additional energy of the input RF wave creates amplification.

BWA can be tuned to the desired frequency by adjusting the helix voltage.

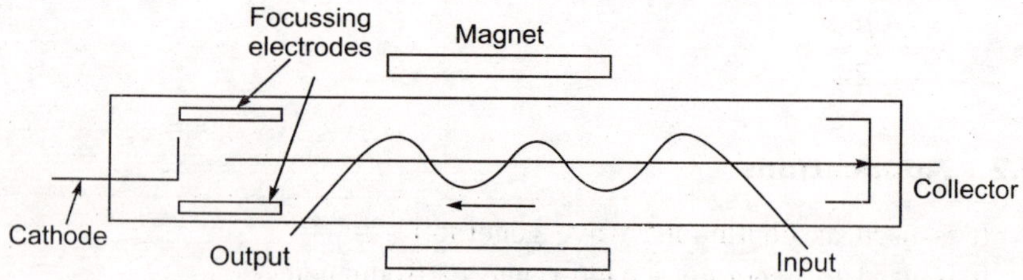

Fig. 2.17 Backward wave amplifier

The velocity of electron beam BWA is given by

$$v_{dc} = \sqrt{\frac{2e}{m}} V_a$$

Here, e = electron charge

m = mass of the electron

V_a = anode voltage

Phase constant of the electron beam is given by

$$\beta_e = \frac{\omega}{V_{dc}}$$

Here, ω_c is the cyclotron angular frequency and is given by

$$\omega_c = \frac{e}{m}\,\beta_m$$

Here, β_m = crossed magnetic flux density

The gain parameter is given by

$$G = \left(\frac{I_a Z_0}{4V_a}\right)^{\frac{1}{3}}$$

I_a = anode current

Z_0 = characteristic impedance

2.11.2 Salient Features

- It has two terminals
- Its frequency of operation depends on helix voltage
- The main principle of operation is velocity modulation
- The velocity modulation is created by interaction between the beam and helix wave.
- Operation of BWA and BWO are similar except that the helix wave depends on input wave.

PROBLEM 2.3 Calculate the cyclotron angular frequency and cut-off voltage for the following parameters of the magnetron.

Anode voltage = 30 kV

Beam current = 30 A

Magnetic flux density = 0.40 Tesla

Radius of cathode = 4 cm

Radius of the vane edge from the centre = 10 cm

Solution The cyclotron angular frequency is given by

$$\omega_c = \frac{e}{m}\,B_m$$

$$\frac{e}{m} = 1.759 \times 10^{11}$$

$$B_m = 0.40 \text{ Tesla}$$

$$\omega_c = 1.759 \times 10^{11} \times 0.40$$

$$\therefore \quad \boxed{\omega_c = 7.036 \times 10^{10} \text{ rad}}$$

The cut-off voltage is given by

$$V_c = \frac{e}{8\,m} B_m^2\, b^2 \left(1 - \frac{a^2}{b^2}\right)^2$$

i.e.
$$V_c = \frac{1}{8} \times 1.759 \times 10^{11} \times (0.4)^2 \times (0.1)^2 \times \left[1 - \left(\frac{4}{10}\right)^2\right]^2$$

$$= 10^{11} \times 35.18 \times 10^{-5} \times 0.36$$

$$\boxed{V_c = 126.6 \times 10^5 \text{ V}}$$

PROBLEM 2.4   The parameters of a two-cavity klystron are given by $V_b = 900$ V, $R_d = 30$ KΩ, $I_b = 20$ mA, $f = 3.2$ GHz, $d = 10^{-3}$ m. Determine (a) electron velocity, (b) transit angle and (c) beam coupling coefficient.

Solution

(a) Electron velocity, $v_e = (0.593 \times 10^6)\sqrt{V_a}$

$$= 0.593 \times 10^6 \sqrt{900}$$

$$\therefore \quad \boxed{v_e = 17.79 \times 10^6 \text{ m/s}}$$

(b) Transit angle, $\quad \theta_t = \omega \dfrac{d}{v_e} = \dfrac{2\pi \times 3.2 \times 10^9 \times 10^{-3}}{17.79 \times 10^6}$

$$\therefore \quad \boxed{\theta_t = 1.13 \text{ radians}}$$

(c) The beam coupling coefficient, $\beta_c = \dfrac{\sin\left(\dfrac{\theta_t}{2}\right)}{\left(\dfrac{\theta_t}{2}\right)}$

$$= \frac{0.53}{0.565}$$

$$\therefore \quad \boxed{\beta_c = 0.946}$$

PROBLEM 2.5   Find the efficiency of the klystron if $I_2 = 28$ mA, $V_2 = 850$ V, $\beta_c = 0.496$, $V_d = 900$ V, $I_b = 26$ mA.

Solution
$$\eta = \frac{\beta_c I_2 V_2}{2 I_b V_b}$$

$$= \frac{0.946 \times 28 \times 10^{-3} \times 850}{2 \, (26 \times 10^{-3}) \times 900}$$

$\therefore$
$$\boxed{\eta = 48.1 \ \%}$$

2.12   POINTS TO REMEMBER

➤ Conventional vacuum tubes are not useful at microwave frequencies.

➤ The limitations of conventional vacuum tubes are the presence of lead inductance, interelectrode capacitance and effect of transit time.

➤ The electron transit angle is defined as $\theta_t = \dfrac{\omega t}{v_0}$.

➤ The transit time across the gap is $\tau_t = \dfrac{d}{v_0}$.

➤ The gain bandwidth product is given by g_m / C.

➤ Two-cavity klystron is popular.

➤ Klystron works on the principle of velocity modulation of electrons.

➤ The efficiency of two-cavity klystron is about 40%.

➤ The power gain of two-cavity klystron is about 30 dB.

➤ In two-cavity klystron the buncher cavity located close to cathode is input-cavity.

➤ In two-cavity klystron catcher cavity is output cavity.

➤ Re-entrant cavity is a cavity in which the metallic boundaries extend into the interior of the cavity.

➤ The velocity of the electron is $\sqrt{\dfrac{2eV_a}{m}}$.

➤ Efficiency of klystron is $\eta = \beta_c \dfrac{I_2 V_2}{I_b V_a}$.

V_a = beam voltage
V_2 = fundamental component of catcher-gap voltage
I_2 = induced current
β_0 = beam coupling coefficient of the catcher gap

➤ BWA is an amplifier

- ➢ BWO is an oscillator
- ➢ TWT uses the slow-wave principle
- ➢ TWT is a broad amplifier
- ➢ TWT has a frequency range of about 0.5 GHz to 95 GHz.
- ➢ TWT has an efficiency of about 5 to 30%.
- ➢ TWT is a low noise RF amplifier.
- ➢ TWT has a long life of about 50,000 hours.
- ➢ TWYSTRON is a combination of klystron and TWT.
- ➢ Magnetron is a cross-field device.
- ➢ Magnetron is an oscillator.
- ➢ The equipment circuit of reflex klystron is

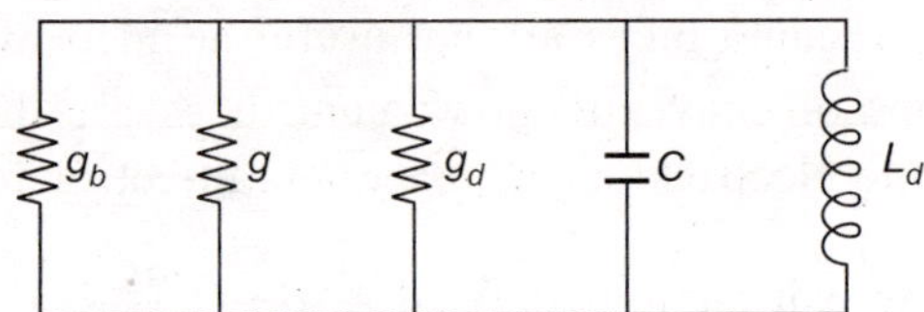

- ➢ The theoretical efficiency of reflex klystron is 22.78%.
- ➢ The axial phase velocity, v_p in TWT is $v_p = v_0(\text{pitch}/2\pi r)$
  v_0 = velocity of light
  r = radius of the helix
- ➢ The gain of TWT is $G = -9.54 + 47.3\,LC(\text{dB})$

- ➢ G = gain parameter = $k\left(\dfrac{I_b}{V_b}\right)^{\frac{1}{3}}$

  I_b = beam current
  V_b = beam voltage
  k = constant
- ➢ L = helix length in wavelength = ℓ/λ_S,
  $\lambda_S = v_p/f$.

- ➢ Hull's cut-off voltage equation in magnetron is $B_c = \dfrac{1}{b}\sqrt{\dfrac{8\,mV_S}{e}}$

  B_c = cut-off magnetic flux density
  b = anode radius
- ➢ The cavities in magnetron have high Q values.
- ➢ The electron velocity in a cavity klystron is

  $v_e = 0.593 \times 10^6 \sqrt{V_a}$ m/sec

  V_a = accelerating voltage

➢ Beam coupling coefficient, $\beta_c = \dfrac{\sin(\theta_t/2)}{(\theta_t/2)}$.

θ_t = gap transit angle

➢ Voltage gain in cavity klystron is $A_v = \dfrac{B_c^2\,\theta_b J_1(X)}{R_b X} R_{sh}$

X = bunching parameter

θ_b = d.c. transit angle between cavities $= \dfrac{\omega L}{v_0}$

$J_1(X)$ = Bessel function of the first order for the argument X.
β_c = beam coupling coefficient
R_b = d.c. beam resistance
L = spacing between two cavities.
R_{sh} = total equivalent shunt resistance of the catcher circuit and load.

➢ Output power delivered to the catcher cavity and load is $p_0 = \dfrac{\beta_c I_2 V_2}{2}$

β_c = beam coupling coefficient
I_2 = current in R_{sh}
V_2 = fundamental component of catcher gap voltage

➢ The depth of velocity modulation in klystron is $x = \beta_c \dfrac{V_{in}}{V_a}$

V_b = beam voltage
V_{in} = amplitude of the signal
β_c = beam coupling coefficient

2.13 MULTIPLE CHOICE QUESTIONS

1. Vacuum triodes are suitable at
   (a) low and high frequencies
   (b) low frequencies
   (c) all frequencies
   (d) high frequencies

2. At high frequencies, conventional vacuum tubes have limitations with the presence of
   (a) transit time effects
   (b) low bandwidth
   (c) high noise
   (d) vacuum

3. The linear beam tube (O type) is
   (a) reflex klystron
   (b) amplitron
   (c) gyrotrons
   (d) magnetron

4. Microwave M type tube is
   (a) klystron
   (b) TWT
   (c) magnetron
   (d) BWA

5. Slow-wave structure is
   (a) klystron
   (b) magnetron
   (c) dematron
   (d) TWT

6. The unit of electron transit angle is
   (a) radian
   (b) radian/m
   (c) radian/sec
   (d) radian-sec

7. The unit of transit time is
   (a) sec/m
   (b) second
   (c) sec/radian
   (d) sec/degree

8. Gain-bandwidth in pentode is
   (a) g_m/C
   (b) C/g_m
   (c) $A\,g_m$
   (d) $r_d\,g_m$

9. Feedback in klystron is by
   (a) using three cavities
   (b) re-entrant
   (c) using 4 cavities
   (d) anode

10. The re-entrant cavities in klystrons can be
    (a) circular
    (b) rectangular
    (c) elliptical
    (d) toroidal

11. The velocity modulation in klystron is represented by
    (a) equivalent circuit
    (b) Applegate diagram
    (c) output voltage
    (d) output power

12. The velocity of electron in klystron in proportional to
    (a) square root of beam voltage
    (b) electron charge
    (c) electron mass
    (d) square root of mass of the electron

13. Beam coupling coefficient in klystron is a function of
    (a) beam voltage
    (b) transit angle
    (c) input signal
    (d) depth of velocity modulation

14. The transit angle is proportional to
    (a) space between the cavities
    (b) v_e
    (c) e
    (d) $1/\omega$

15. The bunching parameter in klystron is proportional to
    (a) beam coupling
    (b) $1/\theta_t$
    (c) V_S
    (d) $1/T_t$

16. The mode number in reflex klystron is
    (a) $N_n = (n + 1/4)$
    (b) $N_n = (n + 3/4)$
    (c) $N_n = n$
    (d) $N_n = (n - 3/4)$

17. The electronic tuning range in reflex klystron is
    (a) the total change in frequency from one end of the mode to the other
    (b) the total change in frequency with repeller voltage
    (c) the total change in frequency with beam voltage
    (d) the total change in frequency with transit angle

18. The output efficiency of reflex klystron is the ratio of
    (a) output RF power to input RF power
    (b) output RF power to input d.c. power
    (c) output d.c. power to input d.c. power
    (d) output d.c. power to RF input power

19. TWYSTRON is a combination of
    (a) klystron and magnetron   (b) klystron and TWT
    (c) klystron and BWA   (d) klystron and BWO

20. Magnetron is
    (a) O – type tube   (b) an oscillator
    (c) an amplifier   (d) a detector

21. Magnetron is
    (a) a low frequency oscillator   (b) a low power device
    (c) a high power device   (d) a low gain amplifier

22. BWO is
    (a) a continuous wave oscillator   (b) a pulse generator
    (c) a square wave generator   (d) an amplifier

23. The electron velocity in reflex klystron is

    (a) $\sqrt{2\dfrac{e}{m}\dfrac{1}{V_b}}$ (b) $\sqrt{2\dfrac{e}{m}V_b}$

    (c) $\sqrt{2\dfrac{m}{e}V_b}$ (d) $\sqrt{\dfrac{1}{2}\dfrac{e}{m}V_b}$

24. TWT is
    (a) M – type tube   (b) is an amplifier
    (c) O – type tube   (d) a cross-field tube

25. The force on the electron by the magnetic field in a magnetron is
    (a) $\boldsymbol{F} = -e\,(\boldsymbol{V} \times \boldsymbol{B})$ (b) $\boldsymbol{F} = (\boldsymbol{V} \times \boldsymbol{B})$
    (c) $\boldsymbol{F} = -e\,(\boldsymbol{B} \times \boldsymbol{V})$ (d) $\boldsymbol{F} = -e\,(\boldsymbol{V} \cdot \boldsymbol{B})$

26. Transit angle in a vacuum tube is

    (a) $\dfrac{\omega d}{v_e}$ (b) $\dfrac{v_e d}{\omega}$

    (c) $\dfrac{\omega}{d v_e}$ (d) $\dfrac{f d}{v_e}$

27. The electron velocity in a tube is
    (a) $0.593 \times 10^6 \, V_b$
    (b) $0.593 \times 10^6 \sqrt{V_b}$
    (c) $0.593 \times \sqrt{V_b}$
    (d) $0.953 \times 10^6 \sqrt{V_b}$

28. The beam coupling coefficient is
    (a) $\dfrac{\sin(T_t/2)}{(T_t/2)}$
    (b) $\dfrac{\theta_t}{2}$
    (c) $\dfrac{\sin(\theta_t/2)}{(\theta_t/2)}$
    (d) $\dfrac{\omega \sin(\theta_t/2)}{(\theta_t/2)}$

29. The bunching parameter of a klystron is
    (a) $\dfrac{\beta_c V_m}{2V_b}\theta_t$
    (b) $\dfrac{V_m}{2V_b}\theta_t$
    (c) $\dfrac{\beta_c V_m}{2V_b}T_t$
    (d) $\dfrac{\beta_c V_m}{2V_b}$

30. The d.c. transit angle between cavities is
    (a) $\dfrac{\omega L}{v_e}$
    (b) $-\dfrac{v_e}{\omega L}$
    (c) $\dfrac{\omega}{v_e}$
    (d) $\dfrac{L}{v_e}$

31. The mutual conductance of a klystron amplifier is ratio of
    (a) beam current to beam voltage
    (b) induced current to input voltage
    (c) beam voltage to beam current
    (d) input voltage to induced current

32. g_m of a klystron is
    (a) $\dfrac{2\beta_c I_b J_1(X)}{V_m}$
    (b) $\dfrac{2 I_b J_1(X)}{V_m}$
    (c) $\dfrac{2\beta_c J_1(X)}{V_m}$
    (d) $2\beta_c I_b J_1(X)$

33. The maximum electronic efficiency of reflex klystron is
    (a) 40 %
    (b) 50 %
    (c) 22.7 %
    (d) 35 %

34. If L is the pitch and v_e is the electron velocity in TWT, the d.c. transit time electron is
    (a) $T_t = L/v_e$
    (b) $T_t = v_e/L$
    (c) $T_t = L\, v_e$
    (d) $T_t = \dfrac{1}{2}L/v_e$

35. The output power gain in dB TWT is
    (a) $-9.54 + 47.3\,LC$
    (b) $9.54 + 47.3\,LC$
    (c) $-9.54 - 47.3\,LC$
    (d) $9.54 - 47.3\,LC$

36. The cut-off voltage in magnetron is

    (a) $\dfrac{e}{m}B^2 b^2\left(1-\dfrac{a^2}{b^2}\right)^2$
    (b) $\dfrac{1}{8}\dfrac{e}{m}B^2 b^2\left(1-\dfrac{a^2}{b^2}\right)^2$

    (c) $\dfrac{1}{8}\dfrac{e}{m}B^2 b^2\left(1-\dfrac{a}{b}\right)$
    (d) $\dfrac{1}{8}\dfrac{e}{m}B^2 b^2\left(1-\dfrac{a}{b}\right)^2$

37. The cyclotron frequency of the circular motion of the electron is

    (a) $\dfrac{eB}{2\pi m}$
    (b) $\dfrac{eB}{m}$

    (c) $\dfrac{eB^2}{2\pi m}$
    (d) $\dfrac{eB}{\pi m}$

38. The electronic efficiency of magnetron is defined as

    (a) $\dfrac{P_{gm}}{P_{dc}}$
    (b) $\dfrac{P_{dc}}{P_{gm}}$

    (c) $\dfrac{P_{gm}}{P_{inRF}}$
    (d) $\dfrac{P_{gm}}{P_{outRF}}$

39. The Hall's cut-off voltage for a linear magnetron is

    (a) $\dfrac{e}{m}B^2 d^2$
    (b) $\dfrac{1}{2}\dfrac{e}{m}B^2 d^2$

    (c) $\dfrac{1}{2}\dfrac{e}{m}Bd$
    (d) $\dfrac{1}{2}\dfrac{e}{m}B^2 d$

40. The gain parameter of Amplitron is

    (a) $\left(\dfrac{I_a Z_0}{4V_a}\right)^{1/3}$
    (b) $\left(\dfrac{I_a Z_0}{4V_a}\right)^{1/2}$

    (c) $\left(\dfrac{I_a Z_0}{V_a}\right)^{1/3}$
    (d) $\left(\dfrac{I_a Z_0}{2V_a}\right)^{1/3}$

2.14 ANSWERS

1. b	2. a	3. a
4. c	5. d	6. a
7. b	8. a	9. b

10. d	11. b	12. a
13. b	14. a	15. a
16. b	17. a	18. b
19. b	20. b	21. c
22. a	23. b	24. c
25. a	26. a	27. b
28. c	29. a	30. a
31. b	32. a	33. c
34. a	35. a	36. b
37. a	38. a	39. b
40. a		

2.15 EXERCISE PROBLEMS

1. In a two-cavity klystron the parameters are, input power = 10 mW, voltage gain = 20 dB, R_{sh} of input cavity = 25 KΩ, R_{sh} of output cavity = 35 KΩ, load resistance = 40 KΩ. Find input voltage, output voltage and the power to the load.

2. The beam voltage, V_b = 250 V, and the beam current, I_b = 15 mA, the signal voltage, V_m = 35 V, are the parameters of a reflex klystron which operates at the mode n = 2. Find input power and efficiency.

3. A reflex klystron operates with V_b = 400 V, R_{sh} = 20 KΩ, f = 9 GHz, L = 10^{-3} m, n = 2, find the repeller voltage and electronic efficiency.

4. In a circular klystron, a = 0.10 m, b = 0.40 m, β = 1.0 mT, V_b = 5 KV, find Hull's cut-off voltage and cut-off magnetic flux density.

5. Find axial phase velocity in a helical TWT if it has a diameter of 3 mm with 40 turns per cm.

6. A TWT operates with the following parameters.

$$V_b = 2.5 \text{ KV}$$
$$I_b = 25 \text{ mA}$$
$$Z_0 = 10 \text{ } \Omega$$
$$\text{circuit length, } L = 50$$
$$f = 9 \text{ GHz}$$

Find gain parameter and power gain.

7. A magnetron operates with the following parameters.

$$\text{Anode voltage} = 25 \text{ kV}$$
$$\text{Beam current} = 25 \text{ A}$$
$$\text{Diameter of cathode} = 8 \text{ cm}$$
$$\text{Radius of vane edge to center} = 8 \text{ cm}$$
$$\text{Magnetic flux density} = 0.34 \text{ Tesla}$$

Find the cyclotron frequency and cut-off voltage.

8. The parameters of a linear magnetron are

$$\text{Anode voltage} = 12 \text{ KV}$$
$$\text{Beam current} = 1.2 \text{ A}$$
$$\text{Magnetic flux density} = 0.15 \text{ Tesla}$$
$$\text{Separation between cathode and anode} = 4 \text{ cm}$$

Find the Hull's cut-off voltage.

Chapter 3

Microwave Semiconductor Devices

The semiconductor devices depend on the relative energy difference between valence and conduction bands for their operation.

3.1 INTRODUCTION

The microwave semiconductor devices are used in

- Amplifiers
- Oscillators
- Frequency multipliers
- Modulators
- Mixers
- Detectors
- Switches etc.

The semiconductor devices are popular as the vacuum tube devices like klystrons, magnetrons and traveling wave tubes are bulky and heavy. Moreover, these tubes require high voltages for their operation and they occupy large space.

The common semiconductors used in microwave devices are Silicon (Si), Gallium Arsenide (GaAs), Indium Phosphate (InP) and Germanium (Ge).

The most popular microwave semiconductor devices are:

- Bipolar transistors
- Field effect transistors
- Bipolar transistors
- Tunnel diodes
- Transferred electron devices
- Gunn diode

- LSA diode
- IMPATT diode
- TRAPATT diode
- BARITT diode
- HEMT devices
- PIN diode

3.2 MICROWAVE BIPOLAR TRANSISTOR

The basic operation of microwave bipolar transistor is the same as that of transistor at low frequencies.

Its operation depends on

- the thickness of the p-n junction depletion region and
- the transit time

3.2.1 Salient Features

- Bipolar transistors are used as amplifiers and oscillators
- They are used in L and S microwave bands.
- They are operated as class C amplifiers.
- Their efficiency is about 50%.
- They are used to get a power gain of about 10 dB.
- All silicon microwave transistors are of npn type as the mobility of electrons is higher than that of holes.
- The reduced active area dimensions and package parasitics are used.
- In microwave transistors emitter strip width and base thickness are made very small. In lithography, the emitter strip width is a submicron and the base thickness is of the order of a few hundred angstroms.

3.3 FIELD EFFECT TRANSISTORS

The operation of FET depends on the conductivity of semiconductor layer of gallium arsenide.

3.3.1 Salient Features

- They can operate in X-band.
- They are used as amplifiers and oscillators.
- They have low noise figure, high efficiency and high input impedance compared to bipolar transistors.
- FETs use mostly GaAs and a metal semiconductor (MES) Schottky junction for gate contact.
- A typical n-channel FET is shown in fig. 3.1.

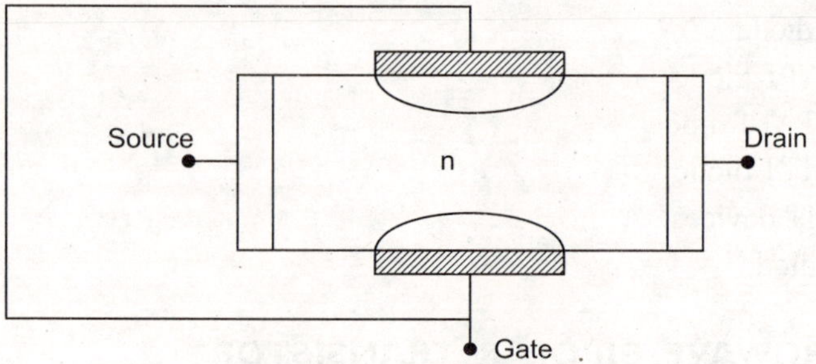

Fig. 3.1 *n*-channel FET

3.4 TUNNEL DIODE

Tunnel diode is a negative resistance semiconductor *p-n* junction device. The influence of negative resistance is to add energy and permit amplification and oscillation in the negative resistance region.

3.4.1 Principle of Operation of Tunnel Diode

It is a negative resistance semiconductor *p-n* junction diode. The negative resistance characteristics are created by the tunnelling effect of the electrons in the junction.

3.4.2 Salient Features of Tunnel Diode

- It differs from ordinary diodes in two ways : 1. In this, *p* and *n* regions are highly doped. 2. The barrier region is very thin.
- The impurity concentration is one part in 10^3.
- It was invented by Esaki in 1958.
- It is a negative resistance device.
- The tunnelling phenomenon is a majority carrier effect.
- It is a low power device.
- It is wide-band device.
- It is a low cost and simple device.
- It requires low voltage power supply.
- It operates in the entire lower half of the microwave spectrum.
- It is represented by $A \bullet\!\!\!-\!\!\!\rhd\!\!\!-\!\!\!\bullet K$
- The junction is thin.
- Its equivalent circuit is in fig. 3.2.

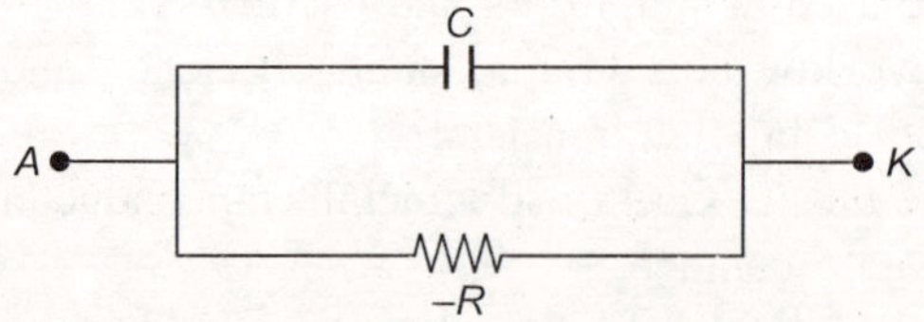

Fig. 3.2 Equivalent circuit of tunnel diode

- In tunnel diode cavity oscillator, mechanical tuning is done by tuning plunger.
- The diode is coupled to the cavity by a probe.
- Loop coupling introduces inductive coupling and hence it is avoided.
- Its volt-ampere characteristic is shown in fig. 3.3.

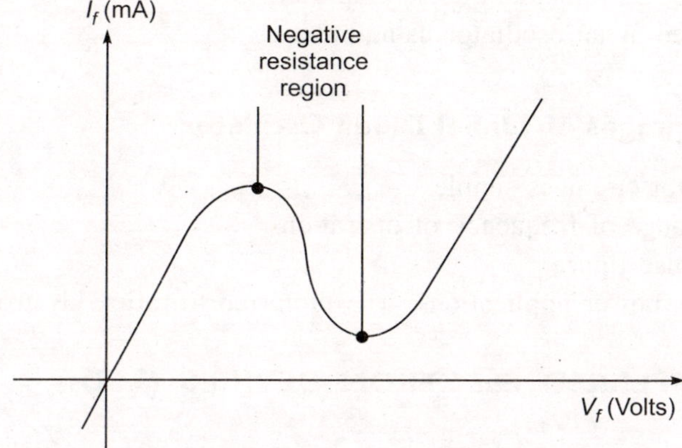

Fig. 3.3 V-I characteristics of tunnel diode

- The property of negative resistances is to add energy.
- The transit time is short.
- It can be used as an amplifier and oscillator.
- It is also called Ekasi diode.
- It can be used as a high speed switch.
- Its transient response is limited only by the shunt capacitance.

Tunnel diode amplifier with a circular as an isolator is shown in fig. 3.4.

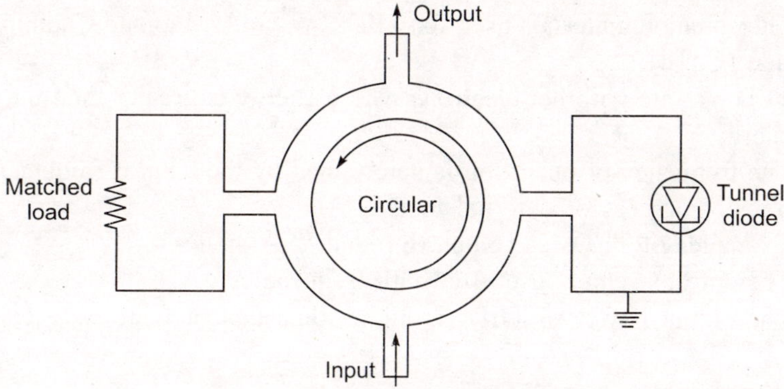

Fig. 3.4 Tunnel diode amplifier

3.4.3 Application of Tunnel Diode

1. A tunnel diode connected to microwave circular is used as negative resistance amplifier.
2. It can be used in oscillator circuits.
3. It is used as high speed switch.
4. It is used as a medium noise amplifier.
5. It is used as an oscillator using cavity.

3.4.4 Advantages of Tunnel Diode Oscillator

➢ Oscillator circuit is simple.

➢ Wide range of frequency of operation.

➢ Low noise figure.

➢ For low power applications, it is preferred to reflex klystron.

3.5 TRANSFERRED ELECTRON DEVICES (TEDs)

TEDs do not have *p-n* junctions. These are negative resistance devices and they are used as amplifiers and oscillators.

3.5.1 Salient Features

- TED is a bulk device and has no junction.
- TED means transferred electron device.
- It is a negative resistance, two-port solid-state device.
- It is a two-terminal semiconductor device.
- It is an active device.
- It is used as a amplifier and oscillator.
- Its operation depends on the bulk negative resistance property of some semiconductor materials.
- The common materials used for TEDs are GaAs, Inp and Cadmium telluride (CdTd).
- TED operate with hot electrons whose energy is greater than the thermal energy.
- The frequency of operation is determined by the natural frequency of the transfer process or external load.
- Examples of TEDs are Gunn, Inp and LSA diodes.
- They are extremely useful at high frequencies.
- The output power of TEDs is higher than that of BJTs.

3.5.2 Applications of TEDs

1. It is used as an oscillator.

2. It is used as an active device.

3. It is used in high frequency applications.

4. It is used for high power generation.

3.6 GUNN DIODES

Gunn diode is a negative resistance device. It is also a Transfer Electron Device (TED). It is used as an amplifier and oscillator.

3.6.1 Salient Features

- Gunn was discovered by Gunn in 1963.
- It is used as an amplifier and oscillator.
- They can be used upto 100 GHz frequency.
- Its power output is about 10 W at lower microwave frequency range.
- It does not have a p-n junction.
- It is an example of transfer electron device (TED).
- It is a wide band device.
- It has low noise characteristics.
- It is a bulk effect device.

- It is represented by

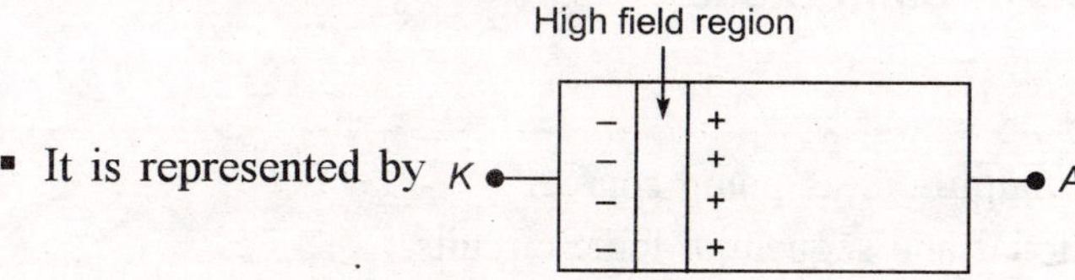

- Its V-I characteristics is in fig. 3.5.

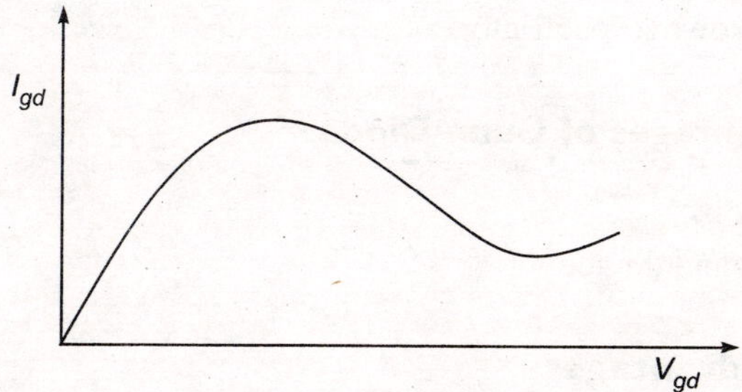

Fig. 3.5 V-I characteristics of gunn diode

3.6.2 Principle of Operation

It operates using the property of negative resistance of some bulk semiconductor materials.

Consider band structure of a typical semiconductor like GaAs (fig. 3.6).

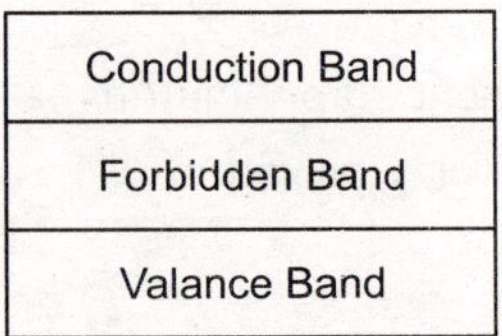

Fig. 3.6 Band structure of GaAs

In GaAs, there exists two energy regions in conduction band. They are known as lower valley and higher valley as shown in fig. 3.7.

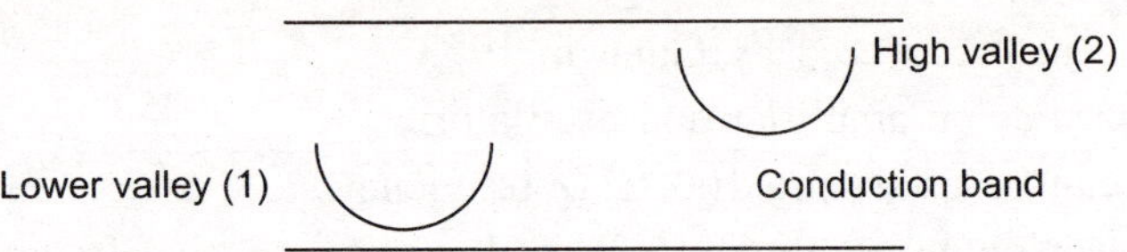

Fig. 3.7 Conduction band of GaAs

There exists energy difference between them. The electron mobility in high valley is smaller than that in lower valley.

3.6.3 Applications of Gunn Diode

It is used

1. in parametric amplifiers as pump source.
2. fast combinational and sequential logic circuits.
3. in different types of radar transmitters.
4. as broadband linear amplifier.
5. as microwave oscillator.

3.6.4 Advantages of Gunn Diode

➢ Low noise
➢ Broadband device

3.6.5 Disadvantages

It is a temperature sensitive device and hence its characteristics change with change in temperature.

3.7 LSA DIODE

LSA means **Limited Space Charge Accumulation Diode.** It is used as an oscillator.

3.7.1   Salient Features

- It is a negative resistance device.
- Its bias voltage is in the center of negative region.
- It is such that it will not permit to create the high field domain.
- Its frequency of oscillation depends only on the external circuit.
- It provides high power output.
- Its *V-I* characteristics is shown in fig. 3.8.

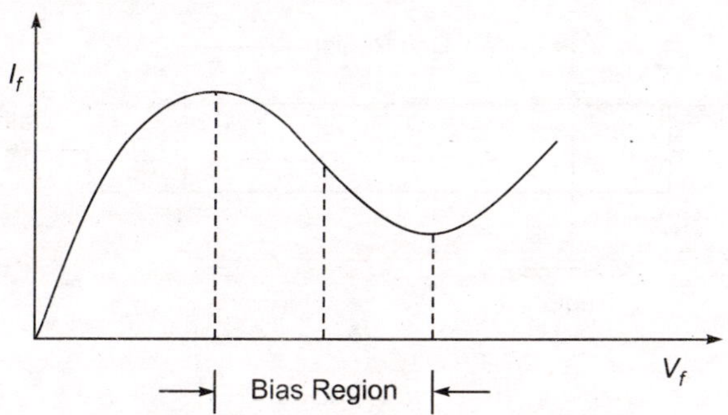

Fig. 3.8   V-I characteristics of LSA diode

3.8   IMPATT DIODE

IMPATT diode means **Impact Ionization Avalanche Transit Time diode.**

> Its operation depends on the reverse breakdown voltage characteristics of *p-n* junction and phase delay of the applied *RF* signal.

3.8.1   IMPATT Diode Mechanism of Oscillation

In a diode resonator circuit, noise voltage excites resonant component. The reverse biased d.c. field and a.c. field makes the diode to swing into and out of the avalanche condition. The carriers drift to the end contacts before the diode swings out of the avalanche region. This happens as the hole drift time is very short. As a result, the a.c. field takes energy from the carriers/d.c. bias source. This process builds the microwave oscillations in the circuit.

The frequency of oscillation is given by

$$f = \frac{v_d}{2\ell}$$

Here, v_d = drift velocity of the holes

ℓ = length of the drift region

3.8.2 Principle of Operation

It is basically a *p-n* junction diode and is reverse biased to breakdown. In this, an avalanche of electron-hole pairs are created by impact ionization in the high field depletion layer of the device. The oscillations are produced by the transit of the carriers through the depletion layer.

The negative resistance of IMPATT diodes is due to the impact ionization and the transit time effects.

The IMPATT diode and avalanche pulse are shown in fig. 3.9.

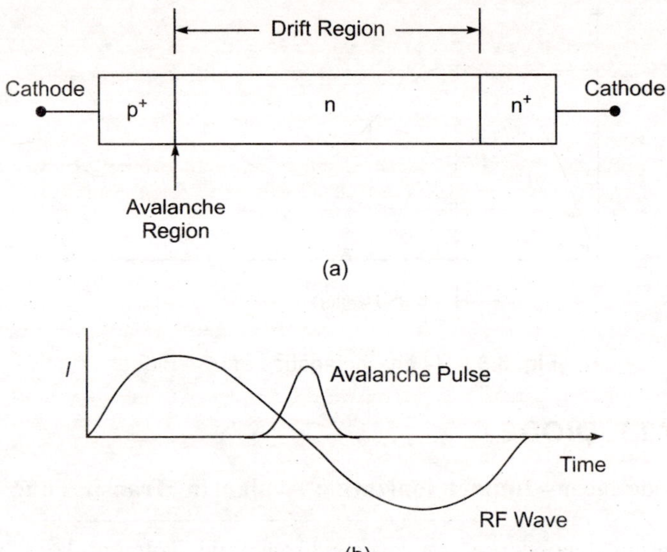

Fig. 3.9 (a) IMPATT diode (b) Avalanche pulse

$$\text{Transit angle, } \theta_t = \omega\tau = \omega\frac{L}{v_d}$$

Here, τ = transit time

$\omega = 2\pi f$

The avalanche resonant frequency is given by

$$\omega_\tau = \sqrt{\frac{2\alpha v_d I}{\epsilon_S A}}$$

Here, A = diode cross section

I = current

ϵ_S = semiconductor permittivity

α = ionization coefficient

$$\alpha = \frac{\partial \alpha}{\partial \in}$$

The IMPATT diode is characterized by an impedance consisting of both resistive and reactive parts. The resistive part R_d is given by

$$R_d = R_p + \frac{2L^2}{v_d \in_a A}\left(\frac{1}{1-\omega^2/\omega_r^2}\right)\left(\frac{1-\cos\theta}{\theta_t}\right)$$

R_p = resistance of inactive region.

In terms of transit time, the frequency of the diode is given by

$$f = \frac{1}{2\tau}$$

Here, τ = transit time

3.8.3 *V-I* Characteristics of IMPATT Diode

The *V-I* characteristics are shown in fig. 3.10.

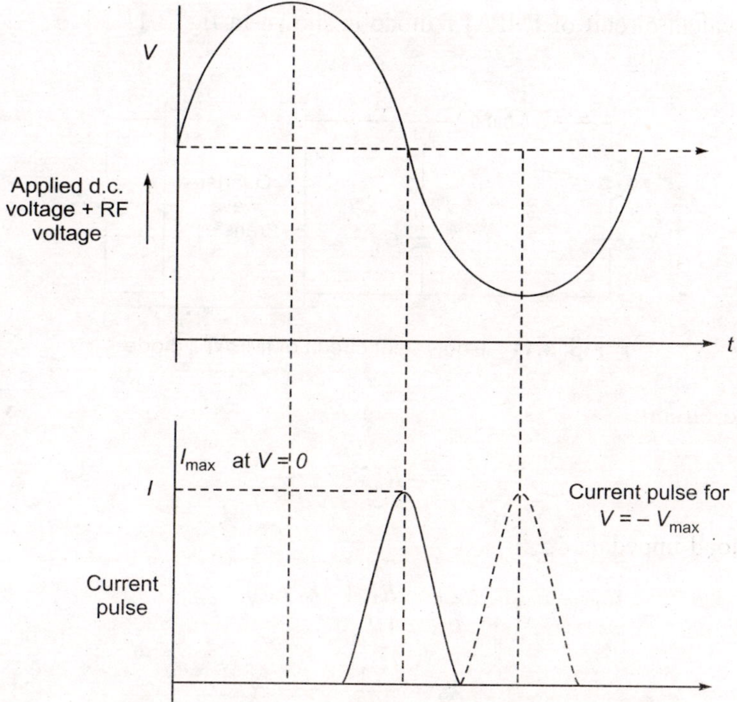

Fig. 3.10   *V-I* characteristics of IMPATT diode

3.8.4 Efficiency of IMPATT Diode

It is given by

$$\eta = \frac{p_{ac}}{p_{dc}} = \left(\frac{V_{ac}}{V_{dc}}\right)\left(\frac{I_{ac}}{I_{dc}}\right)$$

3.8.5 Salient Features

- It is an avalanche transit time device.
- It can be made with germanium, silicon and gallium arsenide.
- It is useful up to 100 GHz.
- It provides highest continuous power compared to other semiconductor devices.
- Its efficiency is less than 30%.
- It is useful at microwave, millimetre and submillimeter wave frequencies.
- IMPATT diodes are popularly used in digital and analog communication and radar systems.
- It can be used as an amplifier and oscillator.

3.8.6 Equivalent Circuit of IMPATT Diode

An equivalent circuit of IMPATT diode is shown in fig. 3.11.

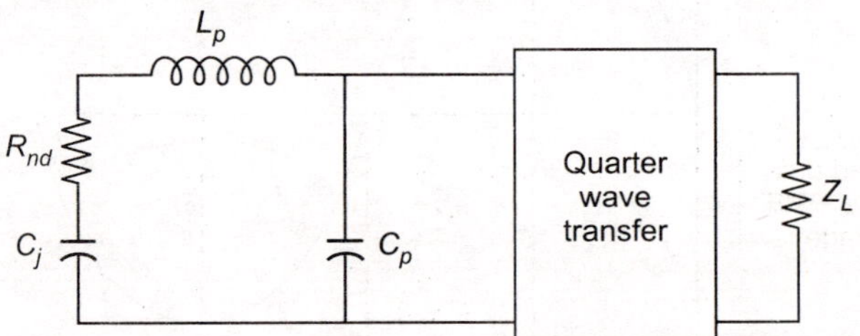

Fig. 3.11   Equivalent circuit of IMPATT diode

In the circuit,

$$Z_d = -R_{nd} + \frac{1}{j\omega C_j}$$

The load impedance

$$Z_L = R_L + jX_L$$
$$R_L = |R_{nd}|$$

$$X_L = \frac{1}{\omega C_j}$$

In the diode mount, the reactance of the circuit is made zero by controlling the inductance. In the circuit

R_{nd} = diode negative resistance
C_j = junction capacitance
L_p = package lead inductance
C_p = capacitance between package terminals
R_L = load resistance

Quarter wave transformer is used to match the low resistance diode with the load.

3.8.7 Applications of IMPATT Diode

It is used

- as microwave oscillator
- as modulated oscillator
- as receiver local oscillator
- as parametric amplifier pump
- in radar reception
- in communication transmission
- as negative resistance amplifier

3.8.8 Advantages of IMPATT Diode

- It is wideband
- Pulse power is high
- Suitable for high frequency use

3.8.9 Disadvantages of IMPATT Diode

- It is a noise device
- Tuning range is not high
- Noise figure is high and is about 80 dB

3.9 TRAPATT DIODE

TRAPATT diode means **Trapped-Plasma Avalanche Trigged Transit diode.**
It has $p^+ n\, n^+$ or $n^+ p\, p^+$ structure. It is useful to produce high microwave power.

3.9.1 Salient Features

- Its structure is similar to IMPATT diode.

- Its structure has $p^+ n\, n^+$ or $n^+ p\, p^+$
- It is a high efficiency diode oscillator.
- Its oscillations depend on the delay in the current caused by avalanche process.
- The diode diameter is about 50 μm for CW operation and it 750 μm at lower frequency for high peak power application.
- It can be operated over a range of 400 MHz to 12 GHz.
- It has an efficiency of 20 – 40 %.
- It has applications in phased array radar systems, proximity fuse sources, altimeters, instrument landing systems and IFF transmitters etc.
- At low frequencies, peak power and efficiencies are high.
- At high frequencies, peak power and efficiency are low.

3.9.2   Principle of Operation

A typical structure of TRAPATT diode is shown in fig. 3.12.

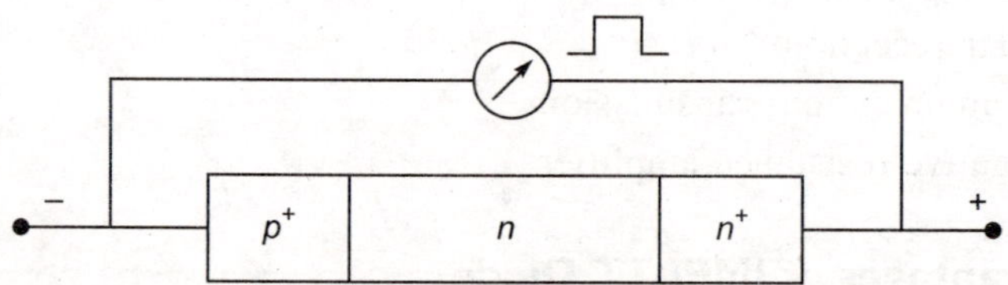

Fig. 3.12   TRAPATT diode

It is operated in the reverse bias. During the application of high reverse voltage, avalanche phenomenon takes place. The depletion layer is filled with a plasma of electrons and holes. These are trapped in low field region behind propagating high field. A voltage pulse is produced and it opposes the applied d.c. voltage. The decreased voltage reduces the current pulse. The decreased current with increased voltage creates a negative resistance region. Basically, TRAPATT oscillators depend on the delay in the current caused by the avalanche process.

The avalanche zone velocity, V_{az} is given by

$$V_{az} = \frac{J}{eN_A}$$

Here, J = current density, A/m^2

e = electron charge = 1.6×10^{-19} C

N_A = doping concentration

The transit time of charge carriers is given by

$$\tau_t = \frac{\ell}{v_S}$$

Here, ℓ = drift region length

v_S = saturated carrier drift velocity

The voltage waveform for TRAPATT diode is shown in fig. 3.13.

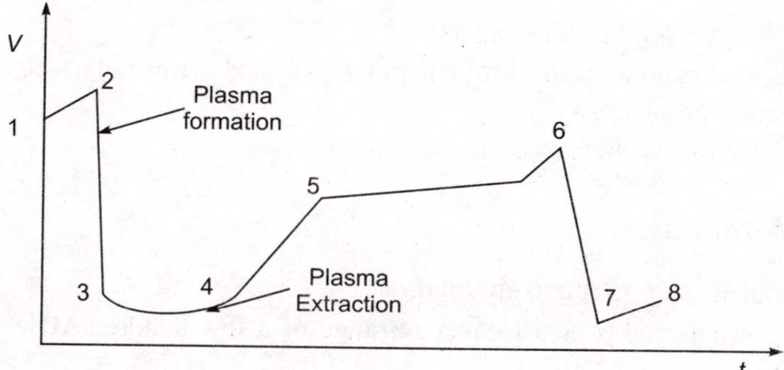

Fig. 3.13   Voltage waveform for TRAPATT diode

The current waveform is shown in fig. 3.14.

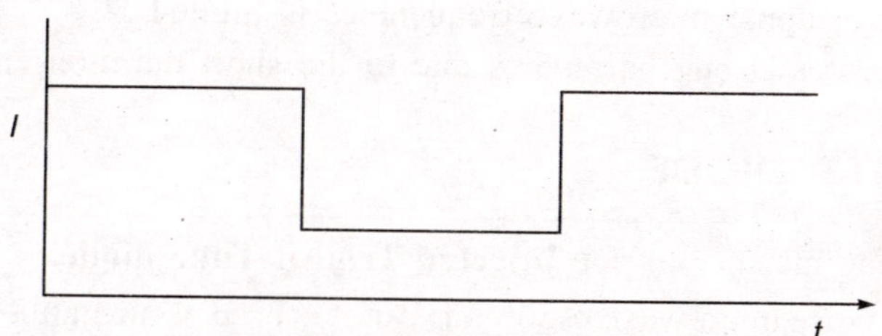

Fig. 3.14   Current waveform

The frequency of oscillation of a TRAPATT diode is

$$f = \frac{1}{2T}$$

Here, $T = 2\left[\dfrac{w}{v_f} + \dfrac{w}{2}\left(\dfrac{1}{v_p} + \dfrac{1}{v_s}\right)\right]$

w = Depletion layer width

v_f = front velocity

v_p = plasma velocity

v_s = saturated drift velocity

The discharge time of plasma is greater than the transit time τ_t of the diode at high field and TRAPATT can be operated at low frequencies. In this diode, the time delay charge carriers during the transit is used to produce the required current phase shift which generates oscillations.

3.9.3 Applications of TRAPATT Diode

It is used

1. in low power Doppler radars.
2. as local oscillators in different radars, phased array radars etc.
3. in ratio altimeters.
4. in microwave beacons and landing systems.

3.9.4 Advantages

- Its efficiency is relatively high.
- It can be used over a frequency range of a few hundred MHz to several GHz.

3.9.5 Disadvantages

- It has high noise figure.
- Its use at upper microwave frequencies is limited.
- It generates strong harmonics due to the short duration current pulse.

3.10 BARITT DIODE

BARITT diode means **Barrier Injected Transit Time diode.**

It is a low noise microwave oscillator. But it is used as an amplifier rather than an oscillator.

3.10.1 Salient Features

- It has $p^+ n p^+$ structure.
- It is a low noise microwave oscillator.
- It is equivalent to a pair of diodes connected back-to-back. One diode is forward biased and 2nd one is reverse biased.
- It can be used upto X-band microwave frequencies.
- It is a narrow band device.
- It is a low efficiency diode.
- It is a low power device.
- Its power output is less than 1 mW.
- Noise figures as low as 9 dB at 6.35 GHz with 15 dB gain are possible.
- BARITT diodes are used as amplifiers rather than oscillators.

3.10.2 Principle of Operation

The structure of BARITT diode is shown in fig. 3.15.

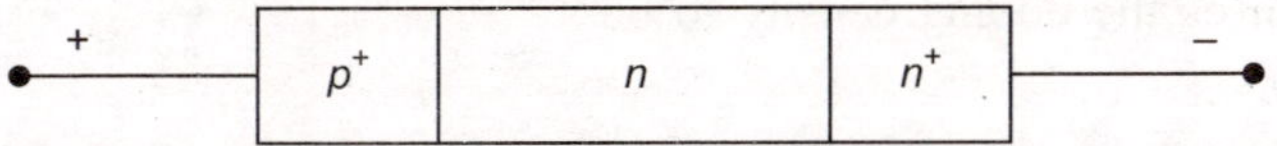

Fig. 3.15 Structure of BARITT diode

It has two semiconductor junctions. One of the junctions is forward biased and the second is reverse biased. If a voltage is applied to the diode, a small voltage is developed across the forward biased junction and the rest of the voltage appears across the reverse biased junction. The carriers have a low unsaturated drift velocity in a major portion of drift region. This makes the carriers to have a large transit time. This in turn produces oscillations. The volt-ampere characteristic of BARITT diode is shown in fig. 3.16.

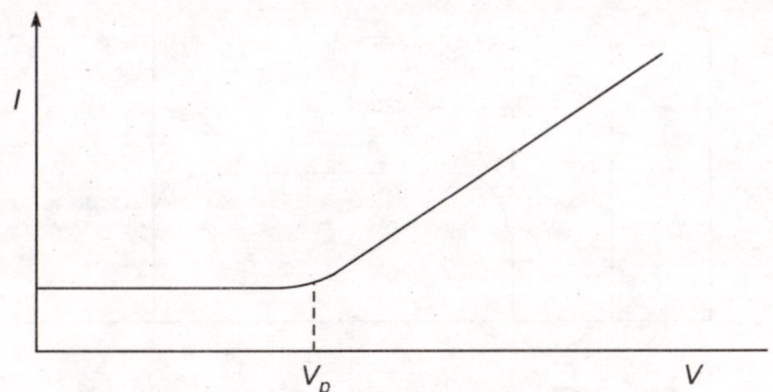

Fig. 3.16 V-I Characteristic of BARITT diode

In fig. 3.16, V_p is known as punch through voltage.

The current increase is due to the magnitude of the critical voltage in the diode and its negative temperature coefficient.

The critical voltage is given by

$$V_c = \frac{qNt}{2\epsilon_s}$$

Here, q = electron charge

N = doping concentration

t = thickness of semiconductor

ϵ_s = permittivity of the semiconductor

In fact, BARITT diodes include p-n-p, p-n-v-p and metal-n-metal.

For a p-n-v-p diode, p-n junction emits holes into v-region in the forward bias. These holes travel into v-region with saturation velocity and they are collected at p contact. It has a negative resistance at transit angles ranging between π and 2π. The optimum transit angle is about 1.6 π.

In these diodes, the drift region is depleted completely to produce punch-through the emitter junction without creating avalanche-breakdown base-collector junction.

This requires the doping density to be

$$E_{\max} = \int_0^W N(x)\,dx$$

The electric field in *p-n-v-p* BARITT diode is given in fig. 3.17.

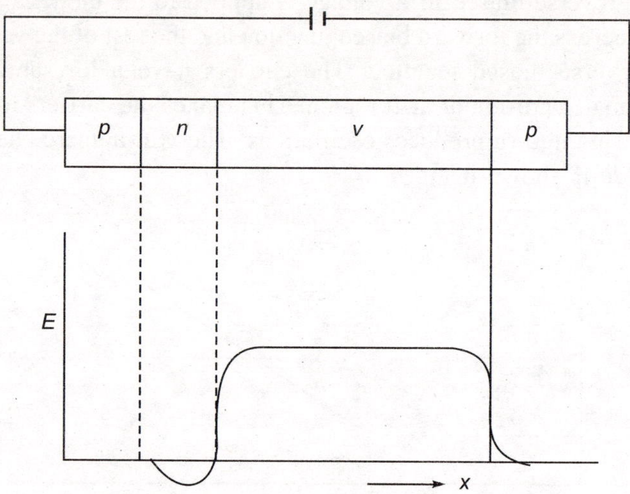

Fig. 3.17 *p-n-v-p* diode and electric field

3.11 PIN DIODE

PIN represents *P*-type semiconductor, intrinsic material and *N*-type semiconductor. That is, a thick intrinsic slab is sandwiched between a thin *P*-semiconductor and a thin *N*-semiconductor (fig. 3.18).

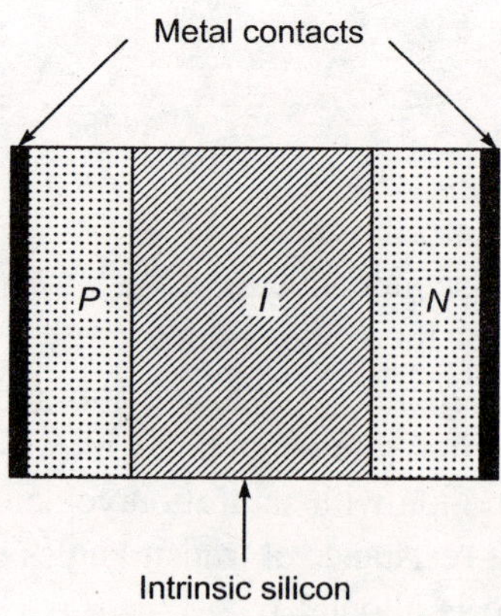

Fig. 3.18 PIN diode

3.11.1   Features of PIN Diode

1. Silicon and Galium Arsenide are used for the construction PIN diode.
2. It can handle high power and exhibits high resistivity of intrinsic region.
3. In practice, intrinsic layer is highly doped N-type semiconductor.

4. Its equivalent circuit is
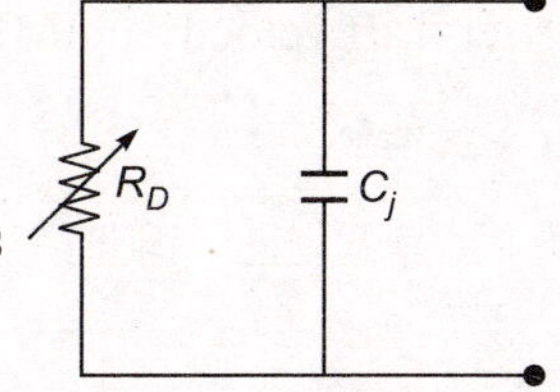

5. The variation R_D of is
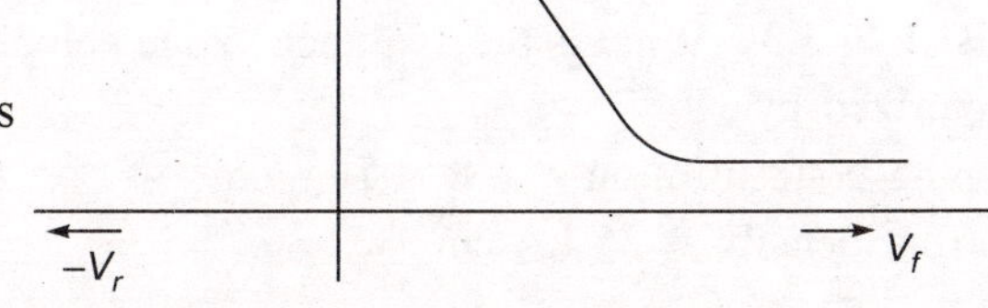

6. It acts as a normal PN diode upto about 100 MHz.
7. At microwave frequencies, it acts as variable resistance.
8. Its forward bias resistance is about 1 to 10 Ω and reverse bias resistance is about 5 to 10 KΩ.
9. It was first proposed by R.N. Hall in 1952 and its potential as a microwave switch was first recognized by Uhlir in 1958.
10. The frequency at which the cross over takes is $f = 0.5 \times t$
    Here, t is provided in specification sheets.

3.11.2   Applications of PIN Diode

1. It is used as a microwave switch.
2. It is used as a modulator.
3. It is used for power limiting.

PROBLEM 3.1   Find the frequency of IMPATT diode with the following parameters.

Carrier drift velocity = 2.2×10^5 m/s

Drift region length = 5 μm

Solution
$$f = \frac{v_d}{2\ell}$$

$$= \frac{2.2 \times 10^5}{2 \times 5 \times 10^{-6}}$$

$$\boxed{f = 22 \text{ GHz}}$$

PROBLEM 3.2   If the carrier drift velocity in IMPATT diode is 3×10^5 m/s and drift region length is 7 µm.

Solution
$$f = \frac{v_d}{2\ell}$$

$$= \frac{3 \times 10^5}{2 \times 7 \times 10^{-6}}$$

$$\boxed{f = 21.4 \text{ GHz}}$$

PROBLEM 3.3   What is the avalanche zone velocity of TRAPATT diode if it is characterized by

Doping concentration $= 1.8 \times 10^{15}/\text{cm}^3$

Current density $= 25$ kA/cm^2

Solution
$$v_{az} = \frac{J}{qN_A}$$

$$= \frac{25 \times 10^3}{1.6 \times 10^{-19} \times 1.8 \times 10^{15}}$$

$$= 0.86 \times 10^6$$

$$\boxed{v_{az} = 86 \times 10^4}$$

PROBLEM 3.4   Find the frequency of Gunn diode oscillator of length 12 µm. The drift velocity in the diode is 2×10^8 m/s.

Solution
$$f = \frac{v_d}{\ell} = \frac{2 \times 10^8}{12 \times 10^{-6}}$$

$$\boxed{f = 16.6 \text{ GHz}}$$

PROBLEM 3.5   The drift length in a Gunn diode is 2.5 µm. Determine the minimum voltage required to operate this Gunn diode.

Solution   Drift length $= 2.5$ µm

The minimum voltage required = the minimum voltage gradient required to start the diode × drift length

$$= 3.3 \times 10^3 \times 2.5 \times 10^{-6}$$

$$\therefore \quad \boxed{V_{min} = 8.25 \text{ mV}}$$

3.12 POINTS TO REMEMBER

- The semiconductors depend on the relative energy differences between valence and conduction bands for their operation.
- The common semiconductors used in microwave devices are Si, GaAs, Inp and Ge.
- The operation of bipolar transistor depends on the thickness of the *p-n* junction depletion region and the transit time.
- In microwave transistor, emitter and base thickness are small.
- FET operation depends on the conductivity of semiconductor layer of GaAs.
- FET is used as amplifier and oscillator.
- FETs use mostly GaAs.
- Tunnel diode is a negative resistance device.
- The impurity concentration in tunnel diode is very high.
- Tunnel diode is also called Esaki diode.
- Tunnel diode is a low power and broadband device.
- The impurity concentration in tunnel diode is very high.
- Tunnel diode is used as amplifier and oscillator.
- Tunnel diode is used as high speed switch.
- TED does not have *p-n* junction.
- TED is a negative resistance device.
- TED is used as an amplified oscillator.
- Gunn diode is a negative resistance device.
- Gunn diode is an oscillator.
- Gunn diode can be used up to 100 GHz.
- Gunn diode is a low noise device.
- Gunn diode is a temperature sensitive device.
- LSA diode means limited space charge accumulation diode.
- LSA diode is an oscillation device.
- IMPATT diode operation depends on the reverse breakdown voltage characteristics of *p-n* junction and phase delay of the applied RF signal.
- IMPATT diode used in digital and analog communication and radar systems.
- IMPATT diode is a noisy device.
- TRAPATT diode produces high power.
- TRAPATT diode is used as local oscillator in transmitter.

> BARITT diode is a low noise oscillator.
> BARITT diode is preferred as an amplifier.
> BARITT diode is a narrow band device.
> BARITT diode is a low power device.

3.13 MULTIPLE CHOICE QUESTIONS

1. The operation of microwave bipolar transistor depends on
   (a) transit time
   (b) p-n junction thickness
   (c) transit time and p-n junction thickness depletion region
   (d) base width
2. Microwave bipolar transistor is operated in
   (a) class A   (b) class B
   (c) class AB   (d) class D
3. The operation of FET depends on
   (a) varying conductivity of semiconductor material
   (b) transit time
   (c) frequency
   (d) pinch off voltage
4. FET is used as
   (a) oscillator   (b) amplitude limiter
   (c) rectifier   (d) switch
5. Tunnel diode is a
   (a) positive resistance device   (b) negative resistance device
   (c) passive device   (d) rectifier
6. Tunnel diode is useful as
   (a) oscillator   (b) rectifier
   (c) low speed switch   (d) clipper
7. Tunnel diode
   (a) high speed switch   (b) clipper
   (c) low gain amplifier   (d) low frequency oscillator
8. In tunnel diode, the tunneling phenomenon is
   (a) minority carrier effect   (b) majority carrier effect
   (c) frequency effect   (d) transit time effect
9. The equivalent circuit of tunnel diode is

(a) 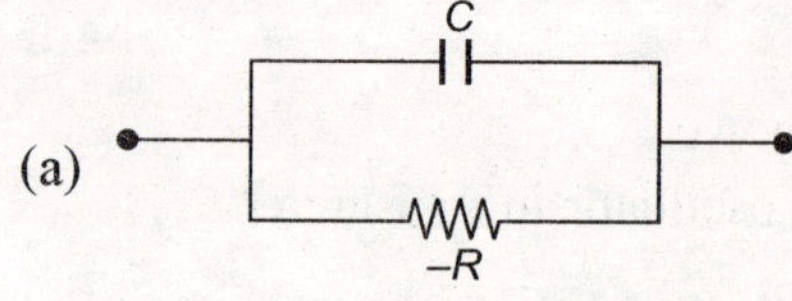

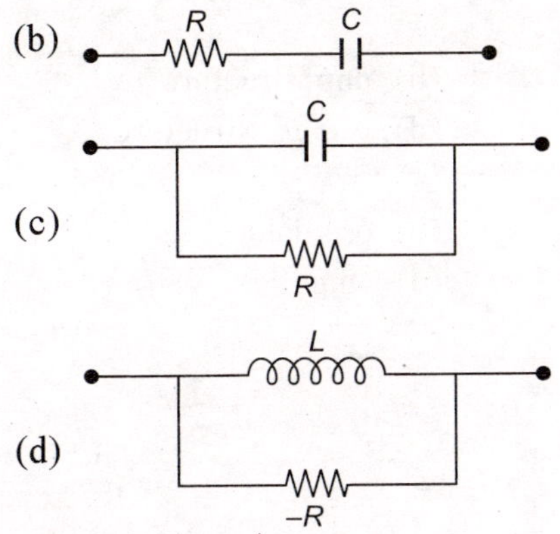

10. TED is a
    (a) FET
    (b) Gunn diode
    (c) tunnel diode
    (d) BARITT diode
11. TED is used as
    (a) switch
    (b) an oscillator
    (c) a clipper
    (d) a rectifier
12. Gunn diode is a
    (a) negative resistance device
    (b) positive resistance device
    (c) high noise device
    (d) low frequency device
13. LSA diode is an example of
    (a) TED
    (b) pn diode
    (c) pnp diode
    (d) pnpn diode
14. IMPATT diode is
    (a) an avalanche transit time device
    (b) a low frequency device
    (c) high efficiency device
    (d) used as switch
15. IMPATT diode is
    (a) a narrow band device
    (b) a wideband device
    (c) a noiseless device
    (d) low frequency device
16. TRAPATT diode has
    (a) p^+nn^+ structure
    (b) n^+p^+n structure
    (c) pnp structure
    (d) npn structure
17. TRAPATT diode is
    (a) high efficiency oscillator
    (b) low efficiency oscillator
    (c) a switch
    (d) a rectifier
18. TRAPATT is used as
    (a) local oscillator in radars
    (b) amplifier in radars
    (c) switch in communication transistor
    (d) low frequency oscillator

19. BARITT diode has
    (a) p^+np^+ structure
    (b) pnp structure
    (c) npn structure
    (d) $p^+n^+p^+$ structure
20. PIN diode is
    (a) phase shifter
    (b) oscillator
    (c) rectifier
    (d) amplifier

3.14 ANSWERS

1. c	2. b	3. a
4. a	5. b	6. a
7. a	8. b	9. a
10. b	11. b	12. a
13. a	14. a	15. b
16. a	17. a	18. a
19. a	20. a	

3.15 EXERCISE PROBLEMS

1. Determine conductivity of the n-type GaAs Gunn diodes if electron density $= 10^{18}$/cm.
   $n_\ell = 10^9$/cm
   $n_{up} = 10^9$/cm
   $T = 300°$ K

2. Determine the electron drift velocity and current density in an n-type GaAs Gunn diode if it has the following parameters.
   Device length $= 12$ μm
   Operating film $= 9$ GHz
   Doping concentration $= 3 \times 10^{14}$/cm^3.

3. Determine the natural frequency of Gunn diode having an active length of 8 μm. The drift velocity in the diode is 2.5×10^7 cm/s.

4. The drift length in a Gunn diode is 3 μm. Find the minimum voltage required to operate Gunn diode. (Note the voltage gradient more than 3.3 kV/m initiates the Gunn diode).

5. The negative resistance of an IMPATT diode is -22 Ω when it is used as a power amplifier, a load resistance of 50 Ω is connected. Determine its power gain.

Chapter 4

Scattering Matrix Parameters

Scattering Matrix describes the relation between the scattered and incident wave amplitudes.

4.1 INTRODUCTION

In microwave circuits, the voltages, currents and impedances are called secondary quantities as they cannot be measured directly. These parameters are also called derived quantities as they are obtained by other directly measurable quantities.

The directly measurable quantities are standing wave ratio, field minimum position and power. In other words, reflection and transmission coefficients are considered to be primary quantities. The reflected or scattered wave amplitudes and phase angles from a junction relative to those of incident waves can be directly measured. The matrix which describes the relationship between the scattered and incident wave amplitudes is known as scattering matrix.

4.2 PROPERTIES OF SCATTERING MATRIX

S-matrix is a

- ➤ square matrix.

  Ex. : $\begin{bmatrix} S_{11} & S_{12} \\ S_{21} & S_{22} \end{bmatrix}$

- ➤ symmetric matrix for a reciprocal junction.

  i.e. $S_{ij} = S_{ji}$

- ➤ unitary matrix

  i.e. $[S]\,[S]^* = I$

  Here, $[S]$ is complex conjugate of $[S]$.

- ➤ $[S]$ matrix satisfies $\displaystyle\sum_{i=1}^{n} S_{ik}\, S_{ij}^* = 0$ for $k \neq j$.

  Here, $j = 1, 2, 3, \ldots n$

  $\qquad k = 1, 2, 3, \ldots n$

> For a lossless junction, $\displaystyle\sum_{i=1}^{n} S_{ik} S_{ik}^{*} = 1$.

4.3 PROOF OF SYMMETRIC PROPERTY

Let $V_m = V_m^{i} + V_m^{r}$

$$I_m = I_m^{i} + I_m^{r} = V_m^{+} - V_m^{-}$$

It is well known that

$$[V] = [V^{i}] + [V^{r}] = [Z][I]$$
$$= [Z][V^{i}] - [Z][V^{r}]$$

In terms of unitary matrix, we have

$$[Z] + [U][V^{r}] = [Z] - [U][V^{r}]$$

Rearranging, we have

$$[V^{r}] = ([Z] + [U])^{-1} ([Z] - [U][V^{i}])$$

As $[V^{r}] = [S][V^{i}]$, we can write

$$[S] = ([Z] + [U])^{-1} ([Z] - [U])$$
$$= ([Z] - [U])([Z] + [U])^{-1} \tag{4.1}$$

The transpose of $[S]$ is

$$[S]_T = ([Z] - [U])_T ([Z] + [U])_T^{-1}$$

Here, $([Z] - [U])_T$ and $([Z] + [U])_T$ are symmetric to each other.

$\therefore \qquad ([Z] - [U])_T = ([Z] - [U])$

$\therefore \qquad S_T = ([Z] - [U]) ([Z] + [U])^{-1} \tag{4.2}$

From the equations (4.1) and (4.2)

$$[S] = [S]_T \quad \text{Hence Proved.}$$

4.4 PROOF OF UNITARY PROPERTY

According to this property,

$$[S][S]^{*} = I \text{ for a lossless network}$$

Assume that the network is lossless.

For this network $\displaystyle\sum_{n=1}^{N} |V_m^{r}|^2 = \sum_{n=1}^{N} |V_n^{r}|^2$

When a single port is excited while the other ports are matched terminated, we have,

$$\sum_{n=1}^{N} S_{nk} |V_k^{i}|^2 = \sum_{n=1}^{N} |V_k^{i}|^2$$

$V_m^i = 0$ except V_k if k^{th} port is excited.

i.e.
$$\sum_{n=1}^{N} |S_{nk}|^2 = 1$$

But
$$\text{LHS} = \sum_{n=1}^{N} S_{nk} \, S_{nk}^*$$

$\therefore$
$$\sum_{n=1}^{N} S_{nk} \, S_{nk}^* = 1$$

As $V_n^i = 0$, $V_k^i \neq 0$ and $V_m^i \neq 0$

We have $\displaystyle\sum_{n=1}^{N} S_{nm} \cdot S_{nk}^* = 0$, $k \neq m$

$\therefore$
$$[S^*][S]^T = [U]$$

4.5 DEFINITION OF SCATTERING MATRIX

The scattering matrix of an m-port junction is a square matrix of a set of elements which relate incident and reflected waves at the ports of the junction. The diagonal elements of the s-matrix represent reflection coefficients and the off-diagonal elements represent the transmission coefficients.

4.6 CHARACTERISTICS OF S-MATRIX

- ➤ It describes any passive microwave component.
- ➤ It exists for linear passive and time-invariant networks.
- ➤ It gives complete information on reflection and transmission coefficients.

4.7 SCATTERING MATRIX OF A TWO-PORT NETWORK

The scattering matrix of a two-port network (fig. 4.1) is defined as

$$S = \begin{bmatrix} s_{11} & s_{12} \\ s_{21} & s_{22} \end{bmatrix}$$

Here, $s_{11} = \dfrac{y_1}{x_1} \bigg/ x_2 = 0$

$s_{21} = \dfrac{y_2}{x_1} \bigg/ x_2 = 0$

$s_{12} = \dfrac{y_1}{x_2} \bigg/ x_1 = 0$

$$s_{22} = \left.\frac{y_2}{x_2}\right/ x_1 = 0$$

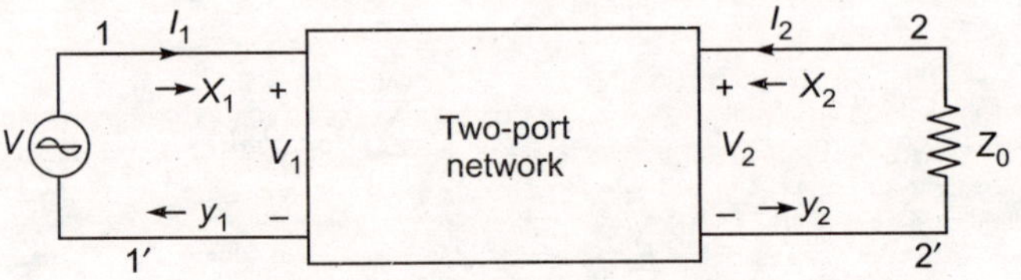

Fig. 4.1   Two-port network

Here, x_1, x_2, y_1 and y_2 are normalized values and $\frac{1}{2} x_i x_i^*$ is input power and $\frac{1}{2} y_i$ y_i^* is the output power at the port i.

For such cases x_1, x_2, y_1 and y_2 are defined by

$$x_1 = \frac{1}{2}\left(\frac{V_1}{\sqrt{z_0}} + \sqrt{z_0}\, I_1\right)$$

$$y_1 = \frac{1}{2}\left(\frac{V_1}{\sqrt{z_0}} - \sqrt{z_0}\, I_1\right)$$

$$x_2 = \frac{1}{2}\left(\frac{V_2}{\sqrt{z_0}} + \sqrt{z_0}\, I_2\right)$$

$$y_2 = \frac{1}{2}\left(\frac{V_2}{\sqrt{z_0}} - \sqrt{z_0}\, I_2\right)$$

Here, $y = Sx$

$$x = \begin{pmatrix} x_1 \\ x_2 \end{pmatrix}$$

$$y = \begin{pmatrix} y_1 \\ y_2 \end{pmatrix}$$

The scattering coefficients can be expressed in terms of eigenvalues of the characteristics equation.

That is,
$$s_{11} = (e_1 + e_2)/2$$
$$s_{21} = (e_1 + e_2)/2$$

For a matched junction, $e_1 = -e_2$

$$s_{11} = 0$$
$$|s_{21}| = 0$$

4.8 SALIENT FEATURES OF *S*-MATRIX

- ➢ Scattering matrix is a square matrix.
- ➢ Its elements relate incident and reflected waves at the terminals of the network.
- ➢ Its main diagonal elements represent reflection coefficient.
- ➢ Its off-diagonal elements represent transmission coefficients.
- ➢ s_{ii}, s_{ij}, s_{ji} of a two-port network represent scattering coefficients.
- ➢ For a matched junction, $s_{ii} = 0$, $|s_{ji}| = 1$.
- ➢ Scattering coefficients are also expressed as

$$s_{11} = s_{22} = -\frac{Y}{Y + 2Y_0}$$

$$s_{21} = s_{12} = \frac{2Y_0}{Y + 2Y_0}$$

Here, Y_0 = characteristic admittance of a transmission line

or
$$s_{11} = s_{22} = \frac{z}{z + 2z_0}$$

$$s_{21} = s_{12} = \frac{2z_0}{z + 2z_0}$$

- ➢ For cascaded circuits, the concept scattering transfer parameters is used. The transfer parameters are expressed in terms of scattering coefficient. That is,

$$T_{11} = s_{12} - \frac{s_{11}\,s_{22}}{s_{21}}$$

$$T_{12} = \frac{s_{11}}{s_{21}}$$

$$T_{21} = -\frac{s_{22}}{s_{21}}$$

$$T_{22} = \frac{1}{s_{21}}$$

- ➢ For a reciprocal network, S matrix is symmetric. i.e.

$$s_{ij} = s_{ji} \quad (i \neq j)$$

➤ For a reciprocal network, transpose of s is equal to s. That is,

$$(s)_t = (s)$$

➤ For a dissipationless junction, scattering marix is unitary. That is,

$$\sum_{m=1}^{M} s_{mn} \cdot s_{mn}^* = 1$$

But for $x_m = 0$ except x_n and x_k

$$\sum_{m=1}^{M} s_{mn} \cdot s_{mk}^* = 0, \quad n \neq k$$

or

$$(s^*)t = (s_1)_t^{-1}$$

➤ S-matrix exhibits phase shift property.
If the reference planes are shifted by electrical phase shifts, the resultant S-matrix is given by

$$(s') = \begin{pmatrix} e^{-j\theta_1} & 0 \\ 0 & e^{-j\theta_2} \end{pmatrix} (s) \begin{pmatrix} e^{-j\theta_1} & 0 \\ 0 & e^{-j\theta_2} \end{pmatrix}$$

Here, $\theta_1 = \beta_1 \ell_1$
$\theta_2 = \beta_2 \ell_2$
ℓ_1, ℓ_2 = path lengths
β_1, β_2 = phase constant

4.9 SCATTERING MATRIX OF MULTI-PORT NETWORK

It is given by

$$S = \begin{bmatrix} s_{11} & s_{12} & \cdots & s_{1N} \\ s_{21} & s_{22} & \cdots & s_{2N} \\ \vdots & \vdots & \cdots & \vdots \\ s_{N1} & s_{N2} & \cdots & s_{NN} \end{bmatrix}$$

Here, $s_{11}, s_{22}, \ldots, s_{NN}$ represent reflection coefficients
$s_{12}, s_{21}, \ldots$ represent transmission coefficients

4.10 LOSSES IN MICROWAVE CIRCUITS

The losses in microwave devices and circuits are expressed in terms of

➤ Return loss
➤ Insertion loss
➤ Transmission loss and
➤ Reflection loss

4.11 RETURN LOSS (RL)

It is defined as

$$RL \equiv \frac{p_i}{p_0}$$

$$RL(\text{dB}) \equiv 10 \log_{10} \frac{p_i}{p_0}$$

4.12 INSERTION LOSS (IL)

It is defined as

$$IL \equiv \frac{p_i}{p_r}$$

$$IL(\text{dB}) = 10 \log_{10} \frac{p_i}{p_r}$$

4.13 TRANSMISSION LOSS (TL)

It is defined as

$$TL \equiv \frac{p_i - p_r}{p_0}$$

$$TL(\text{dB}) = 10 \log_{10} \left(\frac{p_i - p_r}{p_0} \right)$$

4.14 REFLECTION LOSS (Γ)

It is defined as

$$\Gamma \equiv \left(\frac{p_i}{p_i - p_r} \right)$$

$$\Gamma(\text{dB}) = 10 \log_{10} \left(\frac{p_i}{p_i - p_r} \right)$$

Here, p_i = input power at port 1

$\quad p_r$ = reflected power at port 1

$\quad p_0$ = input power at port 2

The above losses are also defined in terms of S-parameters.

$$\text{Return loss } (RL) = 20 \log \frac{1}{|s_{11}|}$$

$$\text{Insertion loss } (IL) = 20 \log \frac{1}{|s_{12}|}$$

$$\text{Transmission loss } (TL) = 10 \log \left(\frac{1 - |s_{11}|^2}{|s_{12}|^2} \right)$$

$$\text{Reflection loss } (\Gamma) = 10 \log \left(\frac{1}{|1 - s_{11}|^2} \right)$$

4.15 IMPEDANCE MATRIX

The impedance matrix for two-port junction is defined as

$$Z \equiv \begin{bmatrix} Z_{11} & Z_{12} \\ Z_{21} & Z_{22} \end{bmatrix}$$

Any two-port junction can be described by the voltage and current relations through impedance matrix.

i.e.
$$V = Z I$$

Here,
$$V = \begin{pmatrix} V_1 \\ V_2 \end{pmatrix}$$

$$I = \begin{pmatrix} I_1 \\ I_2 \end{pmatrix}$$

$$z_{11} = \frac{V_1}{I_1} \bigg/ I_2 = 0$$

$$z_{12} = \frac{V_1}{I_2} \bigg/ I_1 = 0$$

$$z_{22} = \frac{V_2}{I_2} \bigg/ I_1 = 0$$

$$z_{21} = \frac{V_2}{I_1} \bigg/ I_2 = 0$$

These are open circuit parameters.

For a symmetric junction,

$$z_{11} = z_{22}$$

and for a reciprocal junction,

$$z_{12} = z_{21}$$

PROBLEM 4.1   For a T-network, express the elements of Z-matrix in terms of the elements of the equivalent circuit.

Solution   For the T-network of fig. 4.2,

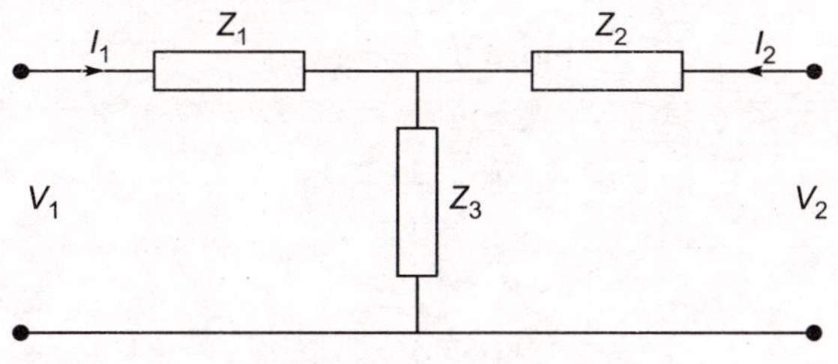

Fig. 4.2 T-network

applying the definitions of z_{11}, z_{12}, z_{21}, z_{22}, we get

$$z_{11} = z_{22} = z_1 + z_3$$
$$z_{12} = z_{21} = z_3$$

4.16   ADMITTANCE MATRIX

It is defined as

$$Y \equiv Z^{-1}$$

or

$$Y \equiv \begin{pmatrix} y_{11} & y_{12} \\ y_{21} & y_{22} \end{pmatrix}$$

Here, $y_{11} = \dfrac{I_1}{V_1} \bigg/ V_2 = 0$

$$y_{12} = \dfrac{I_1}{V_2} \bigg/ V_1 = 0$$

$$y_{21} = \dfrac{I_2}{V_1} \bigg/ V_2 = 0$$

$$y_{22} = \dfrac{I_2}{V_2} \bigg/ V_1 = 0$$

represent short circuit parameters.

PROBLEM 4.2 Obtain short circuit admittance parameters for a π-network.

Solution A typical π-network is

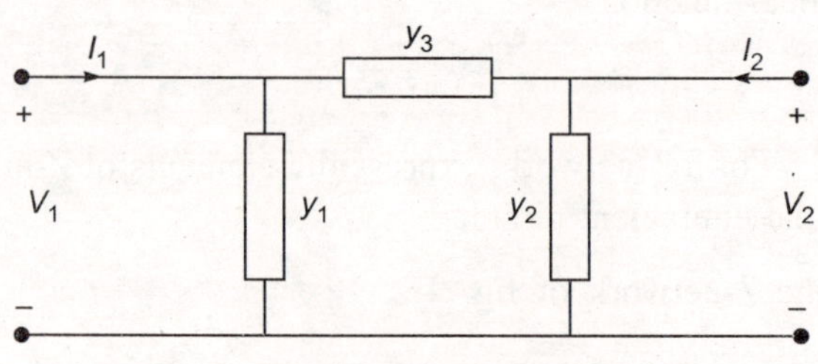

Fig. 4.3 π-network

$$y_{11} = y_1 + y_3$$
$$y_{21} = y_3$$
$$y_{12} = y_{21}$$
$$y_{11} = y_{22}$$

and

4.17 SUMMARY OF S, Z AND Y MATRICS

Parameter	S-matrix	Z-matrix	Y-matrix
type of matrix	square	square	square
Information given by the matrix	It relates the incident and reflected waves at the terminals of network	It relates voltages and currents of junction	It relates currents and voltages of a junction
Example	$\begin{bmatrix} s_{11} & s_{12} \\ s_{21} & s_{22} \end{bmatrix}$	$\begin{bmatrix} z_{11} & z_{12} \\ z_{21} & z_{22} \end{bmatrix}$	$\begin{bmatrix} y_{11} & y_{12} \\ y_{21} & y_{22} \end{bmatrix}$
Properties	It is symmetric, & unitary. For a reciprocal network $s_{12} = s_{21}$. For a symmetric network, $s_{11} = s_{22}$	For a reciprocal network $z_{12} = z_{21}$. For a symmetric network $z_{11} = z_{22}$	For a reciprocal network, $y_{12} = y_{21}$. For a symmetric network $y_{11} = y_{22}$
Units of element	no units	ohms	Mhos
No. of matrix elements	Equal	Equal	Equal

4.18 SHUNT ELEMENT IN A TRANSMISSION LINE

Consider fig. 4.4.

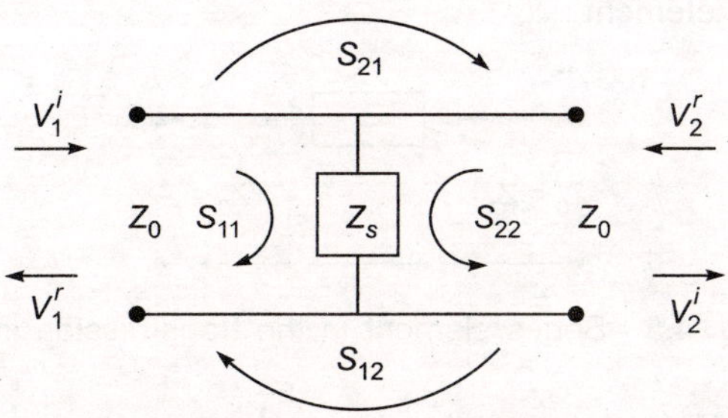

Fig. 4.4 Shunt element in a transmission line

Here, Z_0 = characteristic impedance

Z_s = shunt impedance

For matched output, $V_2^r = 0$.

S_{11} = reflection coefficient on the input side.

$$\frac{Y_0 - Y_i}{Y_0 + Y_i} = \frac{Y_0 - Y_0 - Y_s}{2Y_0 + Y_s}$$

$$\therefore \qquad S_{11} = \frac{-Y_s}{2Y_0 + Y_s}$$

As the above network is symmetrical,

$$S_{22} = S_{11} = \frac{-Y_s}{2Y_0 + Y_s}$$

We know $S_{12} = \sqrt{1 - |S_{11}|^2}$

and

$$V_1^i + V_1^r = V_2^i$$
$$= V_1^i(1 + S_{11})$$

But $V_2^i = S_{21} V_1^i$

$$\therefore \qquad S_{21} = 1 + S_{11} = S_{12}$$

$$= \frac{2Y_0}{2Y_0 + Y_s}$$

$\therefore$ S-matrix for a shunt element in the transmission line is

$$S = \begin{bmatrix} -\dfrac{Y_0}{2Y_0 + Y_s} & \dfrac{2Y_0}{2Y_0 + Y_s} \\[3mm] \dfrac{2Y_0}{2Y_0 + Y_s} & \dfrac{Y_s}{2Y_0 + Y_s} \end{bmatrix}$$

4.19 S-MATRIX OF SERIES ELEMENT IN THE TRANSMISSION LINE

Consider the series element

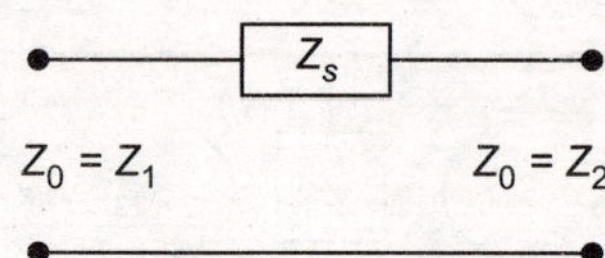

Fig. 4.5 Series element in the transmission line

Let V_1^i, V_1^r, V_2^r, V_2^i be transmission line voltages.

For matched output, we can write

$$\frac{V_1^r}{V_1^i} = \frac{V_{1n}^r}{V_{1n}^i}$$

Here, V_{1n}^r and V_{1n}^i are normalized values.

$$\therefore \quad S_{11} = \frac{Z_i - Z_1}{Z_i + Z_1} = \frac{Z_2 - Z_1 + Z_s}{Z_2 + Z_1 + Z_s}$$

For matched input line,

$$S_{22} = \frac{Z_1 - Z_2 + Z_s}{Z_2 + Z_1 + Z_s}$$

For matched output, we can writ

$$V_1 = V_1^i + V_1^r$$
$$= V_1^i(1 + S_{11})$$
$$I_1 = Y_1(V_1^i - V_1^r) = Y_1 V_1^i(1 - S_{11})$$

and

$$I_2 = -I_2^r = -I_1$$
$$= Y_1 V_1^i(1 - S_{11})$$
$$I_2^i = Y_2 V_2^i$$

i.e.

$$Y_2 V_2^i = Y_1 V_1^i(1 - S_{11})$$

$$\therefore \quad S_{21} = \frac{V_{2n}^i}{V_{1n}^i} = \left(\frac{Y_2}{Y_1}\right)^{1/2} \frac{V_2^i}{V_1^i}$$

$$= \left(\frac{Z_2}{Z_1}\right)^{\frac{1}{2}}$$

$$\therefore \quad S_{21} = S_{12} = \frac{2Z_i}{Z_1 + Z_2 + Z_s}$$

$$= \frac{2\sqrt{Z_1 Z_2}}{Z_1 + Z_2 + Z_s}$$

$\therefore$ S-matrix for series element is

$$S = \begin{bmatrix} \dfrac{Z_2 - Z_1 + Z_s}{Z_2 + Z_1 + Z_s} & \dfrac{2Z_1}{Z_1 + Z_2 + Z_s} \\[2ex] \dfrac{2Z_2}{Z_1 + Z_2 + Z_s} & \dfrac{Z_1 - Z_2 + Z_s}{Z_1 + Z_2 + Z_s} \end{bmatrix}$$

PROBLEM 4.3 The guide wavelength is 4.82 cm and the distance between twice minima points is 0.7 cm. Find VSWR.

Solution

$$\lambda_g = 4.82 \text{ cm}$$
$$d_1 - d_2 = 0.7 \text{ cm}$$

$$\text{VSWR} = \frac{\lambda_g}{\pi (d_1 - d_2)}$$

$$= \frac{4.82}{\pi \times 0.7}$$

$\therefore$
$$\text{VSWR} = 2.193$$

PROBLEM 4.4 Find the scattering matrix of an inductor whose insertion loss is 0.3 dB and an isolation of 40 dB. Assume that the points are well matched.

Solution

$$\text{Insertion loss} = 0.3 \text{ dB}$$
$$= -20 \log |s_{21}|$$

or
$$s_{21} = 10^{-03/20}$$
$$\text{Isolation} = 40 \text{ dB} = -20 \log |s_{12}|$$
$$s_{12} = 10^{-40/20} = 10^{-2}$$
$$s_{11} = s_{22} = 0$$

$$S = \begin{bmatrix} 0 & 10^{-2} \\ 10^{-0.015} & 0 \end{bmatrix}$$

$$S = \begin{bmatrix} 0 & 0.01 \\ 0.966 & 0 \end{bmatrix}$$

PROBLEM 4.5 The measured value of VSWR of a waveguide component is 2.5. The distance measured between twice minima is 0.4 cm. Find guide wavelength.

Solution
$$d_1 - d_2 = 0.4 \text{ cm}$$

$$\text{VSWR} = \frac{\lambda_g}{\pi(d_1 - d_2)}$$

or

$$\lambda_g = \text{VSWR} \times \pi(d_1 - d_2)$$
$$= 2.5 \times \pi \times 0.4$$

$$\boxed{\lambda_g = 3.14 \ \text{cm}}$$

4.20 POINTS TO REMEMBER

- Scattering matrix describes any passive microwave components.
- Scattering matrix is a square matrix.
- Scattering matrix is defined as

$$S = \begin{bmatrix} s_{11} & s_{12} \\ s_{21} & s_{22} \end{bmatrix}$$

- For a matched junction, $s_{ii} = 0$, $|s_{ij}| = 1$.
- The diagonal elements of scattering matrix represents reflection coefficients.
- The off diagonal elements of S represents transmission coefficients.
- For a reciprocal network, S matrix is symmetric.
- S-matrix exhibits shift propagation.

- Return loss is defined as $RL = 10 \ \log_{10} \dfrac{p_i}{p_r}$.

- Insertion loss is defined as $IL = 10 \ \log_{10} \dfrac{(p_i - p_r)}{p_0}$.

- Reflection loss is defined as $\Gamma \equiv 10 \ \log_{10} \left(\dfrac{p_i}{p_i - p_r} \right)$.

- Return loss $= 20 \ \log \dfrac{1}{|s_{11}|}$.

- Insertion loss $= 20 \ \log \dfrac{1}{|s_{12}|}$.

- Transmission loss $= 10 \ \log \dfrac{1 - |s_{11}|^2}{|s_{12}|^2}$.

- Reflection loss $= 10 \ \log \dfrac{1}{1 - |s_{11}|^2}$.

➢ The impedance matrix of a two-port junction is $Z = \begin{pmatrix} z_{11} & z_{12} \\ z_{21} & z_{22} \end{pmatrix}$.

➢ For a symmetric junction, $z_{11} = z_{22}$.
➢ For a reciprocal junction, $z_{12} = z_{21}$.

➢ The admittance matrix of a two-port junction is $Y = \begin{pmatrix} y_{11} & y_{12} \\ y_{21} & y_{22} \end{pmatrix}$.

4.21 MULTIPLE CHOICE QUESTIONS

1. S-matrix is
   (a) a square matrix
   (b) a rectangular matrix
   (c) always 2×2 matrix
   (d) always 3×3 matrix
2. s_{11} of S-matrix represents
   (a) reflection coefficient
   (b) transmission coefficient
   (c) voltage gain
   (d) power gain
3. S-matrix exists for
   (a) two port only
   (b) 3 port only
   (c) multiport network
   (d) 4 port only
4. For matched junction,
   (a) $s_{11} \neq 0$
   (b) $s_{11} = 0$
   (c) $s_{21} = 1$
   (d) $|s_{21}| = 0$
5. For a reciprocity network
   (a) $s_{ij} \neq s_{ji}$
   (b) $s_{ij} = s_{ji}$
   (c) $s_{ji} = s_{ij}$
   (d) $s_{ii} = s_{ji}$
6. For reciprocal network transpose of s is
   (a) $(s)_t = s$
   (b) $(s)_t \neq s$
   (c) $(s)_t = s^{-1}$
   (d) $(s)_t = 0$
7. The units of return loss is
   (a) nil
   (b) watts
   (c) volts
   (d) ohms
8. Insertion loss has
   (a) no units
   (b) units of watts
   (c) units of ohms
   (d) units of volts
9. Return loss is

   (a) $20 \log_{10} \dfrac{1}{|s_{21}|}$
   (b) $20 \log_{10} \dfrac{1}{|s_{11}|}$

   (c) $20 \log_{10} |s_{11}|$
   (d) $20 \log_{10} |s_{21}|$
10. For symmetric junction,
    (a) $z_{11} = z_{21}$
    (b) $z_{11} = z_{22}$
    (c) $z_{11} = z_{21}$
    (d) $z_{11} \neq z_{22}$

4.22 ANSWERS

1. a	2. a	3. c
4. c	5. b	6. a
7. a	8. a	9. b
10. b		

4.23 EXERCISE PROBLEMS

1. For a lossless two port network, prove that

$$\sum_{m=1}^{M} S_{mn} S_{mk}^{*} = \sum_{m=1}^{M} S_{mn}^{*} S_{mk} = 0.$$

2. A two-port network consists of $j10$ as series element and $j20$ as shunt element. Prove that $|s_{11}|^2 + |s_{12}|^2 = 1$.

3. Prove for a matched two-port junction, $s_{11} = 0$ and $|s_{21}| = 1$.

4. If S-matrix of a two-port network is $[S] = \begin{bmatrix} 1 & 1 \\ 1 & -1 \end{bmatrix}$, find the roots of the characteristic equation.

5. The scattering matrix of a two-port network is $S = \begin{bmatrix} 0 & 1 \\ 1 & 0 \end{bmatrix}$, find its impedance matrix.

6. Determine scattering matrix of a 3 dB directional coupler.

Chapter 5

Microwave Passive Components

Microwave passive components are dissipative in nature.

5.1 COMMON PASSIVE COMPONENTS

The common passive components used at microwave frequencies are

- Two-wire lines
- Coaxial lines
- Rectangular waveguides
- Circular waveguides
- Cavity resonators
- Ridge waveguides
- Matched loads
- Corners
- Bends
- Twists
- Circulators
- Isolators
- Attenuators
- Waveguide directional couplers
- Tee junctions
- H-plane Tee
- E-plane Tee
- Hybrid Tee junctions
- Magic Tee
- Ferrite devices
- Phase shifters
- Hybrid rings
- Choke joints
- Flanges
- Tuning screws and posts
- Strip lines
- Microstrips
- Connectors
- Coaxial sections
- Coaxial loads
- Impedance changing devices
- Slotted coaxial lines
- Coaxial cavities
- Slotted coaxial lines
- Probes
- Holes and slots
- Waveguide terminations
- Resonators
- Transition
- Switches

5.2 TWO-WIRE LINES

Two-wire lines are used to transfer power from one place to another. They are used at lower end of microwave range of frequencies.

A typical two-wire line is shown in fig. 5.1.

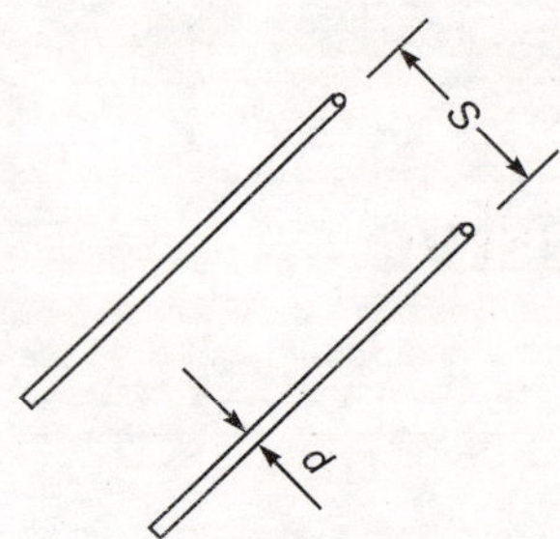

Fig. 5.1 Two-wire transmission line

The two wires have identical dimensions. Their spacing is uniform and the diameter is the same for both of them.

5.2.1 Salient Features of Two-wire Lines

➤ Energy transfer takes place by electric and magnetic fields.
➤ The electric fields carry half of the energy and magnetic fields carry another half.
➤ The electromagnetic energy is distributed around the wires.
➤ Propagation is by transverse electromagnetic (TEM) waves. TEM waves have zero cut-off frequency. They travel with the velocity of light in free space.

➤ The lines have the characteristic impedance given by $Z_0 = \sqrt{\dfrac{L}{C}}$

L is distributed series inductance
C is the distributed shunt capacitance

It is also given by $Z_0 = \dfrac{276}{\sqrt{\epsilon_r}} \log_{10} \dfrac{2s}{d}$

s = spacing between the two lines
d = diameter of the conductance
ϵ_r = relative permittivity of the medium between the wires.
➤ They are not useful at microwave frequencies due to the radiation loss and high attenuation.
➤ The wave penetration in the wires is more at low frequencies and it is less at high frequencies.
➤ Skin depth is inversely proportional to frequency.
➤ At high frequencies, the cross-section of the wire is small.
➤ At high frequencies, attenuation is high and hence power handling capacity of the lines is small.

➤ The time taken for a wave to propagate through a loss less line of length

is $t = \sqrt{LC}$, sec.

This is valid when the resistance and conductance are ignored.

➤ The equivalent circuit is a distributed network given by

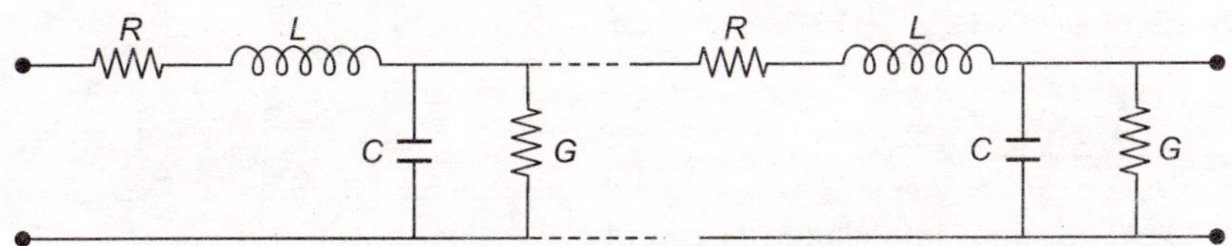

➤ The velocity of propagation of wave along the line is $v \approx \dfrac{1}{\sqrt{LC}}$, m/s

➤ The impedance of the line at any point is $Z = \sqrt{\dfrac{L}{C}}$, Ω

➤ When the transmission line is terminated in a load impedance of z_L, the reflection of the incident wave takes place. Then the reflection coefficient

is given by $\rho \equiv \dfrac{V_r}{V_i} = \dfrac{z_L - z_0}{z_L + z_0}$

V_r = reflected wave

V_i = incident wave

➤ If the load impedance is equal to characteristic impedance, z_0 the reflection coefficient is zero.

i.e. $\rho = 0$ if $z_L = z_0$

➤ If the load is open ($z_L = \infty$), $\rho = 1$.

➤ If the load is short ($z_L = 0$), $\rho = 1$.

➤ If the load is purely reactive, $|\rho| = 1$.

➤ When the load is not terminated in its characteristic impedance, standing waves on the line exist due to reflection.

➤ The presence of standing waves is represented by voltage standing wave

ratio (VSWR). It is defined as $\text{VSWR}\,(S) = \dfrac{V_{max}}{V_{min}}$.

➤ S in terms of reflection coefficient is $S = \dfrac{1 - |\rho|}{1 - |\rho|}$.

➤ $S = 1$ if $|\rho| = 0$.

➤ $S = \infty$ if $|\rho| = 1$.

➤ $1 \leq S \leq \infty$ for $0 \leq |\rho| \leq 1$.

➤ The line has minimum impedance when the voltage is minimum and the current is maximum.

➤ $z_{\min} = \dfrac{z_0}{\text{VSWR}}$

➤ $\text{VSWR} = \dfrac{z_L}{z_0}$ if $z_L > z_0$

➤ $\text{VSWR} = \dfrac{z_0}{z_L}$ if $z_0 > z_L$

➤ The input impedance of the line

$$z_i = z_0 \frac{z_L - jz_0 \tan \beta\ell}{z_0 + jz_L \tan \beta\ell}$$

$$\beta = \frac{2\pi}{\lambda_\ell}, \ \lambda_\ell = \text{wavelength on the line.}$$

➤ The input impedance of half-wave transmission line is $z_i = z_L$.

➤ The input impedance of quarter wave transmission line is $z_i = \dfrac{z_0^2}{z_{\text{output}}}$ or

$$z_0 = \sqrt{z_i\, z_{\text{output}}}$$

5.2.2   Advantages of Two-wire Lines

➤ Low cost
➤ Easy manufacturing
➤ Attenuation is small at low frequencies

5.2.3   Disadvantages

➤ Reduction losses are high
➤ Attenuation is high at high frequencies
➤ Power handling capacity is small at high frequencies.

5.2.4   Applications

They are

➤ To transfer energy from one point to another point
➤ To serve as reactive circuit elements or tuned circuits
➤ As stubs to match the load impedance
➤ As impedance transformers
➤ To match the feed line to the antenna to radiate the available power.

PROBLEM 5.1 Find z_0 of a two-wire transmission line if $L = 1$ mH/km and $C = 0.25$ μF/km.

Solution
$$z_0 = \sqrt{\frac{L}{C}} = \sqrt{\frac{1 \times 10^3}{0.25 \times 10^6}}$$

$$z_0 = \sqrt{4 \times 10^3} \ \Omega$$

$$\boxed{z_0 = 63.24 \ \Omega}$$

PROBLEM 5.2 Determine z_0 of a transmission line if the spacing, $s = 0.49$ cm and the diameter of the wire is 0.1 cm. Assume $\epsilon_r = 1$.

Solution
$$z_0 = \frac{276}{\sqrt{\epsilon_r}} \ \log \frac{2s}{d}$$

$$= 276 \ \log_{10} \frac{2 \times 0.49}{0.1}$$

$$= 276 \ \log_{10} (9.8)$$

$$= 276 \times 0.991$$

$$\boxed{z_0 = 273.51}$$

5.3 COAXIAL LINES

Coaxial line consists of an inner conductor and an outer conductor. A typical coaxial line is shown in fig. 5.2.

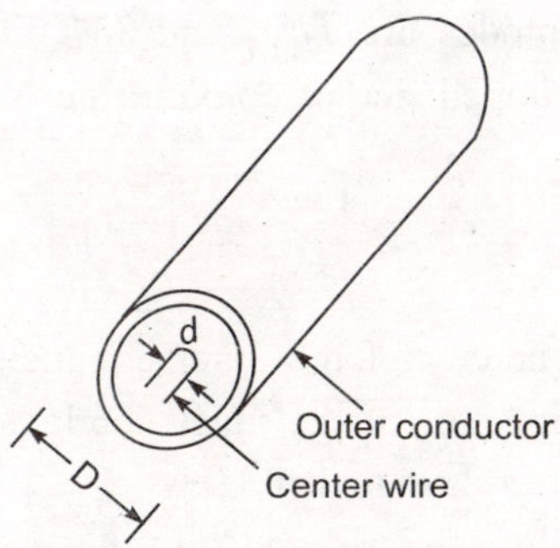

Fig. 5.2 Coaxial cable

These are used to transfer energy from one point to another. It also supports TEM wave. TEM wave is called principal or dominant wave mode. It has no cut-off frequency.

5.3.1 Salient Features of Coaxial Lines

1. TEM wave propagates in coaxial lines
2. There is no cut-off frequency for TEM
3. For a dielectric filled coaxial line, the wavelength is $\lambda = \dfrac{v_0}{\sqrt{\epsilon_r}\, f}$
4. Higher order modes exist in coaxial line if the wavelength is close to the physical dimensions of the line.
5. The lowest wavelength that will propagate is called cut-off wavelength.
6. The cut-off wavelength for *TE* modes

$$\lambda_{\text{cut-off}} = \left(\frac{\pi}{n} \frac{(d + D)}{2} \right)$$

$$= \frac{\pi\,(d + D)}{2n}$$

Here, $n = 1, 2, 3, \ldots,$
$\qquad$ = Number of half-wavelengths around the circle
$\qquad n = 1$ for principal mode

7. The cut-off wavelength of higher order *TM* modes is given by

$$\lambda_{\text{cut-off}} = \frac{1}{m}\,(D - d)$$

$m = 1, 2, 3, \ldots,$
$\qquad$ = Number of half-wavelengths spacing between the conductors.

8. The high orders are expressed in the form of TE_{mn}, TM_{mn}.
9. The lowest order modes are $TE_{1,1}$ and $TM_{0,1}$
10. The characteristic impedance of coaxial line is given by

$$Z_0 = \frac{138}{\sqrt{\epsilon_r}}\,\log\left(\frac{D}{d}\right)$$

Here, D = inner diameter of the outer conductor
$\qquad d$ = outer diameter of the inner conductor

11. Coaxial lines are useful up to 100 GHz.
12. They can be flexible, semigrid and rigid.
13. The use of coaxial cable is limited by operating frequency.
14. If f increases, the diameter of the line must be reduced to eliminate higher mode propagation.
15. The small sized coaxial line can handle more power.
16. They can be interfaced with waveguides.

17. The characteristic impedance, frequency range, power handling, voltage breakdown, applications, attenuation, pulse characteristics and environment describe the coaxial lines.

18. Attenuation of the line depends on conductive loss due to conductor size. It increases with frequency.

19. Attenuation due to conductor losses (α_c) is given by

$$\alpha_c = \frac{1.382}{\sigma \delta z_0} \left(\frac{1}{d} + \frac{1}{D} \right) \text{ dB/m}$$

20. Attenuation due to dielectric losses (α_d) is given by

$$\alpha_d = \frac{27.3\sqrt{\epsilon_r}}{\lambda_0} \tan \delta \text{ dB/m}$$

Here, δ = skin depth

σ = conductivity of the material forming the inner and outer conductor

λ_0 = free space wavelength

$\tan \delta$ = loss tangent

21. The standard value of z_0 is 50 Ω.

22. TEM mode in a coaxial line is shown in fig. 5.3.

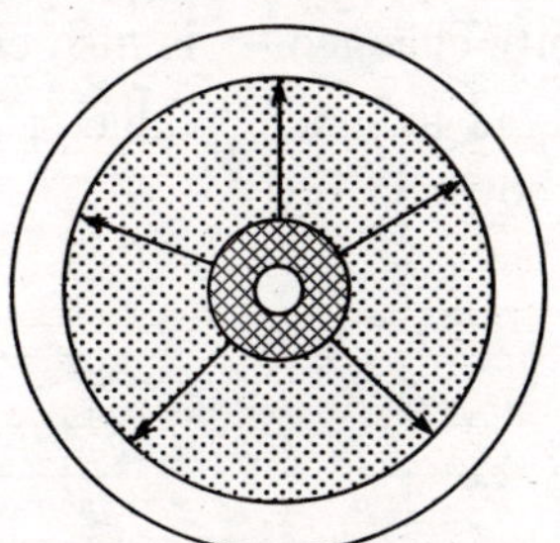

Fig. 5.3   TEM mode

23. The distributed constants of coaxial line are given by

$$L = 0.46 \log_{10} \left(\frac{D}{d} \right) \text{ µH/m}$$

$$C = \frac{241 \epsilon}{\log_{10} \left(\dfrac{D}{d} \right)} \text{ pF/m}$$

$$R = 4.14 \times 10^{-6} \sqrt{f} \left(\frac{1}{D} + \frac{1}{d} \right) \text{ } \Omega/\text{m}$$

24. The wavelength in a coaxial line is $\lambda_{c\ell} = \dfrac{\lambda_0}{\epsilon}$

25. The velocity of propagation is $v_d = \dfrac{v_0}{\sqrt{\epsilon}}$.

26. The velocity of propagation is the same as phase velocity in coaxial line.

27. The electric field inside a point between inner and outer conductor is

$$E = \frac{V}{r \ell n\left(\dfrac{D}{a}\right)} \ \text{volt/m}$$

Here, r is in meters.

28. The inner conductor is supported by shorted circuited stubs. A short circuited line which is one quarter-wavelength long. This line offers infinite impedance.

29. Inner conductor is also supported by dielectric beads.

30. Flexible coaxial cables use flexible dielectric as the support for the inner conductor.

31. The effect of discontinuities if the inner or the outer conductors of a coaxial line is abruptly changed, z_0 is also changed.

32. A thin diaphragm fixed to inner conductor or to the outer conductor is capacitive. This is shown in fig. 5.4.

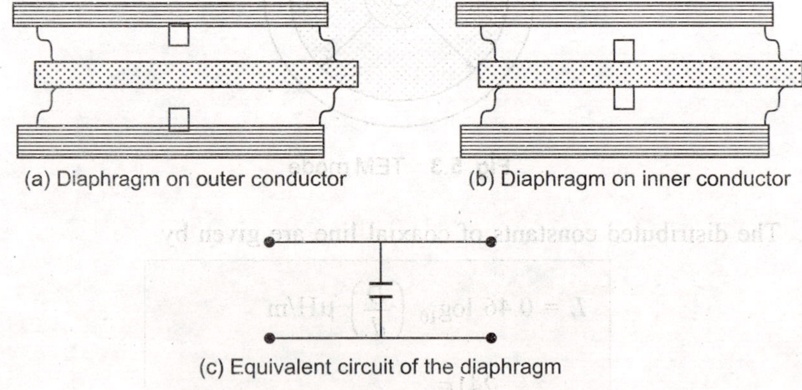

(a) Diaphragm on outer conductor   (b) Diaphragm on inner conductor

(c) Equivalent circuit of the diaphragm

Fig. 5.4 Diaphragm

33. The corners in a rigid coaxial line have the same cross-section and smooth curves.

5.3.2 Applications of Coaxial Lines

1. Coaxial lines can be used up to a frequency of 100 GHz.
2. They are useful for applications where negligible attenuation is required.
3. They are useful to interface waveguides.
4. They are useful for promoting characteristic impedance of 50, 75, 95 and 125 Ω.

5.3.3 Disadvantages of Coaxial Lines

1. The power transmitting capability is limited.
2. At high frequencies, attenuation is more due to low skin depth.

PROBLEM 5.3   What is the wavelength for an operating frequency of 8 GHz in a coaxial line when the dielectric material between the coaxial line when the dielectric material between the conductors has $\epsilon_r = 2.25$.

Solution

$$\text{We have } \lambda = \frac{v_0}{\sqrt{\epsilon_r}\, f}$$

$$= \frac{3 \times 10^8}{\sqrt{2.25} \times 8 \times 10^8}$$

$$= \frac{3}{1.5 \times 12} = \frac{3}{12}$$

$$\boxed{\lambda = 0.25 \text{ m}}$$

PROBLEM 5.4   The radius of the outer conductor is 2.6 mm and the inner conductor has a radius of 0.8 mm in a coaxial cable. Find the frequency for the air dielectric. Also find the highest frequency of possible operation in order to eliminate the lowest mode propagation.

Solution   We have $\lambda_c = \dfrac{\pi}{2n}(d + D)$

For the lowest mode, $n = 1$.

$$\therefore \qquad \lambda_c/_{n=1} = \frac{\pi}{2}(d + D)$$

$$= \frac{\pi}{2}(2.6 + 0.8)$$

$$= \frac{\pi}{2}(3.4)$$

$$\lambda_c = 5.314 \text{ mm}$$

$$\therefore \quad f_c = \frac{3 \times 10^{11}}{\lambda_c}$$

$$= \frac{3 \times 10^{10}}{\lambda}$$

$$= \frac{3 \times 10^{10}}{5.314} = 0.564 \times 10^{10}$$

$$\boxed{f_c = 0.564 \times 10^{10} \text{ Hz}}$$

PROBLEM 5.5 Find z_0 of a coaxial cable if $\epsilon_r = 2.25$, $D/d = 2.25$.

Solution

$$z_0 = \frac{138}{\sqrt{\epsilon_r}} \log \frac{D}{d}$$

$$= \frac{138}{\sqrt{2.25}} \log 2.25 \ \Omega$$

$$= \frac{138}{1.5} \log 2.25$$

$$\boxed{z_0 = 32.384 \ \Omega}$$

PROBLEM 5.6 A coaxial cable has an attenuation is 0.28 dB/m at an operating frequency. Find the output power of 50 m cable if an input of 0.4 W is provided to the line.

Solution The attenuation of the cable

$$\alpha = 28 \text{ dB/m}$$

The attenuation of 50 m cable is

$$= 0.28 \times 50 = 14.0 \text{ dB}$$

The input power, $p_i = 0.4$ W

$$\text{We have } 10 \log \frac{p_i}{p_0} = 14$$

$$\therefore \quad \frac{p_i}{p_0} = 1.4$$

or

$$p_0 = 1.4 \times 0.04$$

$$\therefore \quad \boxed{p_0 = 1.6 \text{ mW}}$$

PROBLEM 5.7 In an open-ended transmission line, the incident voltage is 20 V and the reflected voltage is 12.5 V. Find the percentage of reflected power.

Solution

$$\text{Incident voltage, } V_i = 20 \text{ V}$$

$$\text{Reflected voltage, } V_r = 12.5 \text{ V}$$

$$\text{Reflected voltage coefficient } \rho = \frac{V_r}{V_i} = \frac{12.5}{20} = 0.625$$

But

$$\rho^2 = \frac{\text{reflected power}}{\text{incident power}}$$

$$= \frac{p_r}{p_i} = (0.625)^2$$

$$\rho = 0.391$$

If $p_i = 1$ W

$$p_r = 0.391 \times 1$$

$$\boxed{\% \ p_r = 39.1\%}$$

$\therefore$

PROBLEM 5.8 In a transmission line $z_L \neq z_0$ and maximum voltage of the standing wave is 5 V and minimum voltage is 3 V. Find VSWR.

Solution

$$\text{By definition, VSWR} = \frac{V_{\max}}{V_{\min}}$$

$$= \frac{5}{3} = 1.67$$

or

$$\text{VSWR } (S) = 20 \log_{10} 1.67$$

$\therefore$

$$\boxed{S = 4.44 \text{ dB}}$$

PROBLEM 5.9 The characteristic impedance of a transmission line is 100 Ω and the load impedance is changed from 50 Ω to 125 Ω. Find the VSWR for each load impedance.

Solution

$$\text{VSWR} = \frac{z_0}{z_L} \quad \text{for} \quad z_0 > z_L$$

$$= \frac{100}{50} = 2$$

$$\text{VSWR} = \frac{z_L}{z_0} \quad \text{for} \quad z_0 < z_L$$

$$= \frac{125}{100}$$

$$\boxed{\text{VSWR} = 1.25}$$

PROBLEM 5.10   If the reflected voltage in a transmission line is 0.37 V and the incident voltage is 1.0 V. Find VSWR.

Solution
$$V_r = 0.37 \text{ V}, \ V_i = 1.0 \text{ V}$$

$$\therefore \qquad \rho = \frac{V_r}{V_i} = \frac{0.37}{1.0} = 0.37$$

$$\text{VSWR} = \frac{1+|\rho|}{1-|\rho|} = \frac{1+0.37}{1-0.37}$$

$$\therefore \qquad \boxed{\text{VSWR} = 2.2}$$

PROBLEM 5.11   If $z_L = 10 \, j\Omega$ and $z_0 = 100 \, \Omega$ in a transmission line, find the magnitude of the reflection coefficient.

Solution
$$z_L = 10j \ \Omega$$
$$z_0 = 100 \ \Omega$$

$$\rho = \frac{z_L - z_0}{z_L + z_0}$$

$$= \frac{10j - 100}{10j + 100}$$

$$|\rho| = \frac{\sqrt{10^2 + 100^2}}{\sqrt{10^2 + 100^2}}$$

$$\therefore \qquad \boxed{|\rho| = 1.0}$$

5.4   RECTANGULAR WAVEGUIDES

5.4.1   Definition of Waveguide

It is a metallic hollow pipe which guides the electromagnetic wave.

5.4.2   Salient Features of a Waveguide

1. It supports *TE* and *TM* waves.
2. It does not support *TEM* waves.

3. The electric and magnetic fields in a rectangular waveguide exists as shown in fig. 5.5.

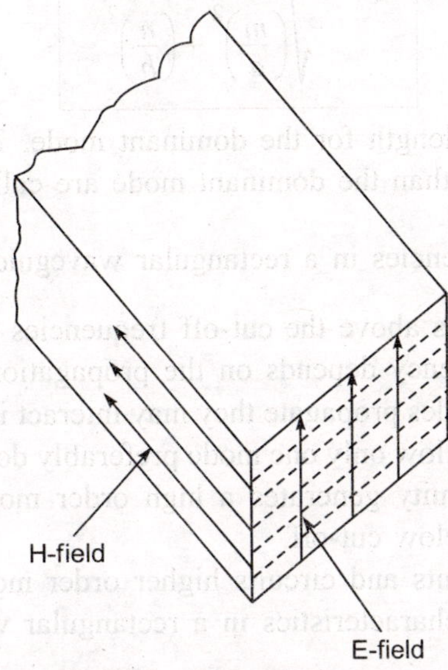

Fig. 5.5   *E* and *H* fields in a waveguide

4. The waves are characterized by a propagation constant given by

$$\gamma_g = \sqrt{\omega^2 \mu \in + \left(\frac{m\pi}{a}\right)^2 + \left(\frac{m\pi}{b}\right)^2}$$

Here, *a* is wide wall dimension

b is narrow wall dimension

m, n represent mode numbers

5. When the wave propagates in the waveguide, it can have infinite types of patterns. These patterns are called modes.

6. The *TE* and *TM* waves are represented by TE_{mn} and TM_{mn}. The first subscript indicates the number of half-wave variations of electric or magnetic field across the wide dimension of the waveguide and second subscript indicates the number of half-wave variations of electric or magnetic field across the narrow wall.

7. If there is no variation across the narrow dimension, the mode is represented by TE_{10} mode. This is called dominant mode.

8. The waveguide dimensions depends on the operating frequency range. For example, *X*-band waveguide has inner dimension of 2.286 cm × 1.016 cm.

9. They are used in transmission lines at microwave frequencies.

10. They can be used as high pass filter radiators and antenna feed elements.

11. The cut-off wavelength in rectangular waveguide is

$$\lambda_c = \frac{2}{\sqrt{\left(\frac{m}{a}\right)^2 + \left(\frac{n}{b}\right)^2}} \cdot$$

12. The cut-off wavelength for the dominant mode, TE_{10} is $\lambda_c = 2a$.
13. The modes other than the dominant mode are called high order modes.

14. The cut-off frequencies in a rectangular waveguide is $f_c = \dfrac{v_0}{2a}$.

15. All the frequencies above the cut-off frequencies are propagated.
16. The cut-off frequency depends on the propagation mode.
17. If two or more modes propagate they may interact in an unwanted manner.
18. It is preferred to allow only one mode preferably dominant mode at a time.
19. When a discontinuity generates a high order mode, the mode dies out quickly as it is below cut-off.
20. In some components and circuits higher order modes are used.
21. The propagation characteristics in a rectangular waveguide are consolidated as

(1) Propagation constant,

$$\gamma_g = \sqrt{\left(\frac{m\pi}{a}\right)^2 + \left(\frac{n\pi}{b}\right)^2 - \omega^2 \mu \in}, \quad \left(\frac{1}{m}\right)$$

$$= \alpha_g \text{ if } \left(\frac{m\pi}{a}\right)^2 + \left(\frac{n\pi}{b}\right)^2 > \omega^2 \mu \in$$

(2) Phase constant,

$$\beta_g = \sqrt{\omega^2 \mu \in - \left(\frac{m\pi}{a}\right)^2 - \left(\frac{n\pi}{b}\right)^2}, \quad \text{(rad/m)}$$

(3) Cut-off frequency,

$$f_c = \frac{1}{2\pi\sqrt{\mu \in}} \sqrt{\left(\frac{m\pi}{a}\right)^2 + \left(\frac{n\pi}{b}\right)^2}, \quad \text{(Hz)}$$

(4) Cut-off wavelength,

$$\lambda_c = \frac{2}{\sqrt{\left(\frac{m}{a}\right)^2 + \left(\frac{n}{b}\right)^2}}, \quad \text{(m)}$$

(5) Phase velocity,

$$v_p = \frac{\omega}{\sqrt{\omega^2 \mu \in - \left(\dfrac{m\pi}{a}\right)^2 - \left(\dfrac{n\pi}{b}\right)^2}}, \ \ (\text{m/s})$$

(6) Guide wavelength,

$$\lambda_g = \frac{2\pi}{\sqrt{\omega^2 \mu \in - \left(\dfrac{m\pi}{a}\right)^2 - \left(\dfrac{n\pi}{b}\right)^2}}, \ \ (\text{m})$$

or

$$\lambda_g = \frac{\lambda}{\left[1 - \left(\dfrac{\lambda}{\lambda_c}\right)^2\right]^{1/2}}, \ \ (\text{m})$$

22. The *TE* waves have the field components E_x, E_y, H_x, H_y, H_z, if z is the direction of propagation.

23. The *TM* waves have the field components E_x, E_y, E_z, H_x and H_y if z is the direction of propagation.

24. The waveguides can be used over a frequency range of 3 GHz to 100 GHz. Its dimensions vary from a few meters to a few centemeters.

25. The waveguides are more efficient compared to coaxial lines. There is no central conductor and hence there are no resistance losses.

26. The power handling capacity depends on the size of the waveguide. The larger waveguide have high power handling capability.

27. The waveguides have different shapes like straight, 90° sections, twisted and flexible waveguides are available. For all these types of waveguides, compatible components like Tees, junctions, couplers, attenuators, cavities and detectors are available.

28. Waveguides can be connected to coaxial line by compatible interface devices.

29. Power loss takes place in the higher order modes.

30. The smallest waveguide for a given frequency is required for principal mode.

31. Small sized waveguides present higher order modes.

5.5 CAVITY RESONATORS

A cavity resonator is a space normally bounded by conducting surfaces in which electromagnetic energy is stored as oscillating fields. The frequency of oscillation is determined by the shape and size of the cavity.

5.5.1 Salient Features

1. A cavity resonator is in the form of closed space and is a resonant circuit.
2. It stores electromagnetic energy in the form of oscillating fields.
3. The resonant frequency depends on the shape and size of the cavity.
4. The cavities are commonly in the shape of rectangular, spherical or cylindrical.
5. The cavities are also designated as re-entrant cavities.
6. The probes, loops and slots are used to remove, electromagnetic energy from the cavities.
7. A typical rectangular cavity resonator is shown in fig. 5.6.

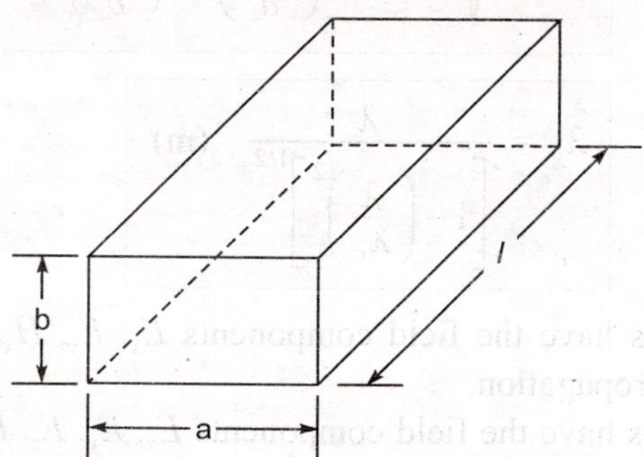

Fig. 5.6 Rectangular cavity resonator

8. The propagation constant in the cavity is given by

$$\gamma_{mn} = \sqrt{\frac{\omega^2}{v^2} - \frac{m^2}{a^2} - \frac{n^2\pi^2}{b^2}}.$$

9. The fields inside the cavity are in the form of TE_{mn} and TM_{mn}.
10. A cavity resonates in an infinite number of modes.
11. The resonant frequency of the rectangular cavity is given by

$$f_r = \frac{v}{2}\sqrt{\frac{m^2}{a^2} + \frac{n^2}{b^2} + \frac{p^2}{\ell^2}}$$

Here, f_r = resonant frequency of the cavity

m = number of half-wavelengths along the 'a' dimension

n = number of half-wavelengths along the 'b' dimension

p = number of half-wavelengths along the dimension

12. Some of the re-entrant cavities are shown in fig. 5.7.

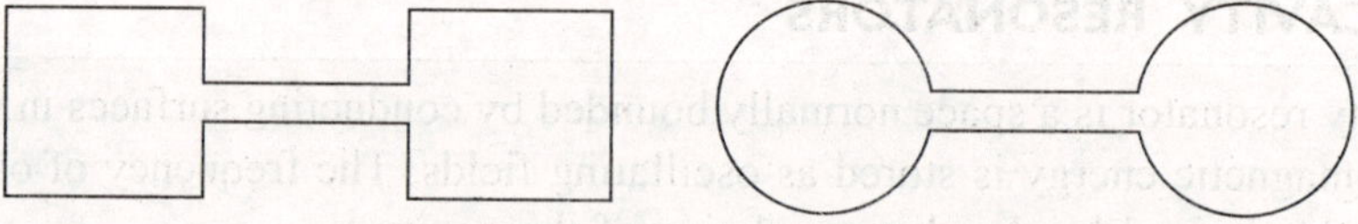

Fig. 5.7 Re-entrant cavities

13. A circular or cylindrical cavity resonator is shown in fig. 5.8.

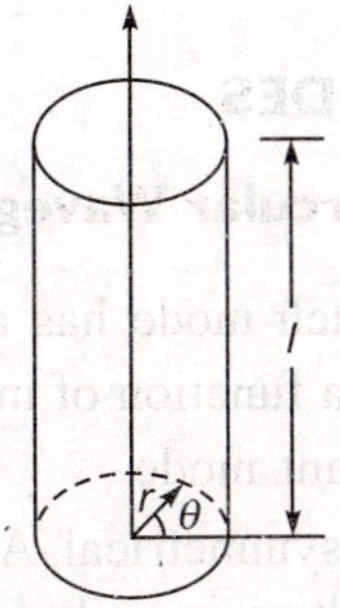

Fig. 5.8 Cylindrical cavity resonator

14. The existing modes in cylindrical cavity resonators are

$$TE_{mnp} \text{ and } TM_{mnp}$$

15. The resonant frequencies are

$$f_r(\text{for } TE_{mnp}) = \frac{v}{2}\left(\frac{p_{mn}^2}{(\pi a)^2} + \frac{p^2}{\ell^2}\right)^{1/2}$$

and

$$f_r(\text{for } TE_{mnp}) = \frac{v}{2}\left(\frac{v_{mn}^2}{(\pi a)^2} + \frac{p^2}{\ell^2}\right)^{1/2}$$

Here, p_{mn} and v_{mn} are the zeros of Bessel function.

p is the number of half cycles of variation of the fields in the z-direction.

v is the velocity of propagation and is equal to $\dfrac{1}{\sqrt{\mu\epsilon}}$

16. The quality factor, Q of the cavity is given by

$$Q = \frac{\text{energy stored per cycle}}{\text{energy loss per cycle}}$$

Q is basically a measure of frequency selectivity of the cavity.

5.5.2 Applications of Cavity Resonators

The cavity resonators are used

1. for the measurement of frequency
2. as filters, in particular selective band pass filters.
3. in microwave sources
4. in klystrons

5. in parametric amplifiers
6. in magnetrons

5.6 CIRCULAR WAVEGUIDES

5.6.1 Salient Features of Circular Waveguides

1. In circular waveguides, each mode has a cut-off wavelength.
2. The cut-off wavelength is a function of inside diameter of the waveguide.
3. In this $TE_{1,1}$ is the dominant mode.
4. $TE_{1,1}$ is not popular as not symmetrical. A bend or discontinuity may twist the mode in the guide resulting in undesired polarization.
5. $TE_{0,1}$ and $TM_{0,1}$ are symmetrical and they will not be changed by twisting in the waveguide.
6. TM_{01} has high cut-off wavelength.
7. The mode TM_{01} is used for rotary joints.
8. It is required to present $TE_{1,1}$ mode from existing.
9. The $TE_{0,1}$ mode has lower attenuation compared to the other modes.

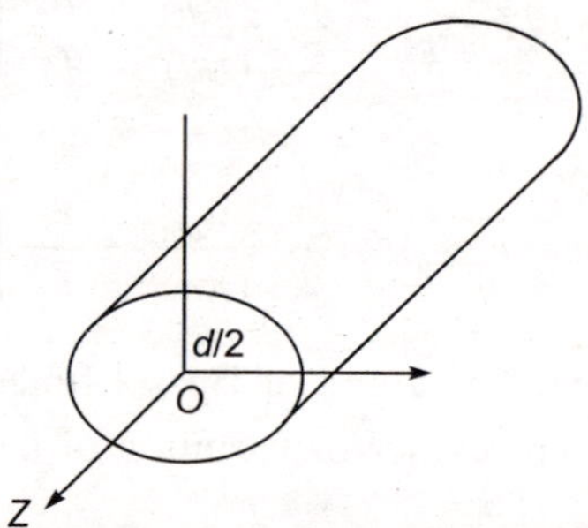

Fig. 5.9 Circular waveguide

10. $TE_{m,n}$ and TE_{mn} exist in circular a waveguide. m represents the number of full wavelengths around the circumference of the inner dimension of the circular waveguide. n represents the number of half-wave lengths across the inner diameter of the circular waveguide.

11. $TE_{1,0}$ in a rectangular waveguide excites $TE_{1,1}$ in a circular waveguide if proper interfacing is done.

12. The cut-off wavelength for $TE_{m,n}$ is given by

$$\lambda_{ce} = \frac{\pi d}{a_{m,n}}$$

Here, d = inner diameter of the waveguide

$a_{m,n}$ = roots of Bessel equation (shown in table 5.1).

Table 5.1 Roots of the Bessel's function (*TE* mode)

$a_{0,1}$	3.821
$a_{1,1}$	1.841
$a_{2,1}$	3.054
$a_{3,1}$	4.201
$a_{0,2}$	7.016
$a_{1,2}$	5.332
$a_{2,2}$	6.706
$a_{3,2}$	8.031

13. The cut-off wavelength for $TM_{m,n}$ is

$$\lambda_{cm} = \frac{\pi d}{a_{m,n}}$$

Here, $a_{m,n}$ = roots of Bessel equation (shown in table. 5.2)

Table 5.2 Roots of the Bessel function (*TM* mode)

$a_{0,1}$	2.405
$a_{1,1}$	3.832
$a_{2,1}$	5.136
$a_{0,2}$	5.520
$a_{1,2}$	7.016
$a_{0,3}$	8.654

14. The circular waveguides are used in rotary joints.
15. The dimensions of the circular waveguides depends on the frequency bands.

5.7 RIDGE WAVEGUIDES

Ridged waveguide is made from rectangular waveguide by adding a rectangular metal bar. The metal bar is added to either the top or bottom of the guide to form a ridge in the metal. The ridge is either single or double.

5.7.1 Salient Features

1. A rectangular metal bar is added at the bottom or at the top wall or at both the walls.
2. Ridge waveguide has low characteristic impedance and high bandwidth.
3. Phase velocity is low in this waveguide.
4. The cut-off frequency is small.
5. Attenuation is high.
6. It is not useful to run for long lengths.
7. A typical ridge waveguide is shown in fig. 5.10.

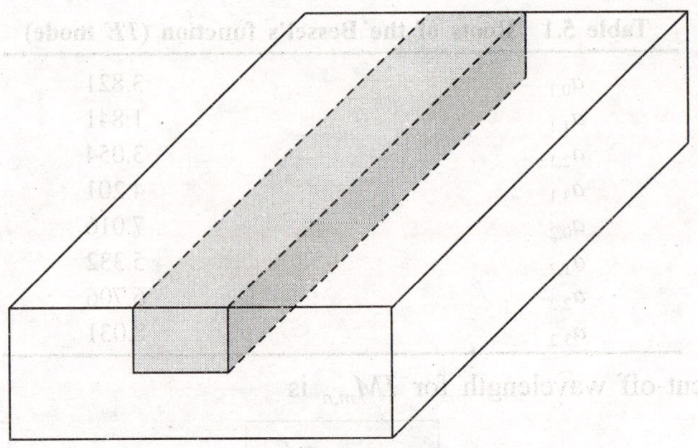

Fig. 5.10   Ridge waveguide using metal bar

8. Ridge waveguide is made in standard sizes.
9. It is compatible with the standard rectangular waveguide.

5.7.2   Advantages of Ridge Waveguide

- High bandwidth
- Its characteristic impedance can be varied over a large range.
- The ridge waveguide is smaller than the rectangular waveguide for the same application.

5.7.3   Disadvantages

- Its use in limited to short lengths.
- It exhibits high attenuation.

5.8   ATTENUATORS

Definition

An attenuator is a device which reduces the power of the signal when the signal passes through it.

Attenuation is the real part of propagation constant and is the reduction of the power and is expressed in positive decibels for convenience. The attenuation of the attenuator is

$$\alpha = 10 \log_{10} \frac{p_i}{p_0}$$

Here, p_i = input power

p_0 = output power

5.8.1   Types of Attenuators

There are

> ➢ Fixed attenuators
> ➢ Step attenuators
> ➢ Continuously variable attenuators

5.8.2   Fixed Attenuators

In these, the attenuation is reduced by a fixed amount. The 3 dB, 10 dB, 20 dB attenuators are available. These attenuators are made of film resistors around a center conductor and disks. The film resistor attenuators are useful at wide range of frequencies as they exhibit constant values. In this, the material depth is less than the skin depth. The attenuators are specified by frequency and attenuation in dB.

The fixed attenuators are also called pads. For example, 6 dB pad means the output power of attenuator is one fourth of the input.

The typical attenuators are shown in fig. 5.11.

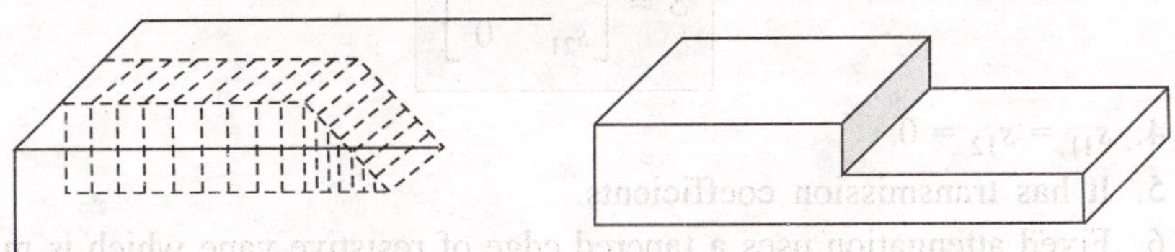

Fig. 5.11   Fixed attenuators

5.8.3   Step Attenuators

These attenuators reduce the power by fixed amounts. They have a number of fixed attenuators and one can be selected for each purpose. Sometimes, fixed attenuators are connected in series in step attenuators. If the net attenuation is required, provision is made to bypass individual attenuators.

5.8.4   Continuously Variable Attenuators

These are of following types

> ➢ Rotary wave attenuator
> ➢ Coaxial line section. In this outer conductors of two sections slide over each other.

The variable attenuators are specified by attenuation, minimum insertion loss, maximum VSWR.

The rotary attenuation is not sensitive and maintains constant phase and hence it is a precision device.

The attenuation is given by

$$\alpha = 2\pi \sqrt{\left(\frac{1}{\lambda_c}\right)^2 - \left(\frac{1}{\lambda}\right)^2}$$

Here, α is attenuation

$\quad\quad \lambda_c$ is cut-off wavelength

$\quad\quad \lambda$ = free space wavelength

5.8.5   Salient Features of a Attenuator

1. It reduces input power given to it.
2. It is a two-port reciprocal network.
3. Its scattering matrix is

$$S = \begin{bmatrix} 0 & s_{12} \\ s_{21} & 0 \end{bmatrix}$$

4. $s_{11} = s_{12} = 0$.
5. It has transmission coefficients.
6. Fixed attenuation uses a tapered edge of resistive vane which is made of lossy material.
7. A variable attenuator consists of a tapered resistive card. The depth of penetration into the waveguide is adjustable.
8. Accuracy of the attenuator increases if the resistive card is replaced by a piece of a metal film with glass coating.
9. The variable attenuator is frequency sensitive and some phase shift is introduced while changing the attenuation.
10. Variable attenuators are constructed by placing lossy resistive card in the waveguide with its surface parallel to the electric field.
11. The variation of attenuation depends on the mount of card insertion.
12. The card is tapered to minimize the reflection.

5.9   CORNERS

The corners are associated with rigid waveguides or coaxial line. They are used for bending upto usually 90°. They may have smooth curves and the well made corners do not produce reflections at any frequency. Typical corners are shown in fig. 5.12.

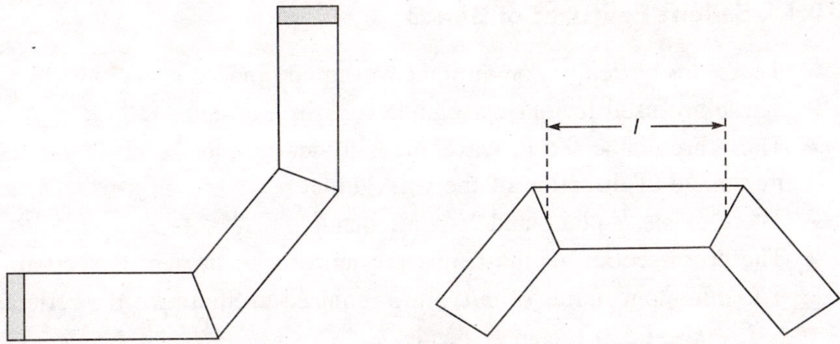

Fig. 5.12 Waveguide corners

➢ These are used to change the direction of the guide through an arbitrary angle.

➢ The length between the continuities equal to odd number of quarter wavelengths, the reflections are minimum, i.e.

$$\ell = (2n + 1)\frac{\lambda_g}{4}$$

Here, λ_g = guide wavelength

$n = 0, 1, 2, \ldots.$

5.10 BENDS

The bends are the bent waveguide pieces. The waveguide bends are used to change the direction of a waveguide to get around obstacles and also to join waveguide runs in a particular system. The bends are of two types.

(i) *E*-plane bend

(ii) *H*-plane bend

These are shown in fig. 5.13.

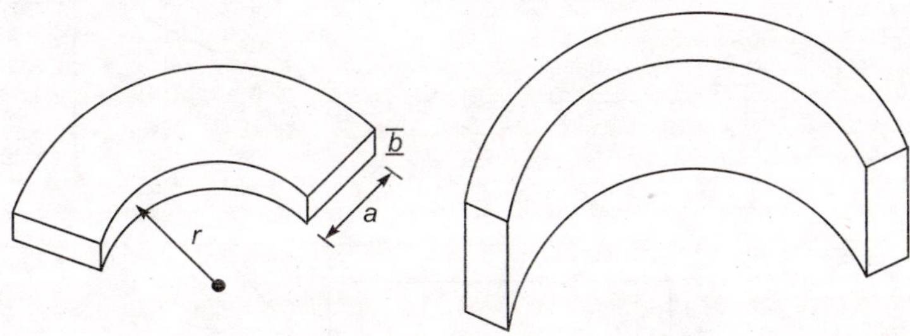

Fig. 5.13 Bends

5.10.1 Salient Features of Bends

➢ These are basically non-straight waveguide pieces.

➢ These are used to join waveguide runs in a system.

➢ These are connected to waveguides to get around the obstacles resulting in change of direction of the waveguide.

➢ There exists E-plane and H-plane bends.

➢ The cross-section of the bends are uniform to present reflection.

➢ The minimum radius of curvature required to minimize the reflection is

$r = 1.5\ a$ for an H bend

$= 1.5\ b$ for E bend

Here, a = broad dimension of the waveguide

b = narrow dimension of the waveguide

➢ The formation of H-plane bend is difficult than the formation of E-plane bend.

➢ The precision bends can be used with minimum VSWR over the entire frequency bend of the waveguide.

➢ Bends are not required in circular waveguides.

➢ When more space is required for the bends, double mitred corners can be used to save some space.

But these are frequency sensitive and hence their use is limited to narrow band applications.

5.11 TWISTS

➢ They are used to change the direction of the guide.

➢ Their required length to minimize the reflections is $\ell = (2n + 1)\dfrac{\lambda_g}{4}$.

➢ The twist can be gradual, step or continuous.

➢ A typical twist is shown in fig. 5.14.

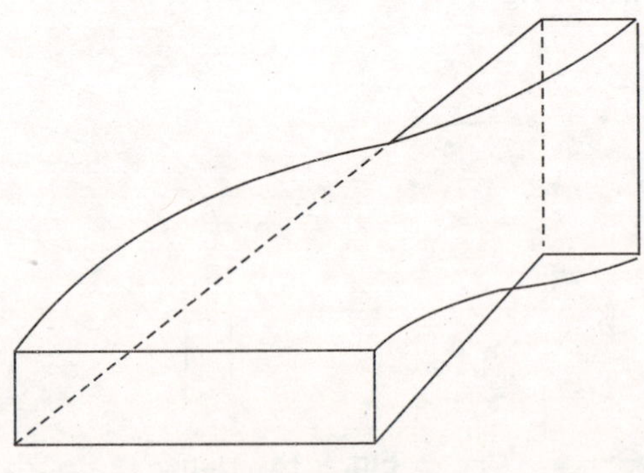

Fig. 5.14 Twist

> ➤ They are also used to change the plane of polarization of propagating wave.
> ➤ The gradual twist changes the plane of polarization gradually or continuously. If VSWR is less than 1.05 if the length of the twist is greater than a few wavelengths.
> ➤ Step twists are useful if the available space is small in the direction of propagation.
> ➤ Multistep twists are used for broadband applications.

5.12 CIRCULATORS

A circulator is a microwave passive multiport device in which the incident wave at port 1 is coupled to port 2 only, incident wave at port 2 is coupled to port only and so on. The ideal circulator is a matched device. It means, when all the ports are terminated in a matched load except one port, the input impedance of the other ports is equal to characteristic impedance of the input line.

A four-port circulator is shown in fig. 5.15.

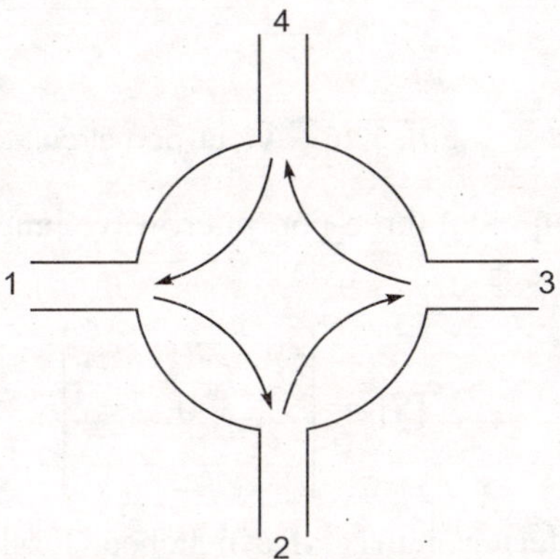

Fig. 5.15 A four-port circulator

A four-port circulator is also constructed using a 3 dB side hole directional coupler and rectangular waveguide and non-reciprocal phase shifters. This is a more compact circulator.

5.12.1 Operation of a Four-Port Circulator

Assume that a wave is incident at port 1 (fig. 5.15). It is split into two equal amplitude in phase waves. These propagate in the side arms of the hybrid junction. These waves reach arm a and b and come out from port 2. The incident wave at port 2 is split into two waves. One wave reaches c with a phase θ and the other wave reaches d with phase of $\phi + \pi$ obtained by gyrator. These waves combine and emerge from port 3. The incident wave at port 3 is split into two equal amplitude waves but with a phase difference of 180°. These waves emerge from port 4.

Similarly, the incident wave at port 4 is split into two equal waves with a phase difference of 180°. The waves combine and emerge from port 1. This is how the circulating property is satisfied.

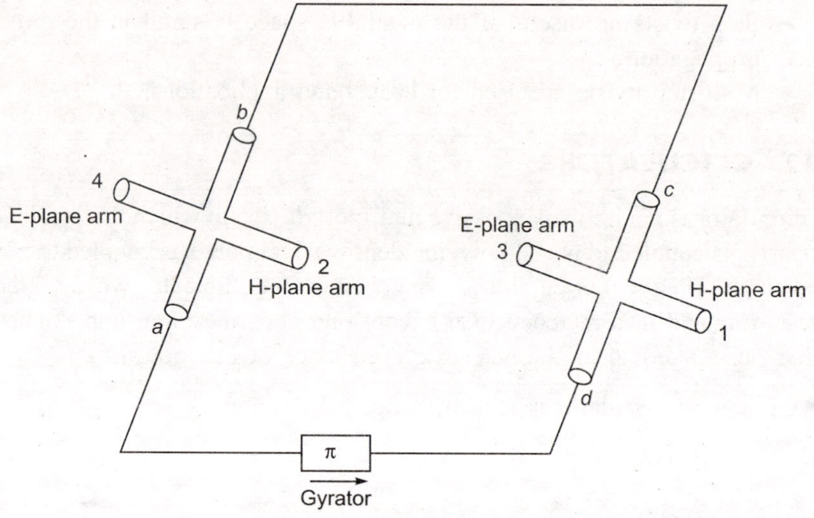

Fig. 5.16 A four-port circulator

A lossless, non-reciprocal three-port microwave junction is a perfect three-port circulator. Its S-matrix is

$$[S] = \begin{bmatrix} 0 & s_{12} & s_{13} \\ s_{21} & 0 & s_{23} \\ s_{31} & s_{32} & 0 \end{bmatrix}$$

However, the scattering matrix of a matched, lossless, non-reciprocal three-port junction is

$$S = \begin{bmatrix} 0 & 0 & s_{13} \\ s_{21} & 0 & 0 \\ 0 & s_{32} & 0 \end{bmatrix}$$

5.12.2 Applications

- It is used to separate the input and output in negative resistive applications.
- It is used to couple a transmitter and receiver to a common antenna.

5.12.3 Salient Features

1. It is usually made of ferrite.
2. It is a mutliport device.

3. In this device, if energy is fed into first port, it goes only to the second port. If energy is fed into second port, it goes only to third port. If energy is fed into the last port, it goes to the first.

4. There exists isolation or attenuation from any port to all ports other than the next to it.

5. The circulators containing three or four ports are the most common.

6. A circulator consists of two non-reciprocal phase shifter waveguide sections mounted side by side with two slots in the walls.

7. Two magic tees are connected into 180° phase shift to create a four-port circulator.

8. Circulator can handle 100 or more kilowatts of average power.

9. Their bandwidth is more than one octane.

5.12.4   S-Matrix of Three-Port Circulator

Its matrix is given by

$$[S] = \begin{bmatrix} s_{11} & s_{12} & s_{13} \\ s_{21} & s_{22} & s_{23} \\ s_{31} & s_{32} & s_{33} \end{bmatrix}$$

As it is perfectly matched junction,

$$s_{11} = s_{22} = s_{33} = 0$$

The circulator is non-reciprocal and hence it is not symmetrical. It means $s_{ij} \neq s_{ji}$

But $[S]$ is unitary.

i.e. $$[S][S] = [1]$$

This gives

$$|s_{12}|^2 + |s_{13}|^2 = 1$$
$$|s_{21}|^2 + |s_{23}|^2 = 1$$
$$|s_{31}|^2 + |s_{32}|^2 = 1 \quad \text{and}$$
$$s_{13}\, s_{23} = 0$$
$$s_{12}\, s_{32} = 0$$
$$s_{21}\, s_{31} = 0$$

Solving the above equations, we get

$$|s_{21}| = 1$$
$$|s_{32}| = 1$$
$$|s_{13}| = 1$$

$$s_{12} = 0$$
$$s_{23} = 0$$
$$s_{31} = 0$$

$$\therefore \quad [S] = \begin{bmatrix} 0 & 0 & s_{13} \\ s_{21} & 0 & 0 \\ 0 & s_{32} & 0 \end{bmatrix}$$

The values s_{13}, s_{21}, s_{32} become 1 when the phase angle of s_{13}, s_{21}, s_{32} are equal zero. This is possible if the locations of terminal planes in the three-port input lines are properly selected.

In this case,

$$[S] = \begin{bmatrix} 0 & 0 & 1 \\ 1 & 0 & 0 \\ 0 & 1 & 0 \end{bmatrix}$$

5.13 TERMINATIONS

Terminations are basically of three types.

1. Matched load
2. Standard mismatches
3. Adjustable shorts.

5.14 MATCHED LOAD

Definition

A matched load is a device which absorbs the incident power completely with no reflections.

It is a one port device. The impedance of matched load is equal to the characteristic impedance of transmission line.

A typical matched load consists of a piece of long resistive card placed in the waveguide parallel to the electric field. The front portion of the card is tapered to avoid discontinuity to the signal. The long cards absorb almost all the incident power. The card length of about one to two guide wavelengths is sufficient to avoid reflections. It produces VSWR of about 1.01.

5.14.1 Salient Features

- It is a type of termination.
- It provides a termination that absorbs all the incident power.

- It is equivalent to terminating the line in its characteristic impedance.
- The matched load for a waveguide is usually a tapered wedge or a slab of lossy material in the waveguide.
- Reflections are eliminated by tapering the lossy material into a wedge.
- The termination is like a lossy tapered transmission line.
- A length of about 1 λ is sufficient to provide a matched load.
- It provides an SWR of less than 1.01.

5.14.2   Tapered Loads

These are used for narrow band, broadband, low power, medium power and high power applications. These tapered loads can be fixed or movable.

Reflections are quite small and SWR is < 1.04.

5.14.3   Applications of Tapered Loads

1. Used in the measurement of reflection in two-port networks.
2. They can be used to measure microwave power. These have 40% bandwidth.

5.14.4   Step Loads

These are useful in applications where it is difficult to taper a load. Its length is smaller than tapered load and usually it is one half of it. But it has low bandwidth. Step loads have 10% bandwidth. SWR is typically less than 1.10. It can be used for broadband applications if two or three section stepped loads.

A typical configuration of step load is shown in fig. 5.17.

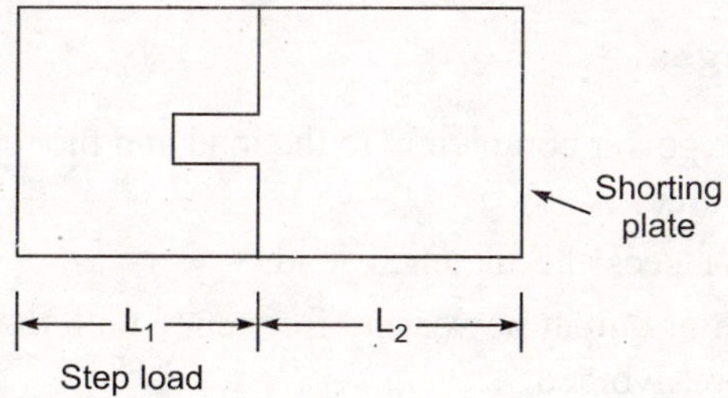

Fig. 5.17   Step load

L_1 and L_2 are chosen so that the reflectors are cancelled to make SWR unity.

5.14.5 Standard Mismatches

- ➢ These are used to calibrate reflections and impedance measuring equipment.
- ➢ Standard mismatches are often used in waveguide equipment
- ➢ These have VSWR of 1.50.

5.14.6 Adjustable Shorts

- ➢ These provide variable reactance in waveguides.
- ➢ It is made of good conducting metal like copper.
- ➢ A typical adjustable short is shown in fig. 5.18.

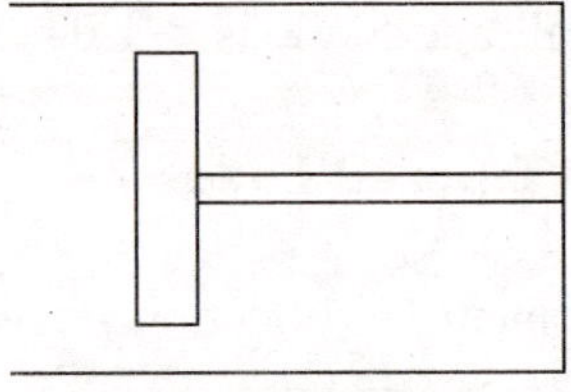

Movable metal plate

Fig. 5.18   Adjustable short

5.15 ISOLATORS

An isolator is a passive device which transmits the wave in the forward direction and does not allow transmission in the reverse direction. It means, it does not exhibit any attenuation in the transmission from port 1 to port 2 and it attenuates totally while transmitting in the reverse direction. It is also called uniline. It behaves like a *p-n* diode. It is used to load network.

5.15.1 Advantages

1. It delivers the power completely to the load and the reflected signals do not reach the source.
2. The generator sees the matched load.
3. The variation of output power and frequency with the variation in the load impedance are avoided.

The common used isolators are

- ▪ Faraday's – rotation isolator
- ▪ Rectangular waveguide resource isolator

A Faraday's rotation isolator is shown in fig. 5.19.

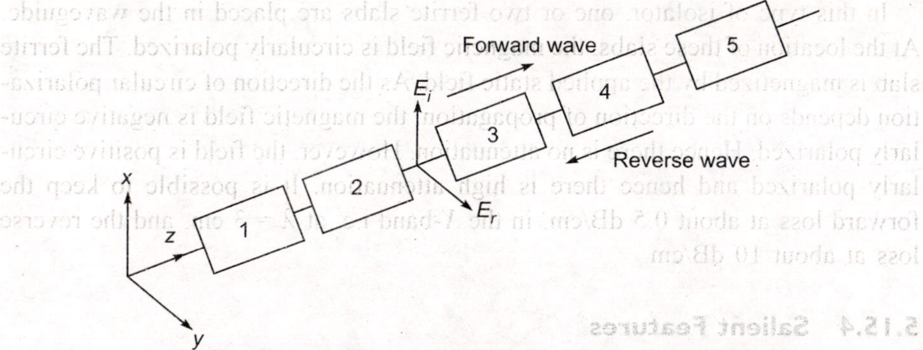

1. Input waveguide, 2. input resistive card 3. Ferrite rod
4. Output resistive card 5. Output waveguide

Fig. 5.19 Faraday's rotation isolator

5.15.2 Operation of Faraday's Isolator

When a wave propagates from port 1 to port 2, its polarization is rotated by 45°
in center clockwise by the waveguide twist section and it is rotated by 45° in
clockwise by the Faraday's rotator.

The wave comes out from port 2 with the correct polarization. But when the
wave coming from port 2 to port 1, its plane of polarization is isolated by 90° and
enter the guide at port 1. Its electric field is parallel to the resistance card and it
is absorbed. However, if the resistance card is placed, the wave will be reflected
from port 1 due to incorrect polarization and hence it cannot propagate in the guide
containing port 1. The resistance card improves the performance of the isolator
and it helps to suppresses the effect of multiple reflections in both directions.

5.15.3 Rectangular Waveguide Resonance Isolator

It is shown in fig. 5.20.

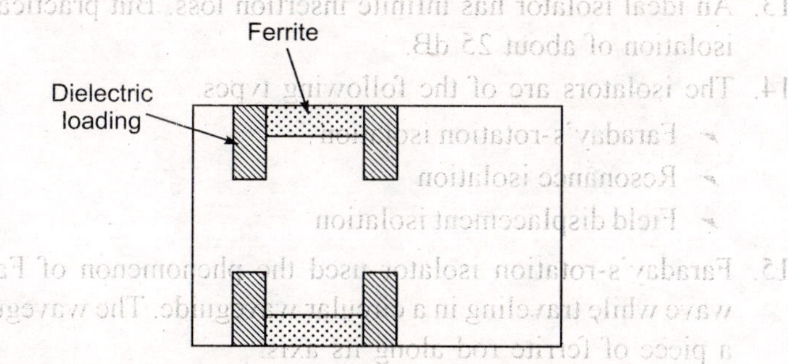

Fig. 5.20 Rectangular waveguide resonance isolator

In this type of isolator, one or two ferrite slabs are placed in the waveguide. At the location of these slabs, the magnetic field is circularly polarized. The ferrite slab is magnetized by the applied static field. As the direction of circular polarization depends on the direction of propagation, the magnetic field is negative circularly polarized. Hence there is no attenuation. However, the field is positive circularly polarized and hence there is high attenuation. It is possible to keep the forward loss at about 0.5 dB/cm, in the X-band i.e. at $\lambda = 3$ cm, and the reverse loss at about 10 dB/cm.

5.15.4 Salient Features

1. It is usually made of ferrite.
2. It is a non-reciprocal two-port device.
3. It passes microwave signals with low loss in the forward direction and absorbs energy in the reverse direction.
4. The measurements are made in both directions.
5. It exhibits low insertion loss in the forward direction and high attenuation in the reverse direction.
6. It is described by back-to-front ratio. The back-to-front ratio is the ratio of reverse attenuation in decibels to the forward insertion loss in decibels.
7. A resonant absorption isolator consists of a section of a rectangular waveguide with ferrite material placed halfway to the center for the waveguide along the axis of the guide.
8. The isolator is popularly known as unilne.
9. Isolator is used to improve the frequency stability in microwave generators like klystrons and magnetrons etc.
10. The isolator is kept between the generator and load to prevent reflected power.
11. It protects the generators from reflected power.
12. An ideal isolator has insertion loss of zero dB. A practical isolator has an insertion loss of about 0.5 dB in the forward direction.
13. An ideal isolator has infinite insertion loss. But practically, it provides an isolation of about 25 dB.
14. The isolators are of the following types.
    - Faraday's-rotation isolation
    - Resonance isolation
    - Field displacement isolation
15. Faraday's-rotation isolator used the phenomenon of Faraday rotation of wave while traveling in a circular waveguide. The waveguide is loaded with a piece of ferrite rod along its axis.
16. It behaves like a diode.

17. It is also called uniline.

18. It is also similar to a gyrator in construction except that it uses a 45° twist section and 45° Faraday's rotation.

19. Its band width is 10 percent.

20. Typical transmission loss is less than 1 dB in the forward direction and it is about 30 dB in the reverse direction.

The input and output resistive cards are at 45° to each other. The forward electric field is perpendicular to the axis of input resistive card and the wave propagates through ferrite rod freely and undergoes 45° rotation without attenuation. When the wave is reflected from the load end, it undergoes 45° rotation ferrite rod and the wave becomes parallel to the axis of input resistive card which absorbs the wave. The static magnetic field is applied to the rod along its axis. In fact, the rotation of the wave is a function of magnetic field, and the dimension of the ferrite rod.

5.16 DIRECTIONAL COUPLERS

5.16.1 Definition

A waveguide directional coupler is a four-port waveguide junction device that samples part of the *EM* wave power through the main waveguide.

5.16.2 Salient Features of Directional Couplers

1. It is a passive four-port device.

2. It consists of a primary guide with ports 1 and 2 and a secondary guide with port 3 and 4.

3. A typical directional coupler is shown in fig. 5.21.

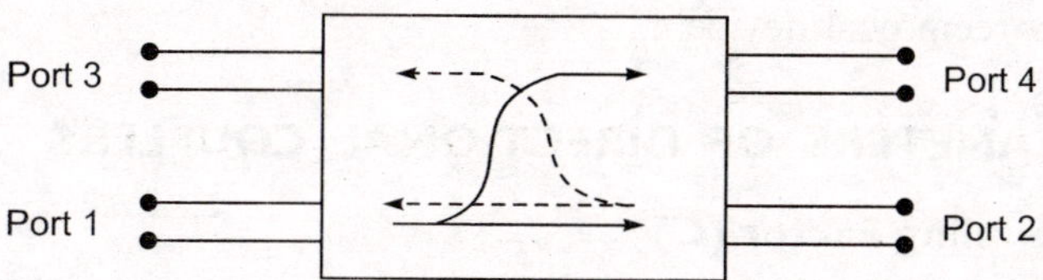

Fig. 5.21   Directional coupler

4. It is made of two connected waveguides. One of the waveguides curves away (fig. 5.22).

5. The waveguides are coupled through holes between them.

6. The directional coupler is said to be consisting of main arm and an auxiliary arm (fig. 5.22).

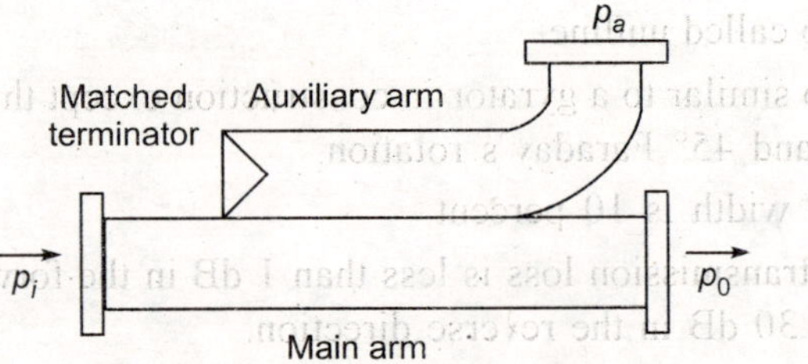

Fig. 5.22 Directional coupler

7. The amount of power coupled to the auxiliary arm depends on the number of holes and their size.

8. The matched termination absorbs reaching it without reflections.

9. The coupler is used to find the power reflection coefficient in a waveguide and to find out incident and reflected power values.

10. The powers at ports 4 and 2 have a phase difference of 90°. Similarly, the powers at port 3 and 1 has a phase difference of 90° when the propagation is in reverse direction.

11. The guides 1-2 and 3-4 are identical. Any one of them can be used as primary and the other acts the auxiliary guide.

12. The directional couplers are described by coupling factor directivity and VSWR.

13. The scattering matrix of the directional coupler is

$$[S] = \begin{bmatrix} 0 & s_{12} & s_{13} & 0 \\ s_{12} & 0 & 0 & s_{24} \\ s_{13} & 0 & 0 & s_{34} \\ 0 & s_{24} & s_{34} & 0 \end{bmatrix}$$

14. It is a reciprocal device.

5.17 PARAMETERS OF DIRECTIONAL COUPLERS

5.17.1 Coupling Factor (C)

It is defined as the ratio of input power and the output power at auxiliary arm. That is,

$$C \equiv 10 \log_{10} \frac{p_i}{p_a}, \text{ dB}$$

Here, p_i = input power to the primary guide

p_a = power output at auxiliary guide.

5.17.2 Directivity (D)

Definition: It is defined as the ratio of power in the auxiliary arm due to power in forward direction to the power at the same port due to power in the reverse direction. That is,

$$D(\text{dB}) = 10 \log_{10} \frac{p_{af}}{p_{ar}}$$

Here, p_{af} = power in the auxiliary arm due to power in forward direction

p_{ar} = power in the auxiliary arm in the reverse direction.

If D is more, the isolation of the auxiliary arm from leakage outputs due to reverse power leakage is more.

5.17.3 Applications of Directional Couplers

1. They are used in the measurement of parameters of antennas like slot coupled Tee junctions, horns, etc.
2. They are used to sample the incoming power.
3. They are widely used in impedance bridges for microwave measurements.
4. They are used for power monitoring.
5. They are used in microwave mixers.
6. They are used as input and output couplers in balanced microwave amplifier circuits.

5.18 SCATTERING MATRIX OF DIRECTIONAL COUPLER

The directional coupler is a four-port and reciprocal device. One output is isolated from input port. All ports are matched. The general S-matrix of a four-port device is

$$S = \begin{bmatrix} s_{11} & s_{12} & s_{13} & s_{14} \\ s_{21} & s_{22} & s_{23} & s_{24} \\ s_{31} & s_{32} & s_{33} & s_{34} \\ s_{41} & s_{42} & s_{43} & s_{44} \end{bmatrix}$$

Assuming

1. one adjacent port is isolated from the input port, we have
$$s_{12} = s_{21} = s_{34} = s_{43} = 0$$
2. all ports are matched, we have
$$s_{11} = s_{22} = s_{33} = s_{44} = 0$$
Moreover, directional coupler is reciprocal.
Hence, $\qquad s_{13} = s_{31}$
$$s_{14} = s_{41}$$

$$s_{23} = s_{32}$$
$$s_{24} = s_{42}$$

Substituting the above conditions, S becomes

$$S = \begin{bmatrix} 0 & 0 & s_{13} & s_{14} \\ 0 & 0 & s_{23} & s_{24} \\ s_{13} & s_{23} & 0 & 0 \\ s_{14} & s_{24} & 0 & 0 \end{bmatrix}$$

From the symmetry of junction, we have

$$s_{13} = s_{24}$$
$$s_{14} = s_{23}$$

$$\therefore \quad S = \begin{bmatrix} 0 & 0 & s_{13} & s_{14} \\ 0 & 0 & s_{14} & s_{13} \\ s_{13} & s_{14} & 0 & 0 \\ s_{14} & s_{13} & 0 & 0 \end{bmatrix}$$

For a lossless network, $|s_{13}|^2 + |s_{14}|^2 = 1$

This is according to the conservation of energy and

$$s_{13}\, s_{14} + s_{13}\, s_{14} = 0$$

A possible solution of this equation is

$$s_{13} = \alpha$$
$$s_{14} = j\beta$$

$$\therefore \quad S = \begin{bmatrix} 0 & 0 & \alpha & j\beta \\ 0 & 0 & j\beta & \alpha \\ \alpha & j\beta & 0 & 0 \\ j\beta & \alpha & 0 & 0 \end{bmatrix}$$

For a directional coupler.

5.19 SCATTERING MATRIX OF 3 dB DIRECTIONAL COUPLER

The scattering of a directional coupler is

$$S = \begin{bmatrix} 0 & 0 & \alpha & j\beta \\ 0 & 0 & j\beta & \alpha \\ \alpha & j\beta & 0 & 0 \\ j\beta & \alpha & 0 & 0 \end{bmatrix}$$

In 3 dB directional coupler, the power is equally divided between ports 3 and 4. Such a symmetric coupler is a hybrid junction. Applying symmetry to such a coupler, we have

$$\alpha = \beta = 0.707.$$

$$\therefore \quad S = \begin{bmatrix} 0 & 0 & 1 & j \\ 0 & 0 & j & 1 \\ 1 & j & 0 & 0 \\ j & \alpha & 0 & 0 \end{bmatrix}$$

5.20 TEE JUNCTIONS

There are basically of 3 types.

> *H*-plane Tee
> *E*-plane Tee
> Hybrid Tee

5.21 *H*-PLANE TEE JUNCTION

In this, a piece of waveguide is attached perpendicular to the narrow wall of a waveguide. (fig. 5.23).

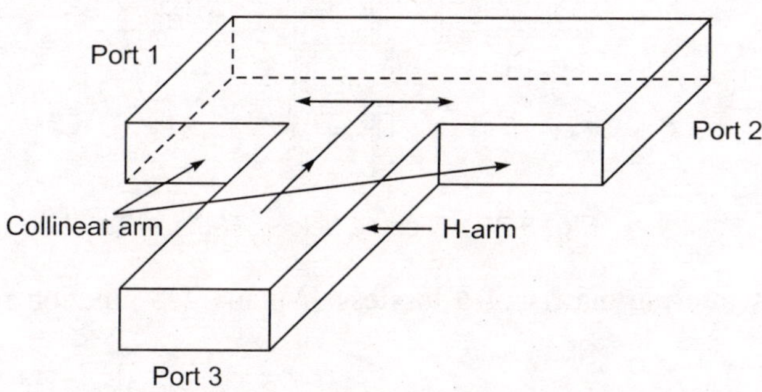

Fig. 5.23 *H*-plane Tee junctions

5.21.1 Salient Features

1. The port 3 is usually the input port.
2. The arms containing port 1 and 2 are in shunt and hence it is also called a shunt Tee.
3. Its equivalent circuit is shown in fig. 5.24.

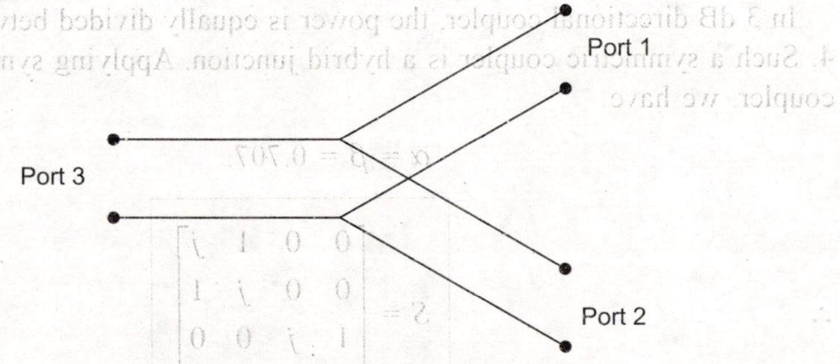

Fig. 5.24   Equivalent circuit of *H*-plane Tee

4. The detailed equivalent circuit consists of a series and shunt reactances with the output arms. These are due to the presence of fringing fields in the junction.
5. The *H*-plane can be matched by Iris of designed length, suitable tapers and transformers etc.
6. In general, it is difficult to match all the three ports. The *H*-arm is usually matched.
7. The Iris matching provides over a small range of frequencies on the other hand, transformers are used to match over a wide band of frequencies.
8. The electric field in *H*-plane is shown in fig 5.25.

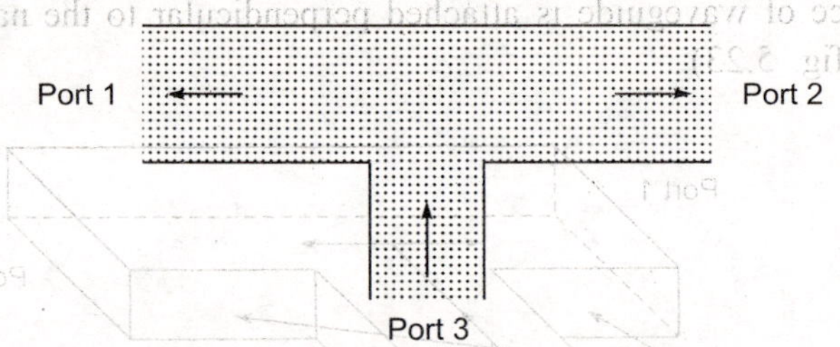

Fig. 5.25   Electric field in *H*-plane Tee

9. The scattering matrix of a lossless *H*-plane Tee junction is

$$S = \begin{bmatrix} \dfrac{1}{2} & -\dfrac{1}{2} & \dfrac{1}{\sqrt{2}} \\[2mm] -\dfrac{1}{2} & \dfrac{1}{2} & \dfrac{1}{\sqrt{2}} \\[2mm] \dfrac{1}{\sqrt{2}} & \dfrac{1}{\sqrt{2}} & 0 \end{bmatrix}$$

10. The input is given at ports 1 and 2, output at port 3 is in phase and additive.
11. If the input is given at port 3, the power is divided equally with ports 1 and 2 in phase but with the same magnitude.

5.21.2 Scattering Matrix of *H*-Plane Tee Junction

H-plane Tee is a three-port junction and hence its scattering matrix is given by

$$S = \begin{bmatrix} s_{11} & s_{12} & s_{13} \\ s_{21} & s_{22} & s_{23} \\ s_{31} & s_{32} & s_{33} \end{bmatrix}$$

From the plane of symmetry, we have

$$s_{13} = s_{23}$$

From the symmetric property,

$$s_{12} = s_{21}$$
$$s_{23} = s_{32}$$
$$s_{13} = s_{31}$$

For a matched junction, $s_{33} = 0$.
Using these values, S becomes

$$[S] = \begin{bmatrix} s_{11} & s_{12} & s_{13} \\ s_{12} & s_{22} & s_{13} \\ s_{13} & s_{13} & 0 \end{bmatrix}$$

According to unitary property,

$$[S][S]^* = [I]$$

or

$$\begin{bmatrix} s_{11} & s_{12} & s_{13} \\ s_{12} & s_{22} & s_{13} \\ s_{13} & s_{13} & 0 \end{bmatrix} \begin{bmatrix} s_{11}^* & s_{12}^* & s_{13}^* \\ s_{12}^* & s_{22}^* & s_{13}^* \\ s_{13}^* & s_{13}^* & 0 \end{bmatrix} = \begin{bmatrix} 1 & 0 & 0 \\ 0 & 1 & 0 \\ 0 & 0 & 1 \end{bmatrix}$$

Taking the product on the LHS, we get

$$s_{11} s_{11}^* + s_{12} s_{12}^* + s_{13} s_{13}^* = 1$$

But $s_{11} s_{11}^* = |s_{11}|^2$

$$s_{12} s_{12}^* = |s_{12}|^2$$
$$s_{13} s_{13}^* = |s_{13}|^2$$

i.e.

$$|s_{11}|^2 + |s_{12}|^2 + |s_{13}|^2 = 1$$
$$|s_{12}|^2 + |s_{22}|^2 + |s_{13}|^2 = 1$$
$$|s_{13}|^2 + |s_{13}|^2 = 1$$
$$s_{13} s_{11}^* + s_{13} s_{12}^* = 0$$

Solving the above equations, we get

$$s_{13} = \frac{1}{\sqrt{2}}$$

$$s_{11} = s_{22}$$

$$s_{11} = -s_{12}$$

$$s_{12} = -\frac{1}{2}$$

$$s_{22} = \frac{1}{2}$$

Using the above values, S becomes

$$[S] = \begin{bmatrix} \dfrac{1}{2} & -\dfrac{1}{2} & \dfrac{1}{\sqrt{2}} \\[2ex] -\dfrac{1}{2} & \dfrac{1}{2} & -\dfrac{1}{\sqrt{2}} \\[2ex] \dfrac{1}{\sqrt{2}} & \dfrac{1}{\sqrt{2}} & 0 \end{bmatrix}$$

5.22 *E*-PLANE TEE JUNCTION

In this a piece of waveguide is attached to the broad wall of the waveguide (fig. 5.26).

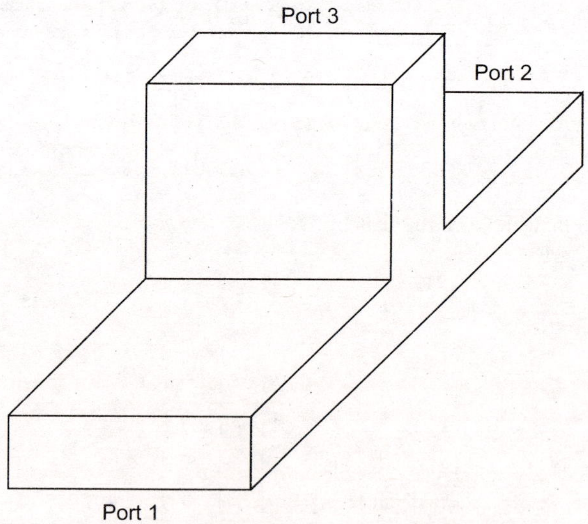

Fig. 5.26 *E*-plane Tee junction

As it similar to an *E*-plane bend, it is called *E*-plane junction.

5.22.1 Salient Features

1. Port 3 is usually the input port.
2. The ports 1 and 2 are in series and hence it is called a series Tee junction.
3. Its equivalent circuit is shown in fig. 5.27.

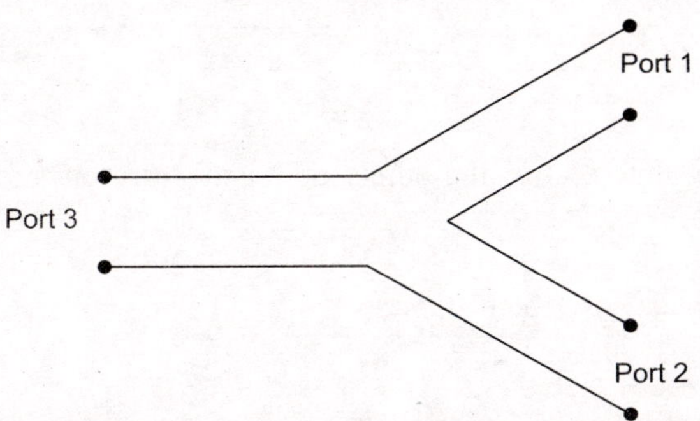

Fig. 5.27 Equivalent circuit of E-plane Tee junction

4. The detailed equivalent circuit consists of series and shunt reactances.
5. This Tee can be matched by the screw turners or Iris or inductive or capacitive windows at the junction.
6. In general, it is difficult to match all the three ports. The E-arm is usually matched.
7. If the input is given at port 3, power is divided equally in ports 1 and 2 with equal magnitude but in opposite phase.
8. The electric field in E-plane is shown in fig. 5.28.

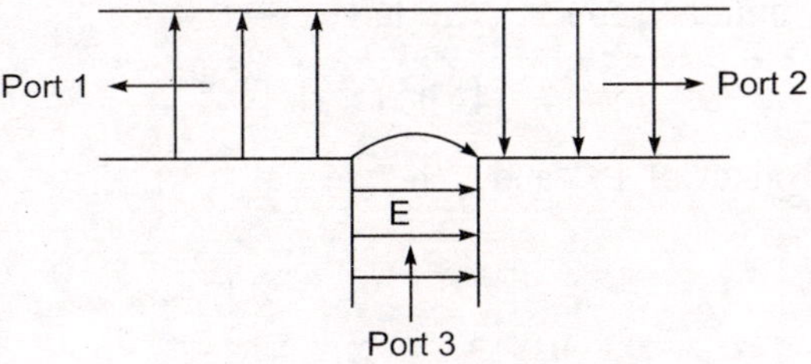

Fig. 5.28 Electric field in E-plane

9. The scattering matrix of loss less E-plane Tee junction is

$$S = \begin{bmatrix} \dfrac{1}{2} & \dfrac{1}{2} & \dfrac{1}{\sqrt{2}} \\[2mm] \dfrac{1}{2} & \dfrac{1}{2} & -\dfrac{1}{\sqrt{2}} \\[2mm] \dfrac{1}{\sqrt{2}} & -\dfrac{1}{\sqrt{2}} & 0 \end{bmatrix}$$

5.22.2 Scattering Matrix of *E*-Plane Tee

In *E*-plane Tee junction, power in port 3 is the difference of two signals entering at 1 and 2 simultaneously. It is a three-port junction and its S-matrix is given by

$$S = \begin{bmatrix} s_{11} & s_{12} & s_{13} \\ s_{21} & s_{22} & s_{23} \\ s_{31} & s_{32} & s_{33} \end{bmatrix}$$

If the power enters at 3, the output at 1 and 2 are out of phase by 180°.

Hence $$s_{23} = -s_{13}.$$

For a matched 3 is matched,

$$s_{33} = 0$$

Due to symmetric property,

$$s_{12} = s_{21}$$
$$s_{13} = s_{31}$$
$$s_{23} = s_{32}$$

Therefore, S becomes

$$S = \begin{bmatrix} s_{11} & s_{12} & s_{13} \\ s_{21} & s_{22} & s_{13} \\ s_{13} & s_{12} & 0 \end{bmatrix}$$

According to unitary property, we have

$$[S][S] = [I]$$

Taking the product of $[S]$ and $[S]^{*}$, we get

$$\begin{bmatrix} s_{11} & s_{12} & s_{13} \\ s_{21} & s_{22} & s_{13} \\ s_{13} & s_{12} & 0 \end{bmatrix} \begin{bmatrix} s_{11}^{*} & s_{12}^{*} & s_{13}^{*} \\ s_{21}^{*} & s_{22}^{*} & s_{13}^{*} \\ s_{13}^{*} & s_{12}^{*} & 0 \end{bmatrix} = \begin{bmatrix} 1 & 0 & 0 \\ 0 & 1 & 0 \\ 0 & 0 & 1 \end{bmatrix}$$

i.e.
$$|s_{11}|^2 + |s_{12}|^2 + |s_{13}|^2 = 1$$
$$|s_{12}|^2 + |s_{22}|^2 + |s_{13}|^2 = 1$$
$$|s_{13}|^2 + |s_{13}|^2 = 1$$
$$s_{13} \cdot s_{11}^{*} - s_{13}^{*} s_{12}^{*} = 0$$

Solving the above equations, we get

$$s_{11} = s_{22}$$
$$s_{13} = 1/\sqrt{2}$$

$$s_{11} = s_{12} = s_{22} = \frac{1}{2}$$

$$\therefore \quad S = \begin{bmatrix} \dfrac{1}{2} & \dfrac{1}{2} & \dfrac{1}{\sqrt{2}} \\[2ex] \dfrac{1}{2} & \dfrac{1}{2} & -\dfrac{1}{\sqrt{2}} \\[2ex] \dfrac{1}{\sqrt{2}} & -\dfrac{1}{\sqrt{2}} & 0 \end{bmatrix}$$

5.23 HYBRID TEE (MAGIC TEE)

It is a combination of E-plane Tee and H-plane Tee.

5.23.1 Salient Features

1. It is a four-port device and it is also called magic Tee because of its unusual characteristics.
2. Its four arms are two side arms, shunt arm and series arm. The shunt arm is called H-arm and series arm is called E-arm. The side arms are called collinear arms.
3. The electric fields in the shunt and series arms are perpendicular to each other and hence they are said to be cross-polarized.
4. There is no coupling between shunt and series arm.
5. In this device, E-arm can see only side arm and similarly H-arm can see only side arm.
6. The magic Tee will be matched if the shunt and series arm are matched.
7. If a signal is given at the shunt arm port, power divides equally into the side arm and they are in phase. There is no coupling to the series arm.
8. If a signal is given to the series arm port, power divides equally into the side arms and they are 180° out of phase. There is no coupling to H-arm.
9. If a signal is given to a side arm, it divides equally into series and shunt arms. There is no coupling to the 2nd side arm.
10. If signals are given to side arms, they will add in phase in H-arm and they add in E-arm with 180° out of phase.
11. H-arm is called sum arm and E-arm is called difference arm.
12. Magic Tee is used to produce sum and difference of signals simultaneously.
13. A typical magic Tee is shown in fig. 5.29.

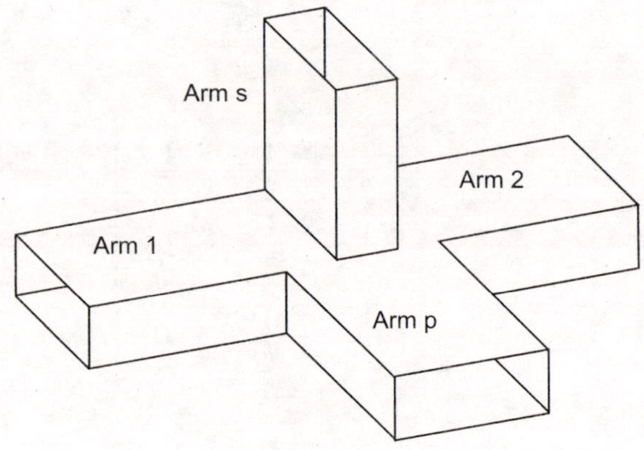

Fig. 5.29   Magic Tee

14. The scattering matrix of magic Tee is

$$
S = \begin{bmatrix}
0 & 0 & s_{13} & s_{14} \\
0 & 0 & s_{23} & s_{24} \\
s_{31} & s_{32} & 0 & 0 \\
s_{41} & s_{42} & 0 & 0
\end{bmatrix}
$$

5.23.2   Applications

The magic Tee is used for

- ➤ mixing
- ➤ duplexing
- ➤ producing sum and difference signals.
- ➤ impedance measurements
- ➤ to couple two transmitters to the antenna without loading each other.

5.23.3   Scattering Matrix of Magic Tee

Magic Tee is a four-port junction. S-matrix is given by

$$
S = \begin{bmatrix}
s_{11} & s_{12} & s_{13} & s_{14} \\
s_{21} & s_{22} & s_{23} & s_{24} \\
s_{31} & s_{32} & s_{33} & s_{34} \\
s_{41} & s_{42} & s_{43} & s_{44}
\end{bmatrix}
$$

As it consists of both H and E-plane Tee arms,

$$s_{23} = s_{13}$$
$$s_{24} = -s_{14} \quad \text{and}$$

as port 3 and 4 are isolated,

$$s_{34} = s_{43} = 0$$

Due to symmetric property

$$s_{12} = s_{21}$$
$$s_{13} = s_{31}$$
$$s_{23} = s_{32}$$
$$s_{34} = s_{43}$$
$$s_{24} = s_{42}$$
$$s_{41} = s_{14}$$

and $s_{33} = s_{44} = 0$ when 3 and 4 are matched

Using the above values, we get

$$S = \begin{bmatrix} s_{11} & s_{12} & s_{13} & s_{14} \\ s_{12} & s_{22} & s_{13} & -s_{14} \\ s_{13} & s_{13} & 0 & 0 \\ s_{14} & -s_{14} & 0 & 0 \end{bmatrix}$$

Using the unitary property,

$$[S][S]^* = [I],$$

we get

$$|s_{11}|^2 + |s_{12}|^2 + |s_{13}|^2 + |s_{14}|^2 = 1$$
$$|s_{12}|^2 + |s_{22}|^2 + |s_{13}|^2 + |s_{14}|^2 = 1$$
$$2|s_{13}|^2 = 1$$
$$2|s_{14}|^2 = 1$$

Solving the above equations, we get

$$s_{13} = \frac{1}{\sqrt{2}}$$

$$s_{14} = -\frac{1}{\sqrt{2}}$$

$$s_{11} = 0$$
$$s_{22} = 0$$

The *S*-matrix of magic tee becomes

$$[S] = \begin{bmatrix} 0 & 0 & \dfrac{1}{\sqrt{2}} & \dfrac{1}{\sqrt{2}} \\[2mm] 0 & 0 & \dfrac{1}{\sqrt{2}} & -\dfrac{1}{\sqrt{2}} \\[2mm] \dfrac{1}{\sqrt{2}} & \dfrac{1}{\sqrt{2}} & 0 & 0 \\[2mm] \dfrac{1}{\sqrt{2}} & -\dfrac{1}{\sqrt{2}} & 0 & 0 \end{bmatrix}$$

5.24 FERRITE DEVICES

Ferrite is a high resistance magnetic material and it consists of mainly ferrite oxide and one or more other metals. Ferrite material is extremely useful at microwave frequencies. Ferrite is made by inserting metallic atoms into iron oxide in place of some of the iron atoms. The addition of zinc atoms is iron oxide forms zinc ferrite (ZnFe2O$_3$). On the other hand, the addition of manganese forms manganese ferrite ($MnFe_2O_3$). Similarly $NiFe_2O_3$, $CuFe_2O_3$, $CdFe_2O_3$ are created.

5.24.1 Salient Features

1. Ferrite is a high resistance magnetic material.
2. Electromagnetic waves pass through ferrites with negligible attenuation.
3. The electromagnetic wave propagation undergoes phase shift and attenuation.
4. The attenuation and phase shift due to ferrites are influenced by the applied d.c. magnetic fields.
5. The ferrites are popularly used in isolators, circulators, phase changers, switches, inductors as core material, TV deflection yokes, miniature antennas, ferrite limiters and ferrite memories. etc.
6. The specific resistivity of ferrites for use at microwave frequencies is of the order of 10^{12} Ω-cm.
7. The typical relative permittivities of ferrites lie in the range of 5 to 20.
8. Ferrites are used at microwave frequencies.

5.24.2 Applications of Ferrites

They are used in

- ➢ isolators
- ➢ circulators
- ➢ phase changers

> switches
> inductors as core material
> TV deflection yokes
> Miniature antennas
> Limiters
> Memories

5.25 PHASE SHIFTERS

The phase is defined as the fraction of time period which has elapsed measured from fixed origin.

The phase shift is defined as change in displacement periodic wave as a function of frequency. Its magnitude is measured in degrees.

Phase shifter is a device which introduces a phase shift between input and output.

5.25.1 Salient Features

1. The phase shifter is obtained by ferrites
2. The maximum number of phase changes take place in the regions of maximum attenuation in a wave guide.
3. The d.c. magnetic field is applied to produce the required phase shift of electromagnetic wave.
4. When the ferrite is placed at the center of the waveguide with the magnet placed outside the waveguide, the phase shift remains the same in both directions.
5. The change of phase shift is obtained by moving the ferrite away from the axis of the waveguide.
6. The non-reciprocal ferrite phase shifters are used in microwave frequencies.
7. In these shifters, the phase design depends on the direction of electromagnetic wave.
8. The common phase shifters are two types : linear and rotary type.
9. In linear type, the phase change is obtained by moving a dielectric slab longitudinally.
10. In rotary type, a half-wave dielectric slab is rotated to obtain the change in phase.

5.26 HYBRID RINGS (RAT RACE COUPLER)

An hybrid ring is called a rat-race coupler hybrid junction is nothing but a 3 dB branch-line coupler.

A 0° or 180° hybrid junction gives a magic Tee.

A hybrid junction is shown in fig. 5.30.

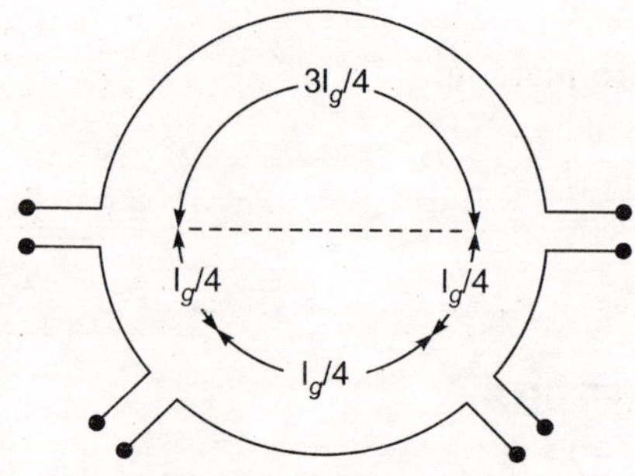

Fig. 5.30 Hybrid ring

5.26.1 Operation

Let a wave be incident at port 1. It splits into two waves and they travel around the ring in opposite direction. These two waves reach in phase port 2 and port 4. But they are out of phase at port 3. The ports 1 and 3 are not coupled and port 2 and 4 are also not coupled as the two ports are apart by $\lambda/2$.

5.26.2 Salient Features

1. It is four-port hybrid junction.
2. The ports are separated by $\lambda_g/4$.
3. The ports are connected either in series or in parallel.
4. The characteristics of hybrid rings are similar to those hybrid Tee.
5. If the input is given to port 1, the phase difference of the two waves propagating in clockwise and anti-clockwise is $180°$. Hence, the signal is absent at port 3.
6. If signal is given at port 2, it will similarly be absent at port 4.
7. For an ideal hybrid ring, the S matrix is

$$S = \begin{bmatrix} 0 & s_{12} & 0 & s_{14} \\ s_{21} & 0 & s_{23} & 0 \\ 0 & s_{32} & 0 & s_{34} \\ s_{41} & 0 & s_{43} & 0 \end{bmatrix}$$

8. In practical hybrid rings, leakage of waves exists at port 3 and 4 for signals fed at 1 and 2 respectively. The zeros appearing in the above S matrix are not exactly zeros for practical rings.

5.27 CHOKE JOINTS

It is a shorted slot that is used to eliminate discontinuities between two waveguide sections.

5.27.1   Salient Features

1. It is a half-wave shorted slot.
2. It is placed in the waveguide section which is terminated with a flange.
3. It is used in rotary joints as it provides some space for rotation. (fig. 5.31).

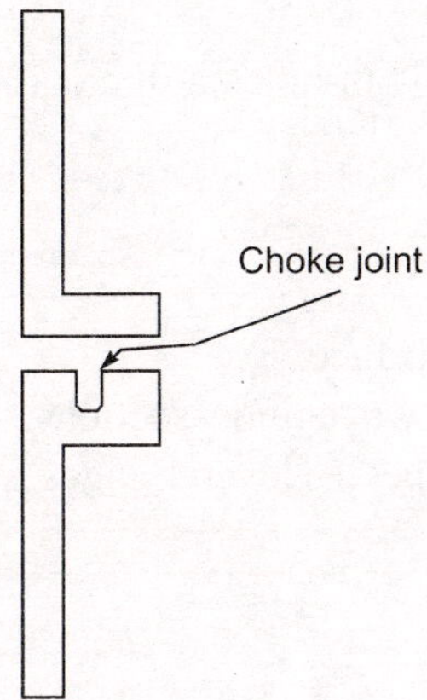

Fig. 5.31   Choke joint

5.28   TUNING SCREWS OR POSTS

5.28.1   Salient Features

1. It is basically metal post of small diameter and it is in the form of a screw.
2. It is inserted in the broad wall of a rectangular waveguide.
3. It forms either an inductor or a shunt capacitance.
4. By tuning, it is possible to change the capacitance or inductance and hence their susceptances.
5. A typical tuning screw is shown in fig. 5.32.

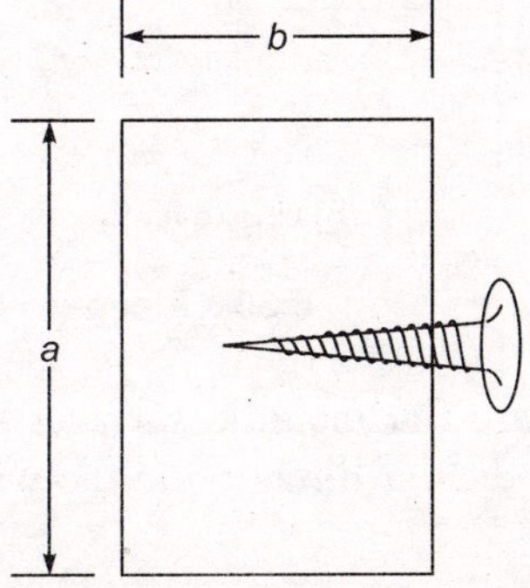

Fig. 5.32   Tuning screw

6. Its susceptance is a function of the depth of the screw inside the guide broad wall dimension and frequency.

7. If the length of the post is less than $\lambda/4$, its susceptance is capacitive and if its length is more than $\lambda/4$, is inductance.

5.29 FLANGES

Flange is a metallic surface with holes at the corner used to connect waveguide sections.

5.29.1 Salient Features

1. Flange is a metallic surface.
2. It is used to connect waveguide sections.
3. Flanges are of two types : Cover flanges and choke flanges as shown in fig. 5.33.

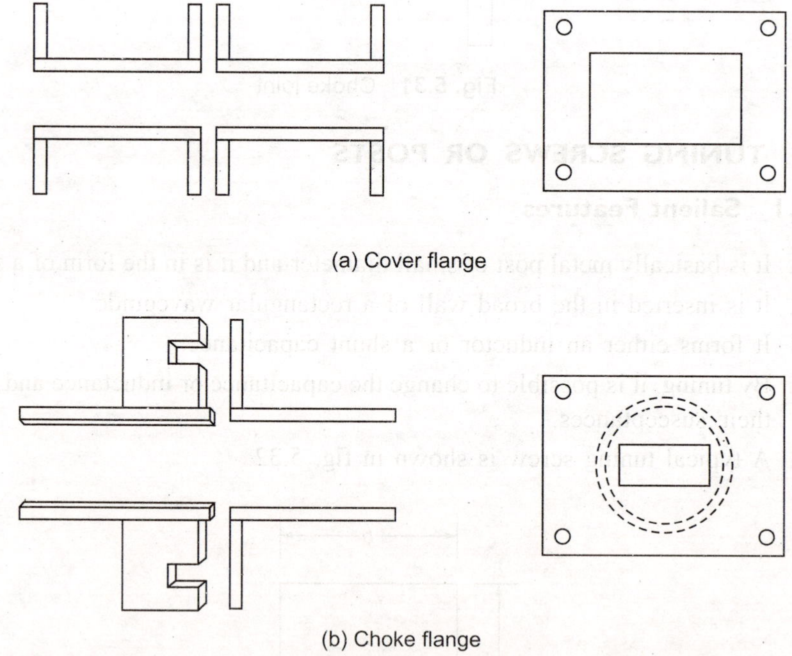

(a) Cover flange

(b) Choke flange

Fig. 5.33 Flanges

4. The cover flanges are flat metallic surfaces connected to the waveguide sections in a perpendicular direction to the wave propagation through the guide.

5. Waveguides are connected by cover flanges with bolts and nuts resulting a continuous surface inside the guide.

6. The cover flanges are fastened and mated well to eliminate any discontinuity between the waveguide sections. Any amount of discontinuity causes voltage breakdown at high power flow.

7. The choke joint is used to provide good electrical connection if proper matching two waveguide is not possible. This type of connection includes both choke and cover flanges.

8. The choke flange has a circular slot and is equivalent to a short circuited line of quarter wave length long.

9. The choke joints are used to separate the two waveguides while maintaining electrical *r-f* connection.

5.30 TRANSITIONS

Transitions are basically tapers from one size coaxial line to another and from one rectangular waveguide to another.

5.30.1 Salient Features of Transitions

1. They have VSWR within limits.
2. They do not add any reflections from the discontinuity.
3. They are basically transducers.
4. They are used to change the impedance.
5. In the case waveguide transition, they are smooth tapers.
6. The waveguides of different sizes are joined with transitions by soldering flanges.
7. Transitions from rectangular to circular waveguides are also used.
8. A typical transition is shown in fig. 5.34.

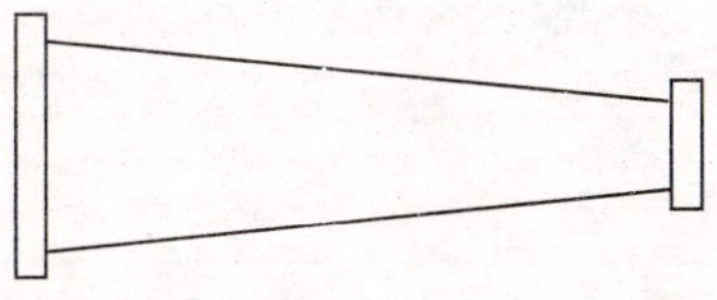

Fig. 5.34 Transition

5.31 MICROWAVE FILTERS

A filter is a network or device which passes a band of frequencies only.

Types of filters

1. Low pass filter
2. High pass filter
3. Band pass filter
4. Band stop filter

Low Pass Filter

A low pass filter at low frequencies is a lumped circuit. A microwave filter consists of a high impedance line for an inductance and a low impedance for a capacitor.

High Pass Filter

At microwave frequencies, waveguide is a high pass filter. In these filters, the transition from the pass band to stop band is gradual.

Band Pass Filter

At microwave frequencies, a resonant cavity is a band pass filter. The high Q of the cavity makes the pass band narrow and vice-versa. With additional cavity, the transition can be made steeper.

Band Stop Filter

It is a combination of a low pass and a band pass filters. In this, the cut-off frequency of low pass filter is made as the lower cut-off of the stop band. At the same time, the lower cut-off of the band pass of the filter is made as the upper cut-off of stop band under consideration.

5.32 POINTS TO REMEMBER

➤ Transmission lines transfer energy from one place to another.

➤ Z_0 of lossless transmission line is $\sqrt{L/C}$.

➤ $Z_0 = \dfrac{276}{\sqrt{\epsilon_r}} \log \left(\dfrac{D}{d} \right)$ for coaxial line.

➤ The standard value of Z_0 for coaxial line is 50 Ω.

➤ VSWR $= \dfrac{V_{max}}{V_{min}}$

➤ $Z_{min} = \dfrac{Z_0}{\text{VSWR}}$

➤ VSWR $= \dfrac{R_0}{Z_L}$ if $R_0 > Z_L$.

➤ VSWR $= \dfrac{Z_L}{R_0}$ if $Z_L > R_0$.

➤ The range of VSWR is 1 and ∞.
➤ Waveguide is a transmission line, radiator, filter.
➤ The cut-off wavelength of dominant in rectangular waveguide is $2a$.
➤ Ridge waveguide has low Z_0 and high bandwidth.

- Attenuation is $10 \log \dfrac{p_i}{p_0}$.
- Attenuator is a device which reduces the power of the signal.
- A matched load is a device which absorbs incident power.
- Corners are used to bend the waveguides up to 90°.
- The waveguide bends are used to change the direction of a waveguide to get around obstacles.
- Twist is used to change the direction of the guide.
- The termination can be matched, open or shorts.
- Isolates are used to protect sources from reflected power.
- Directional couplers are four-port waveguide junctions that sample a part of EM wave through the main guide.
- The S-matrix of directional coupler is $[S] = \begin{bmatrix} 0 & p & 0 & jq \\ p & 0 & jq & 0 \\ 0 & jq & 0 & p \\ jq & 0 & p & 0 \end{bmatrix}$.
- The H-plane Tee junction is a three-port device and its equivalent circuit is

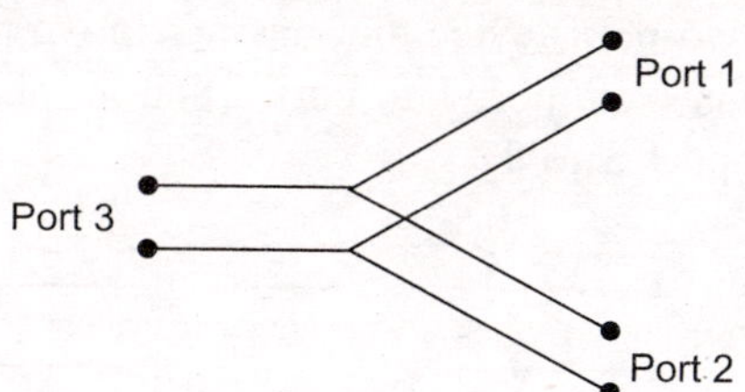

- The E-plane Tee junction is a three-port device and its equivalent

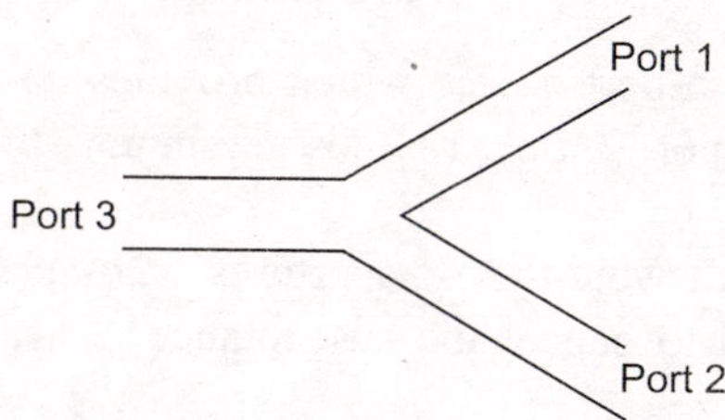

- The S-matrix of H-plane Tee junction is $S = \begin{bmatrix} \dfrac{1}{2} & -\dfrac{1}{2} & \dfrac{1}{\sqrt{2}} \\ -\dfrac{1}{2} & \dfrac{1}{2} & \dfrac{1}{\sqrt{2}} \\ \dfrac{1}{\sqrt{2}} & \dfrac{1}{\sqrt{2}} & 0 \end{bmatrix}$

> The S-matrix of E-plane Tee junction is $S = \begin{bmatrix} \dfrac{1}{2} & \dfrac{1}{2} & \dfrac{1}{\sqrt{2}} \\ \dfrac{1}{2} & \dfrac{1}{2} & -\dfrac{1}{\sqrt{2}} \\ \dfrac{1}{\sqrt{2}} & -\dfrac{1}{\sqrt{2}} & 0 \end{bmatrix}$.

> The S-matrix of rat race junction is $S = \begin{bmatrix} 0 & s_{12} & 0 & s_{14} \\ s_{21} & 0 & s_{23} & 0 \\ 0 & s_{32} & 0 & s_{34} \\ s_{41} & 0 & s_{43} & 0 \end{bmatrix}$.

> The rat race junction is also a hybrid junction.

> Isolation is defined as $I = 10 \log_{10} \dfrac{p_i}{p_b}$ dB.

> p_i = incident power
> p_b = back power
> Gyrator is a two-port device that has a relative phase difference of 180° for transmission from port 1 to port 2 and no phase difference for transmission from port 2 to 1.

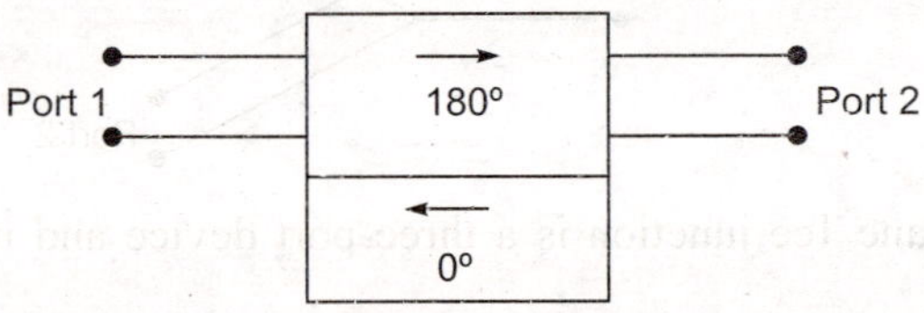

> Isolator is a two-port device which provides no attenuation for transmission from 1 to port 2, but provides maximum attenuation for transmission from port 2 to 1.
> Irises are called windows or apertures or diaphragm or obstacles.
> Irises are used to cancel the susceptance of waveguide.

> The equivalent circuit of iris   is

> The equivalent circuit of iris   is

> The posts are metallic cylindrical pieces.

➤ The posts are introduced in the broad wall of the waveguides to cancel out reactance port of the waveguide.

➤ The equivalent circuit of the post 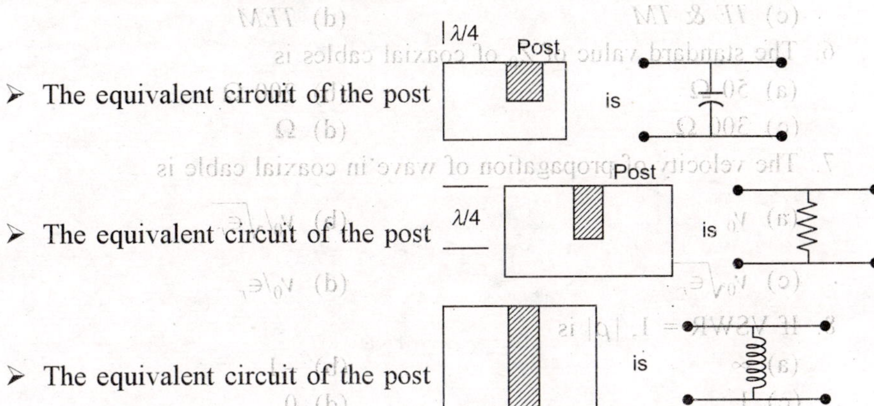 is

➤ The equivalent circuit of the post is

➤ The equivalent circuit of the post is

➤ The advantage of the post over an iris is that it is adjustable.
➤ An adjustable port is called a screw.
➤ A probe is a short antenna if placed correctly, the required modes are produced.
➤ The coupling loop placed at the centre of the shorted and of the waveguide produces dominant mode.
➤ Circulator is a four-port device in which each port is connected only to the next clockwise port.

5.33 MULTIPLE CHOICE QUESTIONS

1. Two-wire lines are used as
   (a) radiator
   (b) transmission lines
   (c) filters
   (d) TEM wave generators
2. The Z_0 of a two-wire transmission line is

   (a) $\sqrt{\dfrac{L}{C}}$
   (b) $\sqrt{LC}$

   (c) $\sqrt{\dfrac{C}{L}}$
   (d) LC

3. The transmission line is used as
   (a) antenna
   (b) stub
   (c) oscillator
   (d) filter
4. The Z_0 of a lossless transmission line for which $L = 2$ mH/km and $C = 0.2$ µf/km is
   (a) 100 Ω
   (b) 200 Ω
   (c) 50 Ω
   (d) 75 Ω

5. Coaxial line supports only
   - (a) *TE*
   - (b) *TM*
   - (c) *TE & TM*
   - (d) *TEM*

6. The standard value of Z_0 of coaxial cables is
   - (a) 50 Ω
   - (b) 200 Ω
   - (c) 300 Ω
   - (d) Ω

7. The velocity of propagation of wave in coaxial cable is
   - (a) v_0
   - (b) $v_0/\sqrt{\epsilon_r}$
   - (c) $v_0\sqrt{\epsilon_r}$
   - (d) v_0/ϵ_r

8. If VSWR = 1, $|\rho|$ is
   - (a) ∞
   - (b) -1
   - (c) 1
   - (d) 0

9. If VSWR = ∞, $|\rho|$ is
   - (a) 1
   - (b) -1
   - (c) 0
   - (d) ∞

10. If $R_0 = 75$ Ω, $Z_L = 50$, VSWR is
    - (a) 2
    - (b) 3
    - (c) 1.5
    - (d) 2.5

11. VSWR is
    - (a) $\dfrac{1-|\rho|}{1+|\rho|}$
    - (b) $\dfrac{1+|\rho|}{1-|\rho|}$
    - (c) $\dfrac{1+\rho}{1-\rho}$
    - (d) $\dfrac{1-\rho}{1+\rho}$

12. If $Z_L = Z_0$, VSWR is
    - (a) 0
    - (b) ∞
    - (c) 10
    - (d) 1

13. If Z_L is reactive, $|\rho|$ is
    - (a) 1
    - (b) -1
    - (c) 0
    - (d) ∞

14. 1 nW is equal to
    - (a) -30 dBm
    - (b) 30 dB
    - (c) 60 dB
    - (d) 90 dB

15. 100 W is equal to
    - (a) 10 dB
    - (b) -10 dB
    - (c) 20 dB
    - (d) 10 dB$_m$

16. 30 dB$_m$ of power is equal to
    - (a) 30 W
    - (b) 1 W
    - (c) 10 W
    - (d) 100 W

17. 100 mW of power is equal to
    (a) 50 dB$_m$
    (b) 10 dB
    (c) 20 dB
    (d) 10 dB$_m$
18. If the broad dimension of rectangular waveguide is 3 cm, the cut-off wavelength of dominant mode is
    (a) 3 cm
    (b) 6 cm
    (c) 9 cm
    (d) 1.5 cm
19. Attenuator
    (a) radiates
    (b) reduces power
    (c) increases power
    (d) matches load
20. Matched load
    (a) absorbs incident power
    (b) transits incident power
    (c) reflects incident power
    (d) scatters incident power
21. Twists exist in
    (a) transmission lines
    (b) waveguides
    (c) filters
    (d) active devices
22. Circulators are created by
    (a) two waveguides
    (b) two magic Tees
    (c) two transmission lines
    (d) circular waveguides
23. Isolator is made of
    (a) non-ferrite
    (b) ferrite
    (c) Si
    (d) Ge
24. Faraday rotation produces
    (a) isolation
    (b) circular properties
    (c) matching properties
    (d) transmission properties
25. Directional coupler produces
    (a) division of power
    (b) amplification
    (c) generator of power
    (d) reduction of power
26. Directional coupler is
    (a) reciprocal device
    (b) non-reciprocal device
    (c) an amplifier
    (d) an oscillator
27. Coupling factor of directional coupler is
    (a) $\dfrac{p_a}{p_i}$
    (b) $\dfrac{p_i}{p_a}$
    (c) $\dfrac{p_r}{p_i}$
    (d) $\dfrac{p_i}{p_r}$
28. Directivity of directional coupler is
    (a) $\dfrac{p_{af}}{p_{ar}}$
    (b) $\dfrac{p_r}{p_i}$
    (c) $\dfrac{p_a}{p_i}$
    (d) $\dfrac{p_i}{p_a}$

29. Magic Tee can produce
    - (a) sum and difference of signals
    - (b) oscillations
    - (c) only sum of signals
    - (d) only difference of signals
30. H-plane Tee junction is
    - (a) a three-port device
    - (b) a two-port device
    - (c) a four-port device
    - (d) one-port device
31. E-plane Tee junction is
    - (a) a three-port device
    - (b) one-port device
    - (c) a four-port device
    - (d) a two-port device
32. Ferrite is a
    - (a) high resistance magnetic material
    - (b) low resistance magnetic material
    - (c) conductor
    - (d) insulator
33. Ferrite is used in
    - (a) isolation
    - (b) amplification
    - (c) oscillation
    - (d) attenuation
34. Phase shift is
    - (a) not a function of frequency
    - (b) a function of frequency
    - (c) obtained from resistance
    - (d) obtained from magnetic materials
35. Circulator is a
    - (a) two-port device
    - (b) three-port device
    - (c) multiport device
    - (d) one-port device
36. Microwave filter consists of
    - (a) one cavity
    - (b) more than one cavity
    - (c) LC lumped components
    - (d) RLC lumped components
37. Probe is
    - (a) a short antenna
    - (b) a filter
    - (c) an amplifier
    - (d) an oscillator
38. The equivalent circuit of the post is
    - (a)
    - (b)
    - (c)
    - (d)

39. Iris is called a
    (a) window   (b) probe
    (c) antenna   (d) filter

40. The directivity of a directional coupler is

    (a) $\dfrac{p_r}{p_b}$ (b) $\dfrac{p_b}{p_f}$

    (c) $\dfrac{p_i}{p_r}$ (d) $\dfrac{p_1}{p_2}$

41. Isolation of directional coupler is

    (a) $\dfrac{p_b}{p_i}$ (b) $\dfrac{p_i}{p_b}$

    (c) $\dfrac{p_r}{p_i}$ (d) $\dfrac{p_1}{p_2}$

42. Isolation has
    (a) no units   (b) unit of power
    (c) unit of voltage   (d) unit of current

43. If the directivity, D of a directional coupler is zero, D in dB is
    (a) zero   (b) infinity
    (c) not defined   (d) $-\infty$

44. Magic Tee is used as
    (a) an amplifier   (b) an oscillator
    (c) an mixer   (a) a filter

45. Impedance can be measured by
    (a) antenna   (b) magic Tee
    (c) iris   (d) post

46. The separation of each post in rat race junction is
    (a) λ_g (b) $\lambda_g/4$
    (c) $\lambda_g/2$ (d) λ_g/g

47. In H-plane Tee junction
    (a) $s_{13} = s_{23}$ (b) $s_{13} \neq s_{23}$
    (c) $s_{33} \neq 0$ (d) $s_{12} = s_{22}$

48. S-matrix is always
    (a) a square matrix   (b) a rectangular matrix
    (c) 2×2 matrix   (d) 3×3 matrix

49. Directivity of directional coupler is always expressed in
    (a) nepers   (b) dB
    (c) without units   (d) watts

50. In Faraday rotation,
    (a) polarization is changed
    (b) plane polarization is changed to elliptical polarization
    (c) circular polarization is changed to elliptical polarization
    (d) no change in polarization exists.

5.34 ANSWERS

1. b	2. a	3. b
4. a	5. d	6. a
7. b	8. d	9. a
10. c	11. a	12. d
13. a	14. a	15. a
16. b	17. a	18. b
19. b	20. a	21. b
22. b	23. b	24. b
25. a	26. a	27. b
28. a	29. a	30. b
31. a	32. a	33. a
34. b	35. c	36. b
37. a	38. c	39. a
40. a	41. b	42. a
43. d	44. c	45. b
46. b	47. a	48. a
49. b	50. b	

5.35 EXERCISE PROBLEMS

1. Determine the characteristic impedance of a coaxial cable if $\epsilon_r = 2.25$, $\dfrac{D}{d} = 3.0$.

2. What is z_0 of a two-wire transmission line if $L = 2$ mH/km, $C = 0.3$ μF/km.

3. Find z_0 of a open-wire transmission line if the spacing, $s = 0.6$ cm and the diameter of the wire is $= 0.16$ cm. Assume $\epsilon_r = 1$.

4. A coaxial cable has an attenuation of 0.2 dB/, at an operating frequency. Determine the output power of 20 m cable if an input of 1.0 W is given to the cable.

5. If a voltage of 30 V is incident in an open-wire lines and the reflected voltage is 22 V. Find the percentage of reflected power.

6. In a transmission line, $Z_L \neq Z_0$ and $V_{max} = 6$ V and V_{max} is 5 V in the standing wave pattern, find VSWR.

Chapter 6

Microwave Transmission Lines

Transmission line is a means of wire communication, a means of energy transferring and also a means of coupling two circuits.

6.1 DEFINITION

A transmission line is defined as a physical structure which guides an electromagnetic wave from place to place.

It is also defined as a device that carries power/information from transmitter to receiver.

It delivers power from a power source to a load and also from component to component in a system.

At low frequencies, transmission line is used for power distribution and at high frequencies, it is used in communication and computer networks.

6.2 FUNCTIONS OF TRANSMISSION LINES

- They convey radio frequency power from one point to another.
- They are used in radio broadcast as transmission channels to carry power from the transmitter to antenna and to carry signals from antenna to receiver.
- They are used as reactive circuit elements or tuned circuits.
- If the length of a transmission line is such that its reactance is infinite, the line can be used as an insulator to support another line.
- The reactive sections of transmission lines are used as impedance matching elements.
- It is used as an impedance transformer.

6.3 EXAMPLES OF TRANSMISSION LINES/MEDIA

1. Free Space
2. Two-wire parallel open lines
3. Twisted pair lines
4. Coaxial lines

5. Parallel plates
6. Rectangular waveguides
7. Cylindrical waveguides
8. Strip lines
9. Microstrip lines
10. Optical fibers

6.4 APPLICATIONS OF TRANSMISSION LINES

The common applications of transmission lines are:

1. They are used in radio broadcast, TV and telephone systems etc.
2. Pair of lines are used for electrical power transmission and distribution system.
3. Twisted pairs and coaxial cables are used in computer networks like internet and Ethernets.
4. The pair of lines used in telephony.
5. Optical fibers are used for long and short distance communications.
6. Waveguides are used in microwave towers to connect transmitter antennas and also to connect antenna to receiver.
7. Free space is used in all wireless communication systems.

6.5 SALIENT FEATURES OF TRANSMISSION LINES

- At low frequencies, power is considered to be delivered to load from source through wire transmission lines.
- At microwave frequencies, power is considered to be delivered through electric and magnetic fields. These fields are guided in waveguides and free space.
- At the very low and medium frequency regions, two-wire lines are used. In these lines, two wires are separated by a fixed distance with an insulator between them.
- In the intermediate range of frequencies, coaxial lines are used.
- At higher frequencies but for short distances, coaxial lines are used.
- Coaxial lines consists of inner and outer conductors separated by dielectric material. They have cylindrical structure.
- At high frequencies waveguides are used.
- Waveguides can be rectangular or cylindrical in shape.
- Microstrips and striplines are used as interconnecting transmission lines between internal components.
- A stripline consists of three conductors separated by a dielectric material.
- A microstrip consists of two conductors separated by a dielectric material.

- The transmission lines can be analyzed by Smith chart or mathematically.
- A signal given to a line of infinite length cannot reach the far end.
- The far end of a line of infinite length has no effect on input end.
- The equivalent circuit of transmission line of unit length is

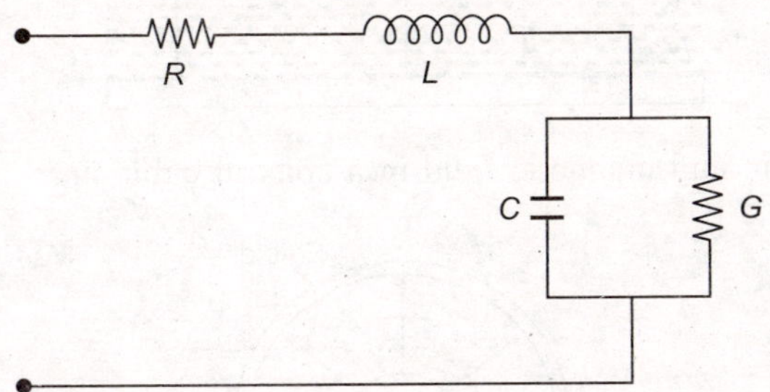

- A transmission line is described by its parameters, namely, series resistance, series inductance, shunt capacitance and shunt conductance.
- The primary constants of a transmission line and R, L, C, G.
- R, L, C, G are called distributed constants.
- The transmission lines are made of conductors. The conductors are described by σ_c, μ_c, ϵ_c.
- The medium between the lines is made of dielectric or insulator and the dielectric is described by σ_d, μ_d and ϵ_d.
- The characteristic impedance of parallel wire transmission line is

$$Z_0 = \sqrt{2/Y} = \sqrt{\frac{R + j\omega L}{G + j\omega C}}.$$

- The propagation constant of the line $\gamma = \sqrt{ZY} = \sqrt{(R + j\omega L)(G + j\omega C)}$.
- The secondary constants of a line are γ and Z_0.
- The condition for a transmission line to be lossless is $R = 0$, $G = 0$.
- The condition for a transmission line to be distortion loss is $\dfrac{R}{L} = \dfrac{G}{C}$.
- The reflection coefficient of a line is $\rho = \dfrac{Z_L - Z_0}{Z_L + Z_0}$.
- The VSWR of a transmission line is $S = \dfrac{1 + |\rho|}{1 - |\rho|}$.
- The reflection coefficient in terms of VSWR is $|\rho| = \dfrac{S - 1}{S + 1}$.
- The transmission lines have different losses : copper losses, dielectric losses and radiation losses.

- The electric and magnetic field in parallel plate transmission lines are

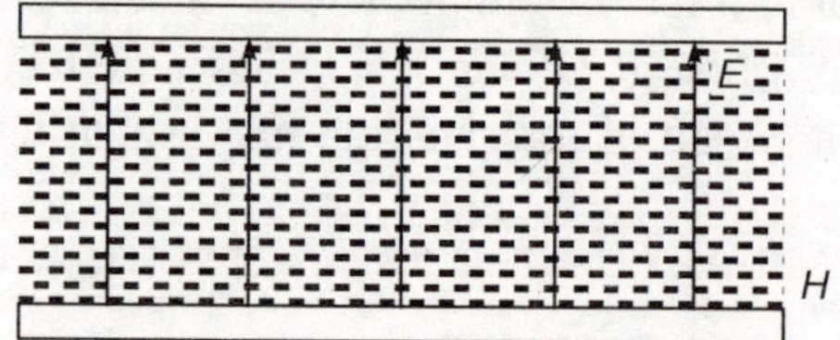

- The electric and magnetic field in a coaxial cable are

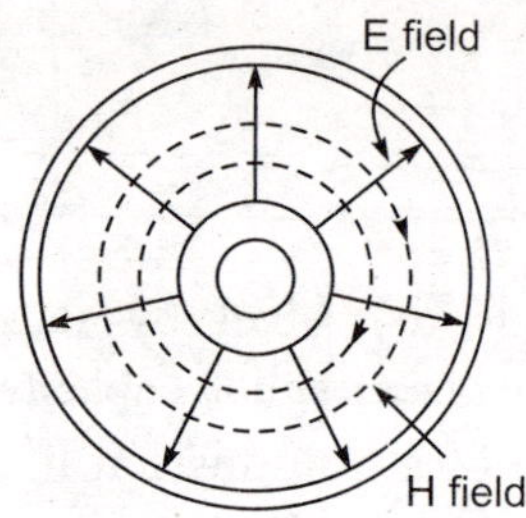

- The equivalent circuit of shorted line of length $\ell < \lambda/4$ is an inductor.
- The equivalent circuit of open line of length, $\ell < \lambda/4$ is capacitor.
- The equivalent circuit of a shorted line of length, $\ell < \lambda/4$ is a tank circuit.
- The equivalent circuit of an open line of length, $\ell = \lambda/4$ is series-resonant circuit.
- The equivalent circuit of shorted line of length $\lambda/4 = \ell = \lambda/2$ is capacitor.
- The equivalent circuit of an open line of length $\lambda/4 = \ell = \lambda/2$ is inductor.
- The equivalent circuit of a shorted line of length, $\ell = \lambda/2$ is series-resonant circuit.
- The equivalent circuit of an open line of length, $\ell = \lambda/2$ is a tank circuit.

6.6 TRANSMISSION MEDIUM – FREE SPACE

The free space is one of best transmission medium for communication between transmitter and receiver.

6.6.1 Characteristics of Free Space

Free space is characterized by the following parameters:

Relative permittivity, $\epsilon_r = 1$
Relative permeability, $\mu_r = 1$
Conductivity, $\sigma = 0$
Characteristic impedance, $\eta = 377 \ \Omega$
Volume charge density, $\rho_v = 0$
Conduction current density, $J = 0$

It supports uniform plane wave. It is truly wide band medium. Ideally, attenuation is zero. Its propagation constant, γ is $j\beta$.

The velocity of propagation and phase velocity of electromagnetic wave in free space are the same.

The velocity of propagation of electromagnetic wave in free space is 3×10^8 m/s.

6.7   TWO-WIRE PARALLEL OPEN LINES

6.7.1   Salient Features of Parallel Open-Wire Transmission Lines

- It consists of two parallel conductors separated by dielectric.
- The dielectric is free space.
- The lines are generally separated by 2 inches and 6 inches.
- *TEM* wave propagates through the transmission lines.
- It is simple in construction.
- It is susceptible to noise pickup.
- As there is no shielding radiation losses exist.
- The open-wire transmission lines are normally operated in the balanced mode.
- These are useful for low frequency transmission.
- Its equivalent circuit is a distributed network and is given in fig. 6.1.

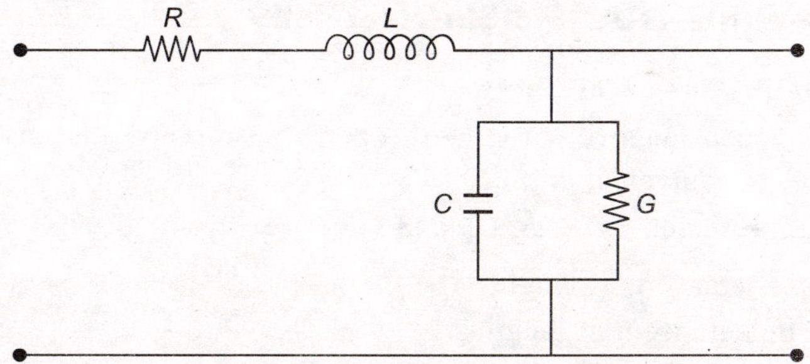

Fig. 6.1   Equivalent circuit of transmission lines

Here, R is loop resistance (Ω/Km)

   L is loop inductance, (H/Km)

   C is shunt capacitance, (F/Km)

   G is shunt conductance, (Mho/Km)

- It is described by the primary, R, L, C, G and secondary constants γ, Z_0, α, β.
- The electric and magnetic fields in the transmission line are

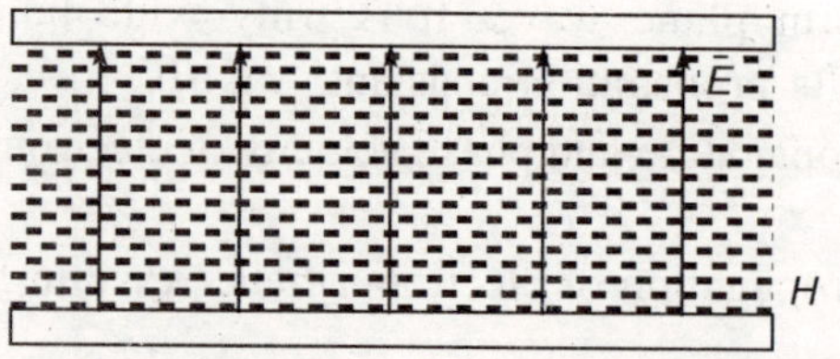

6.7.2   The Structure of Open-Wire Transmission Line

The structure of open-wire transmission line is shown in fig. 6.2 and 6.3.

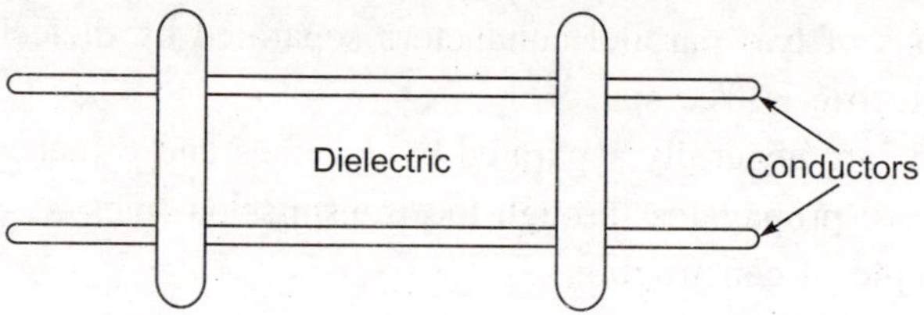

Fig. 6.2   Open-wire transmission line

Fig. 6.3   Cross-section of open-wire parallel transmission line

6.8   DEFINITIONS OF THE PARAMETERS OF TWO-WIRE TRANSMISSION LINES

Definition of Primary Constants

The primary constants of a two-wire open transmission line are distributed parameters of its equivalent circuits.

They are represented by R, L, C and G.

$R \equiv$ Loop resistance/unit length

$L \equiv$ Loop inductance/unit length

$C \equiv$ Shunt capacitance/unit length

$G \equiv$ Shunt conductance/unit length

The transmission lines consist of two conductors separated by dielectric.

If
$\quad a_c$ = radius of the conductors

$\quad d$ = spacing between the wires

$\quad \sigma_c$ = conductivity of the conductors

$\quad \epsilon_d$ = dielectric constant of dielectric material

$\quad \mu_c$ = permeability of the conductor

$\quad\quad = \mu_0\, \mu_{rc}$

$\quad \sigma_d$ = conductivity of dielectric material

$$R_{dc} = \text{dc conductor resistance}$$

$$= \frac{Z}{\pi \sigma_c a_c} \text{ Mho/m for } f < 10 \text{ KHz}$$

$$R_{ac} = \text{ac conductor resistance}$$

$$= \frac{1}{\pi \sigma_c a_c \delta} \ \Omega/\text{m for } f > 10 \text{ KHz}$$

$$L = \left[\mu_{rc} + 9.21 \log_{10} \frac{d}{a_c} \right] \times 10^{-7} \text{ H/m}$$

$$C = \frac{\pi \in_d}{\ell n \left(\dfrac{d}{a_c} \right)} \ \text{F/m}$$

$$G = \frac{C}{\in_d} \sigma_d \text{ Mho/m}$$

If L_i = internal inductance due to internal flux linkages in the conductors, it is given by

$$L_i = \frac{R_{ac}}{2\pi f} \text{ H/m for } f > 10 \text{ KHz}$$

$$= \frac{\mu_0}{2\pi} \text{ H/m for } f < 10 \text{ KHz}$$

If L_e = external inductance due to flux linkages with the flux external to the wire, it is given by

$$L_e = \frac{\mu_d}{\pi} \times \frac{1}{\ell n \left(\dfrac{d}{a_c} \right)} \text{ for } d \gg a_c$$

6.9 DEFINITIONS OF SECONDARY CONSTANTS

From the equivalent circuit of the transmission line, series impedance is defined as

$$Z \equiv R + j\omega L$$

And shunt admittance is defined as

$$Y = G + j\omega C$$

The secondary constants of the line, propagation constant, γ and characteristic impedance, Z_0 are defined as below.

6.10 PROPAGATION CONSTANT

Definition

$$\gamma_\ell \equiv \sqrt{\text{Series Impedance} \times \text{Shunt Admittance}}$$

$$= \sqrt{ZY} \;\; (\text{m}^{-1})$$

$$\gamma_\ell \equiv \ell_n\left(\frac{I_s}{I_L}\right) = \ell_n\left(\frac{V_s}{V_L}\right) \;\text{nepers}$$

$$\gamma_\ell \equiv 20 \log_{10}\left(\frac{I_s}{I_L}\right) = \log_{10}\left(\frac{V_s}{V_L}\right)\text{dB}$$

$$\gamma_\ell \equiv \alpha + j\beta$$

Here, I_S = current at sending end

$\quad\;\; I_L$ = current at load end

$\quad\;\; V_S$ = voltage at sending end

$\quad\;\; V_L$ = voltage at load end

$\quad\;\; \alpha_\ell$ = attenuation constant, dB/m

$\quad\;\; \beta_\ell$ = phase constant, rad/m

6.11 CHARACTERISTIC IMPEDANCE

The characteristic impedance of a transmission line is a special property of the line. It is defined as the ratio of voltage to current at any point on the line. That is,

$$Z_0 \equiv \frac{V_f}{I_f} \;\text{ and also}$$

$$Z_0 \equiv \sqrt{\frac{Z}{Y}}$$

6.11.1 Salient Features of Characteristic Impedance

- Z_0 does not depend on the length of the line, frequency of operation, ohmic value of the load resistance.
- Z_0 depends on the distributed constants, R, L, G and C.
- $Z_0 \approx \sqrt{L/C}$.
- Z_0 of a two-wire open air transmission line is $Z_0 = 276 \log \dfrac{s}{r}$.

Here, s = spacing between wires

$\quad\;\; r$ = radius of conductors.

PROBLEM 6.1   A transmission line has $R = 0$, $G = 0$, $L = 110$ nH/m, $C = 20$ pF/m. Find Z_0 for (a) 10 meters line and (b) 500 meter line.

Solution

(a)
$$Z_0 = \sqrt{\frac{L}{C}} = \sqrt{\frac{110 \times 10^{-9}}{20 \times 10^{-12}} \times \frac{10}{10}}$$

$$= \sqrt{5500} = 74.16$$

$\therefore$ $\boxed{Z_0 = 74.16 \ \Omega}$

(b)
$$Z_0 = \sqrt{\frac{L}{C}} = \sqrt{\frac{110}{20} \times \frac{10^{-9}}{10^{-12}} \times \frac{500}{500}}$$

$\therefore$ $\boxed{Z_0 = 74.16 \ \Omega}$

PROBLEM 6.2   A two-wire open air line consists of two conductors which are separated by 300 mm. The wire diameter is 3.0 mm. Find the characteristic impedamce.

Solution   Spacing, $\qquad\qquad\qquad s = 300$ mm

Radius of the wires, $\qquad\qquad r = \dfrac{3.0}{2} = 1.5$ mm

$\therefore$
$$Z_0 = 276 \log \frac{s}{r}$$

$$= 276 \log \frac{300}{1.5}$$

$$= 276 \log 200$$

$$= 276 \times 2.3$$

$\therefore$ $\boxed{Z_0 = 635 \ \Omega}$

6.11.2   *RF* Lines and Parameters

For *RF* lines, $\qquad\qquad\qquad \omega L \gg R$

$$\qquad\qquad\qquad \omega C \gg G$$

6.11.3   Propagation Constant

$$\gamma_c = \sqrt{ZY}$$

$$= \sqrt{(R + j\omega L)(G + j\omega C)}$$

As $\omega L \gg R$, $\omega C \gg G$

$$= j\omega\sqrt{LC}\,\sqrt{\left(1+\frac{R}{j\omega L}\right)\left(1+\frac{G}{j\omega C}\right)}$$

$$\approx j\omega\sqrt{LC}\left(1+\frac{R}{2j\omega L}\right)\left(1+\frac{G}{2j\omega C}\right)$$

$$= j\omega\sqrt{LC}\left[1+\frac{G}{j2\omega C}+\frac{R}{j2\omega L}-\frac{RG}{4\omega^2 LC}\right]$$

$$\approx j\omega\sqrt{LC}+\frac{G}{2}\sqrt{\frac{L}{C}}+\frac{R}{2}\sqrt{\frac{C}{L}}$$

$$\therefore \quad \gamma_\ell = \frac{R}{2}\sqrt{\frac{C}{L}}+\frac{G}{2}\sqrt{\frac{L}{C}}+j\omega\sqrt{LC}$$

$$\therefore \quad \boxed{\alpha_\ell = \frac{1}{2}\left[R\sqrt{\frac{C}{2}}+G\sqrt{\frac{L}{C}}\right]\quad\text{and}}$$

$$\boxed{\beta_\ell = \omega\sqrt{LC}}$$

6.11.4 Salient Features of *RF* Lines

- For *RF* lines, $\omega L \gg R,\ \omega C \gg G$

- $\gamma_\ell = \dfrac{R}{2}\sqrt{\dfrac{C}{L}}+\dfrac{G}{2}\sqrt{\dfrac{L}{C}}+j\omega\sqrt{LC}$

- $\alpha_\ell = \dfrac{1}{2}\left[R\sqrt{\dfrac{C}{L}}+G\sqrt{\dfrac{L}{C}}\right]$

- $\beta_\ell = \omega\sqrt{LC}$

- $Z_i = Z_0\left[\dfrac{Z_L+jZ_0\tan\beta_\ell}{Z_0+jZ_L\tan\beta_\ell}\right]$

- $v_\ell = \dfrac{1}{\sqrt{LC}}$

6.11.5 Attenuation and Phase Constant in a Transmission Line

$$\gamma_\ell = \sqrt{\frac{1}{2}\left[(RG-\omega^2 LC)+\sqrt{(R^2+\omega^2 L^2)(G^2+\omega^2 C^2)}\right]}$$

$$\beta_\ell = \sqrt{\frac{1}{2}(\omega^2 LC - RG) + \sqrt{(R^2 + \omega^2 L^2)(G^2 + \omega^2 C^2)}}$$

Proof:

$$\gamma_\ell = \sqrt{ZY}$$

$$= \sqrt{(R + j\omega L)(G + j\omega C)}$$

$$= \alpha_\ell + j\beta_\ell$$

$$\therefore \quad \gamma_\ell^2 = \alpha_\ell^2 - \beta_\ell^2 + 2j\alpha_\ell \beta_\ell$$

$$= RG - \omega^2 LC + j\omega RC + j\omega LG$$

$$\alpha_\ell^2 = \beta_\ell^2 = RG - \omega^2 LC \tag{1}$$

We have

$$|\gamma_\ell| = \alpha_\ell^2 + \beta_\ell^2$$

$$\therefore \quad \alpha_\ell^2 + \beta_\ell^2 = \sqrt{(R^2 + \omega^2 L^2)(G^2 + \omega^2 C^2)} \tag{2}$$

From (1) and (2), we have

$$2\alpha_\ell^2 = (RG - \omega^2 LC) + \sqrt{(R^2 + \omega^2 L^2)(G^2 + \omega^2 C^2)}$$

$$\therefore \quad \alpha_\ell = \sqrt{\frac{1}{2}\left[(RG - \omega^2 LC) + \sqrt{(R^2 + \omega^2 L^2)(G + \omega^2 C^2)}\right]} \tag{3}$$

Subtracting (1) from (2), we get

$$2\beta_\ell^2 = \sqrt{(R^2 + \omega^2 L^2)(G^2 + \omega^2 C^2)} - (RG - \omega^2 LC)$$

or

$$\beta_\ell = \sqrt{\frac{1}{2}\left[(\omega^2 LC - RG) + \sqrt{(R^2 + \omega^2 L^2)(G^2 + \omega^2 C^2)}\right]}$$

Hence proved.

6.12 LOSSLESS LINES AND PARAMETERS

6.12.1 Definition of Lossless Line

A line is said to be lossless if $R = 0$ and $G = 0$.

6.12.2 Propagation Constant, γ

$$\gamma_\ell = \sqrt{ZY}$$

$$= \sqrt{(R + j\omega L)(G + j\omega C)}$$

As $R = 0$, $G = 0$

$$\gamma_\ell = \sqrt{(j\omega L)(j\omega C)}$$

$\therefore$ $\qquad\qquad\qquad \gamma_\ell = j\omega\sqrt{LC}$ for lossless line

i.e. $\qquad\qquad\qquad \alpha_\ell = 0,\ \beta_\ell = \omega\sqrt{LC}$

6.12.3 Characteristic Impedance

$$Z_0 = \sqrt{\frac{Z}{Y}}$$

$$= \sqrt{\frac{R + j\omega L}{G + j\omega C}}$$

As $R = 0,\ G = 0$

$$Z_0 = \sqrt{\frac{L}{C}}$$

Velocity of propagation in a lossless line is

$$v_\ell = \frac{\omega}{\beta_\ell} = \frac{\omega}{\omega\sqrt{LC}} = \frac{1}{\sqrt{LC}}$$

6.12.4 Salient Features of Lossless Transmission Lines

- $\gamma_\ell = j\omega\sqrt{LC}$
- $\alpha_\ell = 0$
- $\beta_\ell = \omega\sqrt{LC}$
- $Z_0 = \dfrac{1}{\sqrt{L/C}}$
- $v_\ell = \dfrac{1}{\sqrt{LC}}$
- $Z_i = Z_0\left[\dfrac{Z_L + jZ_0\tan\beta_\ell}{Z_0 + jZ_L\tan\beta_\ell}\right]$
- For matched load, $Z_L = Z_0,\ Z_i = Z_0$

6.13 DISTORTIONLESS LINES AND PARAMETERS

A line is said to be distortionless if $\dfrac{R}{L} = \dfrac{G}{C}$.

6.13.1 Propagation Constant

$$\gamma_\ell = \sqrt{(R + j\omega L)(G + j\omega C)}$$

$$= \sqrt{RG}\sqrt{\left(1 + \frac{j\omega L}{R}\right)\left(1 + \frac{j\omega C}{G}\right)}$$

$$= \sqrt{RG}\sqrt{\left(1 + j\omega\frac{C}{G}\right)\left(1 + \frac{j\omega C}{G}\right)}$$

$$= \sqrt{RG}\left(1 + \frac{j\omega C}{G}\right)$$

$$= \sqrt{RG} + \frac{j\omega C}{G}\sqrt{RG}$$

$$= \sqrt{RG} + j\omega C\sqrt{\frac{GL}{CG}}$$

$$\therefore \quad \gamma_\ell = \sqrt{RG} + j\omega\sqrt{LC}$$

$$\therefore \quad \alpha_\ell = \sqrt{RG}$$

$$\therefore \quad \beta_\ell = \omega\sqrt{LC}$$

6.13.2 Characteristic Impedance

$$Z_0 = \sqrt{\frac{Z}{Y}}$$

$$= \sqrt{\frac{R + j\omega L}{G + j\omega C}}$$

$$= \sqrt{\frac{R(1 + j\omega L/R)}{G(1 + j\omega C/G)}}$$

$$= \sqrt{\frac{R}{G}} = \sqrt{\frac{L}{C}}$$

i.e. $\qquad Z_0 = R_0$ [As reactance is zero]

$$v_\ell = \frac{\omega}{\beta_\ell} = \frac{\omega}{\omega\sqrt{LC}}$$

$$\therefore \qquad v_\ell = \frac{1}{\sqrt{LC}}$$

6.13.3   Salient Features of Distortionless Lines

> The condition for no distortion in the line is $\dfrac{R}{L} = \dfrac{G}{C}$.

- $\gamma_\ell = \sqrt{RG} + j\omega\sqrt{LC}$
- $\alpha_\ell = \sqrt{RG}$
- $\beta_\ell = j\omega\sqrt{LC}$
- $Z_0 = \sqrt{\dfrac{R}{G}} = \sqrt{\dfrac{L}{C}}$
- $Z_0 = R_0$
- $X_0 = 0$
- $v_\ell = \dfrac{1}{\sqrt{LC}}$

6.14   INPUT IMPEDANCE OF A TRANSMISSION LINE

6.14.1   Definition

Input impedance of a line is defined as the ratio of voltage and current at the sending end.

$$Z_i = \frac{V_G}{I_S}$$

Z_i in terms of γ, ℓ, Z_G is given by

$$Z_i = \frac{V_L \cos h\,\gamma\ell + Z_0\, I_L \sin h\,\gamma\ell}{I_L \cos h\,\gamma\ell + \dfrac{V_L}{Z_0}\sin h\,\gamma\ell}$$

Proof:
Consider a transmission line with load Z_L (fig. 6.4.)

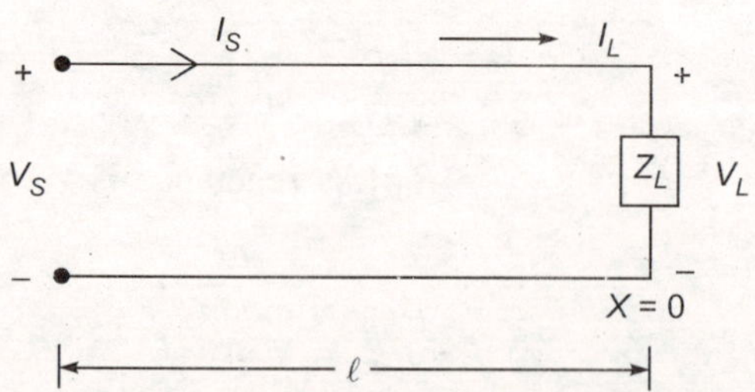

Fig. 6.4   Transmission line with load Z_L

Assume

V = voltage across arbitrary points

I = current at the arbitrary point

From fig. 6.4, we can write

$$\frac{dV}{dx} = -(R + j\omega L)I$$

$$\frac{dI}{dx} = -(G + j\omega C)V$$

Differentiating with respect to x,

$$\frac{d^2V}{dx^2} = -(G + j\omega C)\frac{dV}{dx}$$

i.e.

$$\frac{d^2V}{dx^2} = -(R + j\omega L) \times -(G + j\omega C)I$$

$$= (R + j\omega L)(G + j\omega C)I$$

$$\frac{d^2V}{dx^2} = \gamma^2 I$$

Similarly,

$$\frac{d^2I}{dx^2} = \gamma^2 V$$

Here, $\gamma = \sqrt{(R + j\omega L)(G + j\omega C)}$

The solutions of (1) and (2) are given by

$$V = k_1 \cos h \ \gamma x + k_2 \sin h \ \gamma x$$
$$I = k_3 \cos h \ \gamma x + k_4 \sin h \ \gamma x$$

k_1, k_2, k_3 and k_4 are constants.

The boundary conditions are

$$V = V_L, I = I_L \text{ at } x = 0$$
$$V = V_S, I = I_S \text{ at } x = \ell$$

applying the above boundary conditions, we get,

$$k_1 = V_L, k_3 = I_L$$

$$k_2 = -\sqrt{\frac{R + j\omega L}{G + j\omega C}} \ I_L$$

$$k_4 = -\sqrt{\frac{G + j\omega C}{R + j\omega L}} \ V_L$$

$$\therefore \qquad V_S = V_L \cos h\ \gamma x - Z_0\ I_L \sin h\ \gamma x$$

$$I_S = I_L \cos h\ \gamma x - \frac{V_L}{Z_0} \sin h\ \gamma x$$

Z_L is at $x = 0$ and if $\ell = -x$, we get

$$V_S = V_L \cos h\ \gamma_\ell + Z_0\ I_L \sin h\ \gamma_\ell$$

$$I_S = I_L \cos h\ \gamma x - \frac{V_L}{Z_0} \sin h\ \gamma_\ell$$

$$\therefore \qquad Z_i = \frac{V_L \cos h\ \gamma\ell + Z_0\ I_L \sin h\ \gamma\ell}{I_L \cos h\ \gamma\ell + \left(\dfrac{V_L}{Z_0}\right) \sin h\ \gamma\ell}$$

Hence proved.

6.14.2   Input Impedance for Shorted Line

For shorted load ($Z_L = 0$), $V_L = 0$

$$(Z_i)_{SC} = Z_0 \tan h\ \gamma\ell$$

6.14.3   Input Impedance for Open Line

For open line ($Z_L = \infty$), $I_L = 0$

$$\therefore \qquad (Z_i)_{OC} = Z_0 \cot h\ \gamma\ell$$

From (3) and (4)

$$\therefore \qquad Z_0 = \sqrt{(Z_i)_{SC}\ (Z_i)_{OC}}$$

6.15   LOADING OF TRANSMISSION LINES

6.15.1   Definition of Loading

A transmission line is said to be loaded if an inductance is introduced in series with the line.

6.15.2   Type of Loading

These are

Patch Loading

In this type of loading, sections of continuously loaded cable separated by sections of unloaded cable are used. The cost of this loading is less.

Continuous Loading

In this type, loading is done by winding a type of iron around the conductor. It is expensive.

Lumped Loading

In this type, loading is introduced at uniform intervals. The loading introduces hysteresis and eddy current losses.

6.16 REFLECTION COEFFICIENT IN TERMS OF Z_L AND Z_0

$$\text{Reflection Coefficient, } \rho \equiv \frac{\text{Reflected Voltage}}{\text{Incident Voltage}}$$

i.e.

$$\rho \equiv \frac{V_r}{V_i}$$

and it is given by $\rho = \dfrac{Z_L - Z_0}{Z_L + Z_0}$

Proof:

We have

$$I_L = I_i - I_r$$

$$Z_0 = \frac{V_i}{I_i}$$

$$= -\frac{V_r}{I_r}$$

$$\therefore \qquad I_L = \frac{V_i}{Z_0} - \frac{V_r}{Z_0}$$

$$= \frac{(V_i - V_r)}{Z_0}$$

But $V_L = V_i + V_r$

$$\therefore \qquad Z_L = \frac{V_L}{I_L} = \frac{V_i + V_r}{V_i - V_r} Z_0$$

$$Z_L V_i - Z_L V_r = Z_0 V_i + Z_0 V_r$$
$$V_i(Z_L - Z_0) = V_r(Z_L + Z_0)$$

Dividing both sides by V_i, we get

$$(Z_L - Z_0) = (Z_L + Z_0)\frac{V_r}{V_i}$$

But

$$\frac{V_r}{V_i} = \rho$$

$\therefore$

$$\rho = \frac{Z_L - Z_0}{Z_L + Z_0}$$

Hence proved.

6.17 STANDING WAVES ON THE TRANSMISSION LINES

Definition

Standing wave is an interference wave which exists when two traveling waves travel in opposite direction in the transmission line at the same time.

When $Z_L = Z_0$, there exists reflected waves on the line. As the incident and reflected waves exists on the line, standing waves of voltage and current appear on the line. At some points, on the line, the two waves will be in phase and will add. At some other points, they are out of phase and will cancel. When they are in phase, pattern maximum and when they are out of phase, pattern minima exist.

The incident, reflected and standing waves are shown in fig. 6.5.

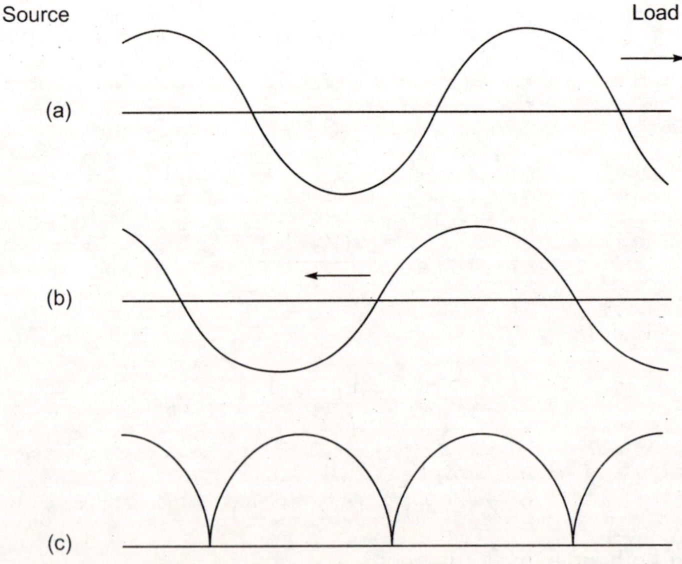

Fig. 6.5   (a) Incident wave (b) Reflected wave (c) Standing wave

6.18 STANDING WAVE RATIO

Definition

The standing wave ratio (SWR) is defined as the ratio of voltage maximum to the voltage minimum.

In the standing wave voltage pattern

i.e.
$$SWR = \frac{V_{max}}{V_{min}}$$

It is also defined as the ratio of current maximum to the current minimum in the standing wave pattern.

i.e.
$$SWR \equiv \frac{I_{max}}{I_{min}}$$

6.19 RELATION BETWEEN STANDING WAVE RATIO AND REFLECTION COEFFICIENT

$$SWR = S = \frac{1+|\rho|}{1-|\rho|}$$

Proof:

$$SWR = \frac{V_{max}}{V_{min}}$$

But $V_{max} = |V_i| + |V_r|$

$$V_{min} = |V_i| - |V_r|$$

$$\therefore \quad SWR = \frac{|V_i|+|V_r|}{|V_i|-|V_r|}$$

$$= \frac{1+\left|\dfrac{V_r}{V_i}\right|}{1-\left|\dfrac{V_r}{V_i}\right|}$$

But $\left|\dfrac{V_r}{V_i}\right| = |\rho|$

$$\therefore \quad SWR = S = \frac{1+|\rho|}{1-|\rho|}$$

Hence proved.

6.19.1 Salient Features of Standing Wave Ratio

- $S = \dfrac{V_{max}}{V_{min}}$

- S ranges between 1 and ∞.
- Minimum value of S is 1 and maximum value of S is ∞.
- If $Z_L = Z_0$, $S = 1$.
- When the line is terminated by purely resistive load,

$$S = \frac{Z_0}{R_L} \text{ if } Z_0 > R_L.$$

$$S = \frac{R_L}{Z_0} \text{ if } R_L > Z_0.$$

- $S = \infty$ if $Z_L = 0$
- $S = \infty$ if $Z_L = \infty$
- $S = 1$ if $\rho = 0$
- $S = \infty$ if $\rho = 1$
- $S = \infty$ if $\rho = -1$

6.20 MATCHED TRANSMISSION LINE

A transmission line is said to be matched if $Z_L = Z_0$. Under this condition, all the incident power is absorbed by the load.

6.21 MISMATCHED TRANSMISSION LINE

A transmission line is said to be mismatched if $Z_L \neq Z_0$. Under this condition, some power is absorbed by the load and the rest is reflected.

6.22 LOSSES IN TRANSMISSION LINES

The losses are of three types.

- Copper losses
- Dielectric losses
- Radiation losses

Copper Losses

These losses exist due to I^2R power loss, skin effect and crystallization. I^2R power exists due to heating in resistance. Copper loss is made small by reducing loop currents and by proper termination. This loss is more for small values of Z_0. For small Z_0, current is high.

For an applied high frequency signal to the transmission line, current exists on the surface of the conductor. This phenomenon is called skin effect. This reduces the cross-sectional area of the conductor with increased frequency. This in turn increases the resistance and hence power losses are increased. Copper losses increase due to ageing, with high temperature, high winds and moisture. The lines become brittle and cracks appear with the bending of the line back and forth. This effect is called crystallization. It increases the resistance of the conductors which increases copper losses.

Dielectric Losses

These losses are due to I^2R dissipation as a result of heating of the solid electric material between the conductors. These losses are proportional to the voltage across the dielectric.

At high frequencies, solid dielectric properties worsen. The transmission lines with air dielectric are useful as the air dielectric loss is very small at high frequencies.

Radiation Losses

These losses are high when the line separation is large as the line acts as an antenna. Radiation losses are high in parallel wire transmission lines.

6.23 SUMMARY OF TRANSMISSION LINE PARAMETERS

The transmission line parameters are presented in table 6.1.

Table 6.1   Transmission line parameters

Parameter	Transmission Line	Lossless Line	Distortionless Line
γ_ℓ	$\sqrt{(R + j\omega L)(R + j\omega C)}$	$j\omega\sqrt{LC}$	$\sqrt{RG} + j\omega\sqrt{LC}$
Z_0	$\sqrt{\dfrac{R + j\omega L}{G + j\omega C}}$	$\sqrt{L/C}$	$\sqrt{L/C}$ or $\sqrt{R/G}$
α_ℓ	$\left[\dfrac{1}{2}\left\{(RG - \omega^2 LC) + \sqrt{(R^2 - \omega^2 L^2)}\sqrt{(G^2 + \omega^2 C^2)}\right\}\right]^{\frac{1}{2}}$	0	$\sqrt{RG}$
β_ℓ	$\left[\dfrac{1}{2}\left\{(\omega^2 LC - RG) + \sqrt{(R^2 - \omega^2 L^2)}\sqrt{(G^2 + \omega^2 C^2)}\right\}\right]^{\frac{1}{2}}$	$\omega\sqrt{LC}$	$\omega\sqrt{LC}$

v_l	$\dfrac{\omega}{\beta_l}$	$\dfrac{1}{\sqrt{LC}}$	$\dfrac{1}{\sqrt{LC}}$
Condition		$\begin{aligned} R &= 0 \\ G &= 0 \end{aligned}$	$\dfrac{R}{L} = \dfrac{G}{C}$
Z_i	$Z_0\left[\dfrac{Z_L + Z_0 \tan h\,\gamma l}{Z_0 + Z_L \tan h\,\gamma l}\right]$	$Z_0\left[\dfrac{Z_L + jZ_0 \tan \beta l}{Z_0 + jZ_L \tan \beta l}\right]$	$R_0\left[\dfrac{Z_L + jR_0 \tan \beta l}{R_0 + jZ_L \tan \beta l}\right]$
Z_i with shorted load	$Z_0 \tan h\,\gamma l$	$jZ_0 \tan \beta l$	$jR_0 \tan \beta l$
Z_i with open circuit load	$Z_0 \cot h\,\gamma l$	$-jZ_0 \cot \beta l$	$-jR_0 \cot \beta l$
Z_i with matched load	Z_0	Z_0	Z_0

6.24 SUMMARY OF DEFINITIONS OF TRANSMISSION LINE PARAMETERS

These are given in table 6.2.

Table 6.2 Definitions of transmission line parameters

Parameters	Definition	Expression	Units
Propagation Constant, γ_l	$\gamma_l = \sqrt{ZY} = \alpha_l + j\beta_l$	$\sqrt{\dfrac{R + j\omega L}{G + j\omega C}}$	m^{-1} or dB/m or nepers/m
Attenuation Constant, α	Real part of propagation constant or it is a constant which indicates the rate at which signal amplitude reduces as it propagates from one point to another point along the line	$\left[\dfrac{1}{2}\left\{(RG - \omega^2 LC) + \sqrt{(R^2 - \omega^2 L^2)}\sqrt{(G^2 + \omega^2 C^2)}\right\}\right]^{\frac{1}{2}}$	m^{-1} or dB/m
Phase Constant, β_l	Imaginary part of γ_l or it is defined as a measure of phase shift of signal as it propagates along the line	$\left[\dfrac{1}{2}\left\{(\omega^2 LC - RG) + \sqrt{(R^2 + \omega^2 L^2)}\sqrt{(G^2 + \omega^2 C^2)}\right\}\right]^{\frac{1}{2}}$	m^{-1} or rad/m
	$Z_0 = \sqrt{Z/Y}$		

Characteristic Impedance, Z_0	$Z_0 \equiv V_f/I_f$ $Z_0 = -V_r/I_r$	$\sqrt{\dfrac{R+j\omega L}{G+j\omega C}}$	Ohm
Input Impedance, Z_i	$Z_i \equiv \dfrac{V_S}{I_S}$	$\left(\dfrac{V_L \cos h\,\gamma\ell + Z_0\,I_L \sin h\,\gamma\ell}{I_L \cos h\,\gamma\ell + \dfrac{V_L}{Z_0}\sin h\,\gamma\ell}\right)$	Ohm
Reflection Coefficient, ρ	$\rho \equiv \dfrac{V_r}{V_i}$ or $\dfrac{I_r}{I_i}$	$\rho = \dfrac{Z_L - Z_0}{Z_L + Z_0}$	No units
Standing Wave Ratio, S	$S \equiv \dfrac{V_{max}}{V_{min}}$ or $\dfrac{I_{max}}{I_{min}}$	$S = \dfrac{1+|\rho|}{1-|\rho|}$	No units

6.25 SMITH CHART

Smith chart is a special chart developed by Philip H. Smith in 1939.

It is a polar plot of the reflection coefficient in terms of normalized impedance, $r + jx$.

The Smith chart gives a relation between reflection coefficient, load impedance and characteristic impedance. It consists of a center line, resistance and reactance circles. The resistance circle has center at $(\rho_r, \rho_i) = \left(\dfrac{r}{1+r}, 0\right)$ with a radius of $\dfrac{1}{1+r}$. The reactance circle has center at $(\rho_r, \rho_i) = \left(1, \dfrac{1}{x}\right)$ with a radius of $\left(\dfrac{1}{x}\right)$.

A typical Smith chart is shown in fig. 6.6.

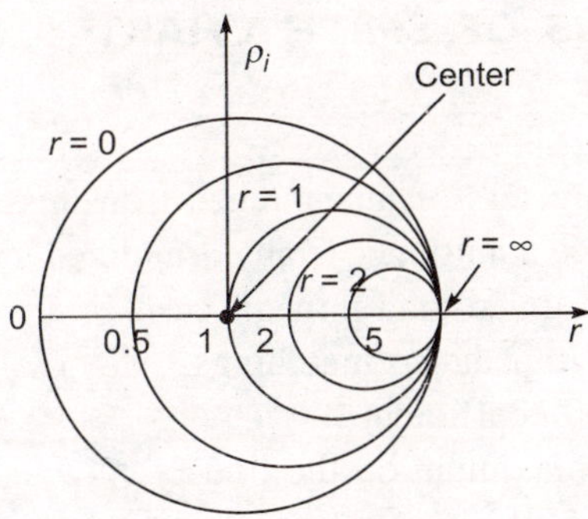

Fig. 6.6 Smith chart

6.25.1 Description of Smith Chart

It is described points-wise.

1. The centers of all r-circles lie on ρ_r-axis where ρ_r is the real part of reflection coefficient. The center straight line is called ρ_r axis.
2. The $r = 0$ circle having unity radius and centered at the origin is the largest.
3. The r-circles become progressively smaller as r increases from 0 to ∞ ending at the point ($\rho_r = 1$, $\rho_i = 1$) for open circuit.
4. All circles pass through the ($\rho_r = 1$, $\rho_i = 0$) point.
5. The centers of all x-circles lie on the $\rho_r = 1$ line.
6. The x-circles for $x > 0$ (inductive reactance) lie above ρ_r-axis and those for $x < 0$ (capacitive reactance) lie below the ρ_r-axis.
7. The $x = 0$ circle becomes ρ_r-axis.
8. x-circle becomes progressively smaller as $|x|$ increases from 0 to ∞ ending at the point ($\rho_r = 1$, $\rho_i = 0$) for open circuit.
9. All x-circles pass through the ($\rho_r = 1$, $\rho_i = 0$) point.
10. The Smith chart consists of two sets of circles or arcs of circles. The complete circles lie on the center line on the chart. These circles correspond to various values of normalized resistance $\left(r = \dfrac{R}{z_0} \right)$ along the line.
11. The arcs of circles on either side of the straight line correspond to various values of normalized line reactance, $\left(jx = \dfrac{jX}{z_0} \right)$.
12. The circles are orthogonal to each other.
13. The perimeter of outer rim of the chart is of $\dfrac{\lambda}{2}$ length.

6.26 APPLICATIONS OF SMITH CHART

It can be used to

1. Find out the parameters of mismatched transmission lines.
2. Find out normalized admittance from normalized impedance or vice-versa.
3. Find VSWR for a given load impedance.
4. Design stubs for impedance matchings.
5. Find out reflection coefficient.
6. Locate a voltage maximum on the line.
7. Find out input impedance of a transmission line.

6.27 STUBS

A stub is a piece of transmission line which can be shorted or open ended.

Salient Features

> 1. It is a transmission line.
> 2. It has pure reactance.
> 3. It is used for impedance matching.
> 4. Single stubs and double stubs are used for impedance matching.

6.28 DESIGN OF SINGLE STUB MATCHING

The design is described in the following steps for readability.

1. Normalize the load impedance with Z_0. i.e. $Z_n = \dfrac{Z_L}{Z_0} = r + jx$.

2. Mark r, and x values on Smith chart. Note the point of intersection of r circle and x-arc.

3. Note the distance of the point from the center of chart.

4. Draw a circle with this distance as radius.

5. Draw a line from this point to the second half of the chart through the center. That is, travel around it through a distance of $\lambda/4$ to find load admittance.

6. The stub is placed parallel to the transmission line and hence admittance parameter is useful.

7. Note the point where the straight line intersects the circle drawn. This point is $y_n = (g + jb)$.

8. Also note the point where the circle cuts the center line of the chart to the right side of the center. It gives VSWR.

9. Note the point nearest to the load. At this point, normalized admittance is $1 \pm jb$. This is the point where $\pm jb$ intersects the $r = 1$ circle. This is also the point at which a stub, signal to tune out $\pm jb$ component will be placed.

10. Measure the distance traveled round the circumference of the chart. This gives the distance of the stub from the load.

11. Find the point by moving clockwise round the perimeter of the chart. At this point, the susceptance tunes out $\pm jb$ susceptance of the line. For example, if the line admittance is $1 + jb$, the required susceptance is $-jb$.

12. Measure the distance in wavelength from $\infty, j\infty$ of the chart of the new point where the susceptance is say $-b$.

6.29 SHORTED AND OPEN STUBS

The typical stubs are shown in fig. 6.7.

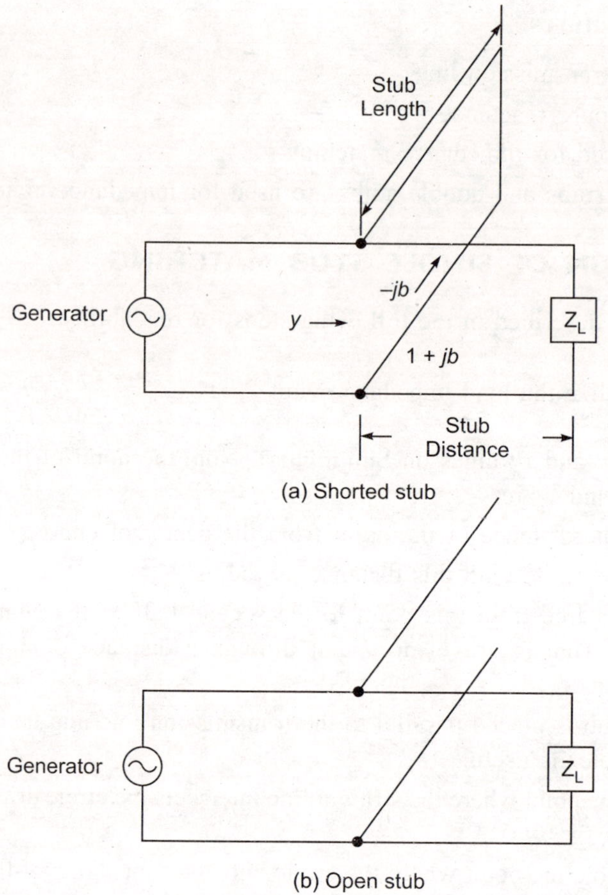

(a) Shorted stub

(b) Open stub

Fig. 6.7 Stubs

PROBLEM 6.3 Determine the input impedance of 50 Ω lossless transmission line of 0.1 λ, if the load is shorted.

Solution

$$Z_L = 0 \ \Omega$$
$$Z_0 = 50 \ \Omega$$

Length of the line = 0.1 λ

The normalized load impedance, $Z_n = 0$

The input impedance of a lossless transmission line is

$$Z_i = Z_0 \left[\frac{Z_L + jZ_0 \tan \beta \ell}{Z_0 + jZ_L \tan \beta \ell} \right]$$

But $Z_L = 0$

$\therefore$

$$Z_i = jZ_0 \tan \beta \ell$$

$$= j50 \, \tan \left(\frac{2\pi}{\lambda} \right) 0.1\lambda$$

$$\therefore \quad \boxed{Z_i = j\, 36.25}$$

PROBLEM 6.4   Determine the input of 50 Ω lossless transmission line of length equal to 0.1 λ if it is terminated in open circuit.

Solution   Length of the line $\ell = 0.1\,\lambda$

$$Z_L = \infty$$
$$Z_0 = 50\ \Omega$$

$$Z_i = Z_0 \left[\frac{Z_L + jZ_0 \tan \beta\ell}{Z_0 + jZ_L \tan \beta\ell} \right]$$

$$= -50\, j \, \cot \beta\ell$$

$$= -50\, j \, \cot \left(\frac{2\pi}{\lambda} \right) (0.1\,\lambda)$$

$$\therefore \quad \boxed{Z_i = -j\, 68.82\ \Omega}$$

PROBLEM 6.5   Find the input impedance of a lossless transmission line if the characteristic impedance, $Z_0 = 100\ \Omega$, length of the line, $\ell = \lambda/3$, load impedance, $Z_L = 150 + j\,60$.

Solution

$$Z_0 = 100\ \Omega$$
$$\ell = \lambda/3,$$
$$Z_L = 150 + j\,60.$$

$$\therefore \quad \tan \beta\ell = \tan(120\,\pi) = -1.225$$

Input Impedance,

$$Z_i = Z_0 \left[\frac{Z_L + jZ_0 \tan \beta\ell}{Z_0 + jZ_L \tan \beta\ell} \right]$$

$$= 100 \left[\frac{150 + j60 + j100(-1.225)}{100 + j(150 + j60)\,(-1.225)} \right]$$

$$= 100 \left[\frac{150 - j62.5}{100 + j183.75 + 73.5} \right]$$

i.e.

$$Z_i = 100 \left[\frac{150 - j62.5}{173.5 - j183.75} \right]$$

$$\therefore \quad \boxed{Z_i = 58.7 + j\, 26.0}$$

PROBLEM 6.6 If the inductance of a transmission line 1.2 µH/m and the capacitance is 12.5 pF/m, determine the time required for the wave to travel 2 m length down the line.

Solution Inductance of the line, $L = 1.2$ µH/m

Capacitance of the line, $C = 12.5$ pF/m

Time of travel for travel

$$2 \text{ m} = \sqrt{LC} \times \text{length of the line}$$

$$= \sqrt{1.2 \times 10^{-6} \times 12.5 \times 10^{-12}} \times 2$$

$\therefore$

$$\boxed{T = 7.745 \text{ nsec.}}$$

PROBLEM 6.7 If a transmission line is characterized by an inductance of 1.5 µH/m and a capacitance of 10 pF, find its characteristic impedance.

Solution

$$L = 1.5 \text{ µH/m}$$

$$C = 10 \text{ pF}$$

$$Z_0 = \sqrt{\frac{L}{C}}$$

$$= \sqrt{\frac{1.5 \times 10^{-6}}{10 \times 10^{-12}}}$$

$$= 0.387 \times 10^3$$

$\therefore$

$$\boxed{Z_0 = 387 \ \Omega}$$

PROBLEM 6.8 A voltage of 50 V is supplied to a transmission line. Find the reflected voltage if the reflection coefficient is 0.25.

Solution

$$V_i = 50 \text{ V}$$

$$\rho = 0.25$$

But $\rho = \dfrac{V_r}{V_i}$

$\therefore$ The reflected voltage,

$$V_r = V_i \rho$$

$$= 50 \times 0.25 = 12.5$$

$\therefore$

$$\boxed{V_r = 12.5}$$

PROBLEM 6.9 The incident voltage on an open-end transmission line 50 V and the reflected voltage is 25 V. Find out the percentage of voltage reflected.

Solution

$$V_i = 50$$

$$V_r = 25$$

$$\rho = \frac{V_r}{V_i} = \frac{25}{50} = 0.5$$

$\therefore$ The percentage of reflected voltage $= \rho \times 100$

i.e.

$$\boxed{\% \; V_r = 0.5 \times 100 = 50\%}$$

PROBLEM 6.10   The maximum voltage of the standing wave on the transmission line due to mismatched load is 50 V and the minimum voltage is 35 V. Find VSWR in dB.

Solution
$$V_{max} = 50 \text{ V}$$
$$V_{min} = 35$$

$$\text{VSWR} = \frac{50}{35} = 1.428$$

$$\text{VSWR in dB} = 20 \log_{10} \text{VSWR}$$
$$= 20 \log_{10} 1.428$$
$$= 3.098$$

$\therefore$
$$\boxed{\text{VSWR} = 3.098 \text{ dB}}$$

PROBLEM 6.11   The characteristic impedance of a transmission line is 75 Ω. Find out the maximum impedance of the line with VSWR of 3.0

Solution
$$Z_0 = 75 \; \Omega$$
$$\text{VSWR} = 3.0$$
$$Z_{max} = Z_0 \times \text{VSWR} = 75 \times 3.0$$

i.e.
$$\boxed{Z_{max} = 225 \; \Omega}$$

PROBLEM 6.12   If the reflection coefficient of transmission line is equal to 0.4, find VSWR.

Solution
$$|\rho| = 0.4$$

The VSWR is given by

$$\text{VSWR} = \frac{1+|\rho|}{1-|\rho|}$$

$$= \frac{1+0.4}{1-0.4}$$

$\therefore$
$$\boxed{\text{VSWR} = 2.33}$$

PROBLEM 6.13   Find out the input impedance at a point which is at a distance of $\lambda/8$ from the shorted load for a transmission line with $Z_0 = 75 \; \Omega$.

Solution
$$Z_L = 0$$

$$\ell = \frac{\lambda}{8}$$

$$\beta = \frac{2\pi}{\lambda}$$

$$Z_0 = 75 \ \Omega$$

$$\beta\ell = \frac{2\pi}{\lambda} \times \frac{\lambda}{8} = 0.785 \text{ rad} = 45°$$

We have

$$Z_i = Z_0 \left[\frac{Z_L + jZ_0 \tan \beta\ell}{Z_0 + jZ_L \tan \beta\ell} \right]$$

$$= 75 \left[\frac{0 + j75 \tan 45°}{75 + j(0) \tan 45°} \right]$$

$$= 75 \times j$$

$$\therefore \qquad \boxed{Z_i = j \ 75 \ \Omega}$$

It means input impedance of the line is inductive.

PROBLEM 6.14 A quarter-wave transformer is used to match a line of $Z_0 = 50 \ \Omega$ to a load impedance of 200 Ω. The frequency of operation is 300 MHz. Find the physical length and characteristic impedance of the transformer.

Solution

$$Z_0 = 50 \ \Omega$$
$$Z_L = 200 \ \Omega$$
$$f = 300 \text{ MHz}$$

$$\lambda = \frac{v_0}{f} = \frac{3 \times 10^8}{3 \times 10^8} = 1 \text{ m}$$

$\therefore$ The length of the transformer $= \dfrac{\lambda}{4} = \dfrac{1}{4} = 0.25$ m

The characteristic impedance of the transformer

$$Z_t = \sqrt{Z_0 \, Z_L}$$

$$= \sqrt{50 \times 200}$$

$$\boxed{Z_t = 100 \ \Omega}$$

PROBLEM 6.15 It is required to keep SWR along a television lead-in line unity. The TV provides a load impedance of 300 Ω to 75 Ω transmission line. If the quarter-wave transformer is connected to the load, find its characteristic impedance.

Solution

$$Z_L = 300 \ \Omega$$
$$Z_0 \text{ of the line} = 75 \ \Omega$$

The characteristic impedance of the transformer is given by

$$Z_t = \sqrt{Z_0 Z_L} = \sqrt{75 \times 300}$$

$\therefore$
$$\boxed{Z_t = 150 \ \Omega}$$

It may be noted that for SWR = 1, the source impedance is equal to the characteristic impedance of the line.

PROBLEM 6.16   A 50 Ω transmission line exhibits a VSWR of 2.0 and the first minima from load is at 0.2 λ. Find out $|\rho|$.

Solution
$$\text{VSWR}, S = 2.0$$
$$Z_0 = 50 \ \Omega$$
$$|\rho| = \frac{S-1}{S+1}$$
$$= \frac{2.0-1}{2.0+1}$$
$$= \frac{1}{3}$$

$\therefore$
$$\boxed{|\rho| = 0.333}$$

PROBLEM 6.17   A 50 Ω shorted line has a length of 0.1 λ and it operates at 500 MHz. Find input impedance of the line.

Solution
$$\ell = 0.1 \ \lambda$$
$$\beta = \frac{2\pi}{\lambda}$$
$$f = 500 \ \text{MHz}$$
$$Z_n = 50$$

$\therefore$
$$\beta\ell = \frac{2\pi}{\lambda} \times 0.1 \ \lambda$$
$$= 0.2\pi = 36°$$

$\therefore$
$$Z_i = jZ_0 \tan 36°$$
$$= j \ 50 \times 0.7265$$

$\therefore$
$$\boxed{Z_i = j \ 36.33 \ \Omega}$$

6.30   APPLICATIONS OF TWO-WIRE OPEN LINES

- They are used in power distribution, telephone and telegraph lines.
- They may be used between a transmitter and antenna or between antenna and receiver.

Advantages

- They are simple in structure.
- Inexpensive.

Disadvantages

- Radiation losses are high.
- Noise is high

6.31 TWISTED PAIR LINES

These transmission lines are twisted and they consist of rubber insulated wires (fig. 6.8).

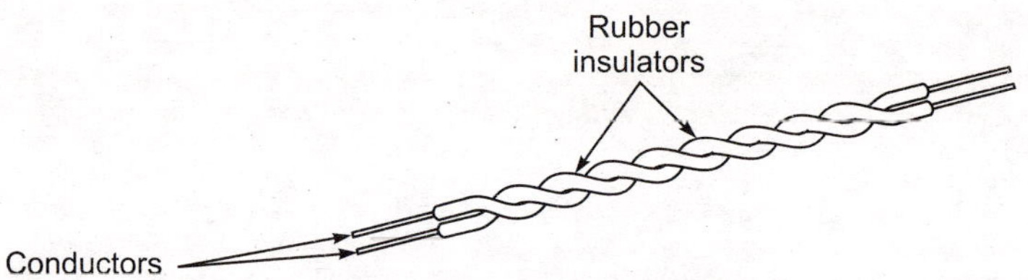

Fig. 6.8 Twisted pair lines

6.31.1 Salient Features

- The wires are twisted
- The wires have rubber insulation
- No spaces are used
- They are flexible
- They exhibit high frequency loss in the rubber insulation
- The losses become high if the line is wet
- They are used only for low frequency applications like ac power cables.

6.32 SHIELDED PAIR LINE

The shield transmission line is shown in fig. 6.9.

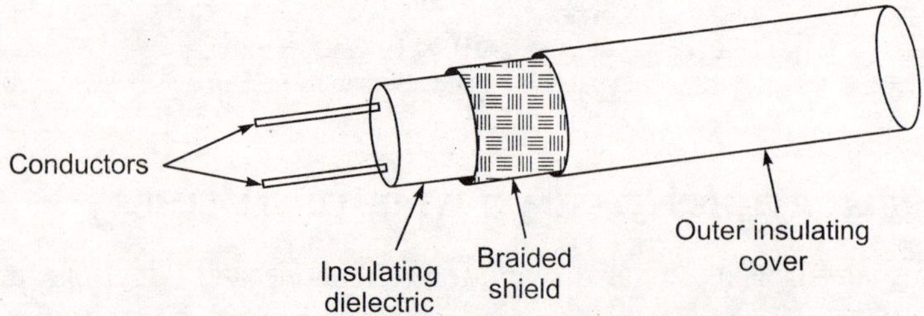

Fig. 6.9 The shielded pair transmission line

6.32.1   Salient Features

- It consists of two parallel wires surrounded by solid, low-loss dielectric. The dielectric is enclosed in a metal braid for shielding. The shielding metal is covered with a rubber to protect the line from moisture.
- This type of line has low radiation losses.
- This has reduced noise pickup.
- This line is best suited for balancing line feeds.
- The electric and leakage losses are high.

Applications

It is the best line to balance line feeds.

Advantages

- Radiation losses are small
- Noise is less

Disadvantages

- Dielectric and leakage losses are high.

6.33   COAXIAL LINES

These are of two types.

- Flexible coaxial lines
- Rigid coaxial lines

A typical coaxial line is shown in fig. 6.10.

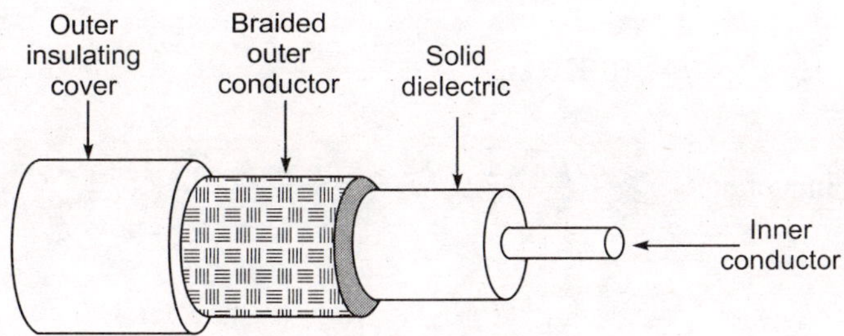

Fig. 6.10   Coaxial lines

6.33.1   Salient Features

- They are unbalanced.
- They consist of inner conductor and grounded outer shield.
- The inner conductor carries RF energy.
- The outer shield presents radiation into free space.

- Radiation losses are minimum.
- Noise pickup is small.
- It supports *TEM* fields.
- Its cut-off frequency is zero.

Applications

- It is used to connect transmitter and antenna or antenna and receiver.
- It is used at high frequencies.

Advantages

- Radiation loss is minimum.
- Noise pickup is less.

Disadvantages

Their use is restricted to unbalanced applications.

6.33.2 Properties of Coaxial Cable

DC resistance, $R_{dc} = \dfrac{1}{\sigma_c \pi}\left[\dfrac{1}{a^2} + \dfrac{1}{t(b+t)}\right]$

AC resistance, $R_{ac} = \dfrac{1}{2\pi \sigma_c \delta}\left[\dfrac{1}{a} + \dfrac{1}{b}\right]$ for $t \gg \delta$

Inter inductance, $L_i = \dfrac{R_{ac}}{2\pi f}$ H/m for $f > 10$ KHz

$L_i = \dfrac{\mu_0}{4\pi}$ H/m for $f < 10$ KHz

External inductance, $L_e = \dfrac{\mu_0}{2\pi}\ell_n\,(b/a)$

$$C = \pi \in_d \left[\frac{2}{\ell n\,(b/a)}\right]$$

$$G = \frac{C}{\in_d}\,\sigma_d\ \text{(mho/m)}$$

$$Z_0 = \frac{138}{\sqrt{\in_r}}\log\frac{r_0}{r_i}$$

a = inner radius

b = outer radius

t = outer thickness

ϵ_d = dielectric constant of the dielectric material

σ_d = conductivity of the dielectric (mho/m)

r_0 = inside diameter of outer conductor

r_i = outer diameter of inner conductor

ϵ_r = relative permittivity

6.34 FIELDS IN COAXIAL CABLE

E and H fields are shown in fig. 6.11.

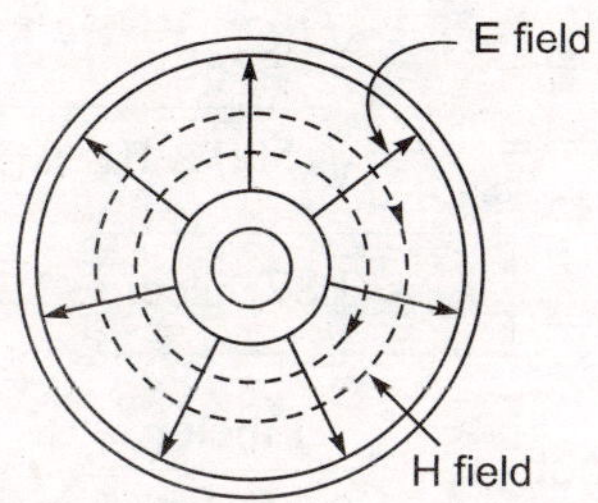

Fig. 6.11 Fields in coaxial cable

Attenuation due to Conductor Losses

$$\alpha_c = \frac{0.691}{\sigma_c \, \delta Z_0}\left[\frac{1}{a} + \frac{1}{b}\right]$$

Attenuation due to dielectric losses

$$\alpha_d = 0.686 \frac{\pi}{\lambda_0}\sqrt{\epsilon_r}\ \tan\delta\ \text{dB/m}$$

Here, δ = skin depth

PROBLEM 6.18 A copper coaxial line has an outside tubing of thickness 2 mm and its outside diameter is 30 mm. The thickness of the inner tubing is 1.0 mm and its outside diameter is 10 mm. Find the characteristic impedance of the line for air dielectric.

Solution Diameter of the outside conductor is $b = 30 - 2 \times 2 = 24$ mm

Diameter of the inner conductor is $a = 10 - 2 \times 1.0 = 8$ mm

$$Z_0 = 138 \log\left(\frac{b}{a}\right)$$

$$= 138 \log\frac{24}{8}$$

$$= 138 \log 3.0$$

$$\therefore \quad \boxed{Z_0 = 65.84 \ \Omega}$$

PROBLEM 6.19 A coaxial has a capacitance of 30 pF/m and an inductance of 500 nH/m. If a signal passing through it, find (a) the time delay for a cable for 1.0 m long (b) propagation velocity (c) propagation delay over a cable length of 10 m.

Solution

(a) The time delay for 1.0 m long cable is

$$t_d = \sqrt{LC}$$

$$= \sqrt{(500 \times 10^{-9})(30 \times 10^{-12})}$$

$$\therefore \quad \boxed{t_d = 3.87 \ \text{nsec.}}$$

(b) Velocity of propagation $v_p = \dfrac{1 \ \text{meter}}{3.87 \times 10^{-9}}$

$$\therefore \quad \boxed{v_p = 2.5839 \times 10^8 \ \text{ms}}$$

(c) The capacitance of 10 m cable is $C = 30 \times 10^{-12} \times 10 = 3 \times 10^{-10}$ F

Inductance of 10 m cable is $L = 30 \times 10^{-12} \times 10 = 5 \times 10^{-6}$ H

$$\therefore \quad L_d = \sqrt{LC} = \sqrt{5 \times 10^{-6} \times 3 \times 10^{-10}}$$

$$\boxed{t_d = 38.7 \ \text{n sec}}$$

PROBLEM 6.20 A coaxial cable has Z_0 of 75 Ω and a capacitance of 70 pF/m. Find its inductance per meter. If the radius of the inner conductor is 0.292 mm and $\epsilon_r = 2.3$, find the radius of the outer conductor.

Solution   We have $\qquad Z_0 = \sqrt{L/C}$

$$C = 70 \ \text{pF/m}$$

$$Z_0 = 75 \ \Omega$$

$$L = Z_0^2 C = 75^2 \times 70 \times 10^{-12}$$

$$\therefore \qquad L = 0.3937 \ \mu\text{H/m}$$

For a coaxial cable, Z_0 is also given by $Z_0 = \dfrac{138}{\sqrt{\epsilon_r}} \log \dfrac{b}{a}$

$$\epsilon_r = 2.3$$

$$a = 0.292 \text{ mm}$$

$$\therefore \quad \log \frac{b}{a} = \frac{Z_0 \sqrt{\epsilon_r}}{138}$$

$$= \frac{75 \times \sqrt{23}}{138}$$

$$= 0.8242$$

$$\frac{b}{a} = \text{antilog} \,(0.8242)$$

$$= 6.671$$

$$b = 0.292 \times 6.671$$

$$\therefore \quad \boxed{b = 1.9479}$$

6.35 PARALLEL PLATES

A typical parallel structure is shown in fig. 6.12.

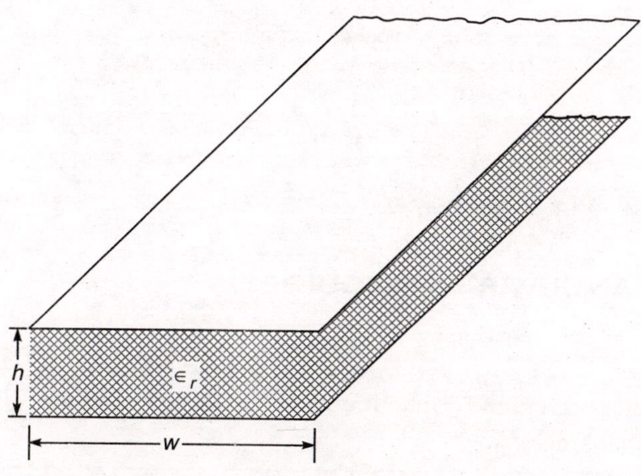

Fig. 6.12 Parallel plate transmission line

6.35.1 Salient Features

- It supports *TEM* wave

- Its $Z_0 = \dfrac{377}{\sqrt{\epsilon_r}} \dfrac{h}{w}$ ohm.

- The capacitance C_ℓ is given by $C_\ell = \dfrac{Cw}{h}$

- The phase velocity, $v_p = \dfrac{v_0}{\sqrt{\in_r}}$ m/s

- Its inductance is given by $L_\ell = \mu_0 \dfrac{h}{w}$

- $R_{dc} = \dfrac{2}{\sigma_C \, wt}$

- $R_{ac} = \dfrac{2}{\pi \sigma_C \, \delta w}$ for $t \gg \delta$

Attenuation factor for *TEM* wave

$$\alpha_p = \frac{1}{\eta_p \, h} \sqrt{\frac{\omega \mu_C}{2\sigma_C}} \;\; \text{neper/m}$$

Attenuation factor for TE wave is

$$\alpha_p = \frac{2m^2 \pi^2 \sqrt{\omega \mu_C / 2\sigma_C}}{\omega \mu_C \, a^2 \sqrt{\omega^2 \mu_C \in_C - \left(\dfrac{n\pi}{a}\right)^2}}$$

Here, $m = 1, 2, 3,$ $\omega = 2\pi f$

6.36 RECTANGULAR WAVEGUIDE

Definition

A rectangular waveguide is a hollow metallic device with four sides closed and two longitudinal sides open.

Applications

It is used as a

- transmission line
- radiator
- high pass filter
- feed element to antennas

6.36.1 Structure of Rectangular Waveguide

It is shown in fig. 6.13.

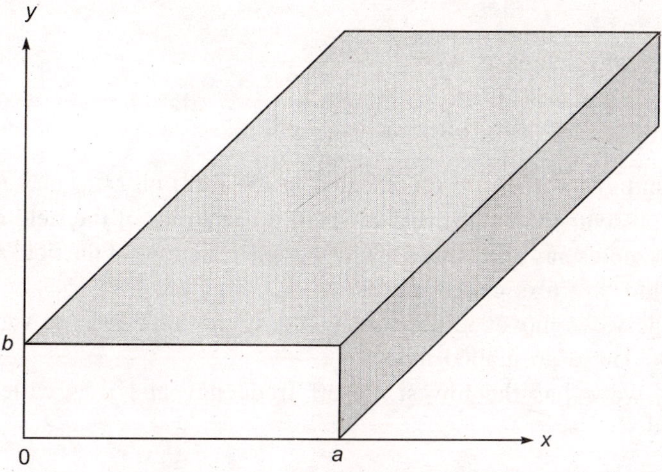

Fig. 6.13 A rectangular waveguide

6.36.2 Salient Features

- It supports only Transverse Electric (TE) and Transverse Magnetic (*TM*) waves.
- It does not support *TEM* wave.
- It is used as a transmission line.
- It is used as a radiator, high pass filter and as a feed element to antennas like parabolic reflectors.
- The power flow is in the form electric and magnetic fields.
- The transmission is by reflection.
- Inside surface is made very smooth.

- The propagation constant in the guide is $\gamma_g = \sqrt{\left(\dfrac{m\pi}{a}\right)^2 + \left(\dfrac{n\pi}{b}\right)^2 - \omega^2 \mu \in}$

- The cut off frequency is $f_c = \dfrac{1}{2\pi\sqrt{\mu \in}} \sqrt{\left(\dfrac{m\pi}{a}\right)^2 + \left(\dfrac{n\pi}{b}\right)^2}$

- The cut-off length is $\lambda_c = \dfrac{2}{\sqrt{\left(\dfrac{m\pi}{a}\right)^2 + \left(\dfrac{n\pi}{b}\right)^2}}$

- Phase velocity is $v_p = \dfrac{\omega}{\sqrt{\omega^2 \mu \in - \left(\dfrac{m\pi}{a}\right)^2 + \left(\dfrac{n\pi}{b}\right)^2}}$

- Guide wavelength is $\lambda_g = \dfrac{\lambda}{\sqrt{1 - \left(\dfrac{\lambda}{\lambda_c}\right)^2}}$

- *TE* and *TM* waves are represented in the form of TE_{mn} and TM_{mn}. Here, m represents the number of half-period variations of the field along x-axis and n represents the number of half-period variations of the field along y-axis.
- m and n are also called modes.
- As the wave moves in the waveguide, it can have infinite variety of patterns. These are called modes.
- TE_{10} wave has the lowest cut-off frequency and it is called dominant mode.

- For TE_{mn} wave, wave impedance is $\eta_{mn} = \dfrac{\eta_0}{\sqrt{1 - \left(\dfrac{f_c}{f}\right)^2}}$

  Here, η_0 = intrinsic impedance of free space.

- Power transmitted in a lossless waveguide is $p_{av} = \dfrac{ab\, E^2}{4\, \eta_{10}^{TE}}$ watts.

- Attenuation due to dielectric loss is $\alpha_d = \dfrac{\omega \mu_d \sigma_d}{2\beta_g}$ nepers/m.

- Attenuation due to wall loss is $\alpha_W = \dfrac{R_{SC}}{\eta}\left(\sqrt{\dfrac{f}{f_c}}\right) \times \dfrac{a + 2b\left(\dfrac{f_c}{f}\right)^2}{ab\sqrt{1 - \left(\dfrac{f_c}{f}\right)^2}}$

- Here, R_{SC} = surface resistance at cut-off frequency of TE_{10} Ω

- $\eta = \sqrt{\dfrac{\mu}{t}}$ Ω

- Total attenuation is $\alpha_t = \alpha_d + \alpha_W$ neper/m

6.37 TRANSVERSE ELECTRIC WAVES (TE WAVES)

6.37.1 Definition

Transverse electric waves are defined as the electromagnetic waves for which there is no component of electric field in the direction of propagation.

If z is the direction of propagation, $E_L = 0$ for *TE* waves.

6.37.2   Field Equations for *TE* Waves

Assuming

1. Free space inside the waveguide
2. The walls are perfectly conducting
3. Direction of propagation of power in z-direction
4. The dimension of the narrow wall is 'b' meters.
5. The dimension of broad wall is 'a' meters.
6. The field vary in z-direction vary as $e^{-\gamma_g z}$.

The field equations are obtained from wave equations, Maxwell's equation and boundary conditions.

The field components for *TE* are obtained as

$$H_z = (K \cos Cx \cos By)e^{(j\omega t - \gamma_g z)}$$

$\therefore$ The field expressions for *TE* waves in complete form are

$$H_z = (K \cos Cx \cos By)\, e^{(j\omega t - \gamma_g z)}$$

$$E_x = \left(\frac{j\omega\mu}{h_g^2} KB \cos Cx \sin By \right) e^{(j\omega t - \gamma_g z)}$$

$$E_y = \left(\frac{-j\omega\mu}{h_g^2} KC \sin Cx \cos By \right) e^{(j\omega t - \gamma_g z)}$$

$$H_x = \left(\frac{\gamma_g}{h_g^2} KC \sin Cx \cos By \right) e^{(j\omega t - \gamma_g z)}$$

$$H_y = \left(\frac{\gamma_g}{h_g^2} KB \sin Cx \cos By \right) e^{(j\omega t - \gamma_g z)}$$

Here, $h_g^2 = \gamma_g^2 + \omega^2 m\epsilon$

$$C = \frac{m\pi}{a},\ B = \frac{n\pi}{b}$$

$$K = A_1 A_3$$

As TE_{10} is popular, its field equations are

$$H_z = K \cos \frac{\pi x}{a} e^{(j\omega t - \gamma_g z)}$$

$$H_x = \frac{\gamma_g a K}{\pi} \sin \frac{\pi x}{a} e^{(j\omega t - \gamma_g z)}$$

$$E_y = \frac{-j\omega\mu a K}{\pi} \sin \frac{\pi x}{a} e^{(j\omega t - \gamma_g z)}$$

$$E_x = 0, \; H_y = 0$$

Here, $h_g^2 = \gamma_g^2 + \omega^2 \mu\epsilon$

$$C = \frac{m\pi}{a}, \; B = \frac{n\pi}{b}$$

$$k = A_1 A_3$$

6.38 DOMINANT MODE

Definition

It is defined as a wave which has the lowest cut-off frequency : It is represented by TE_{10}.

The field patterns of TE_{10} are shown in fig. 6.14.

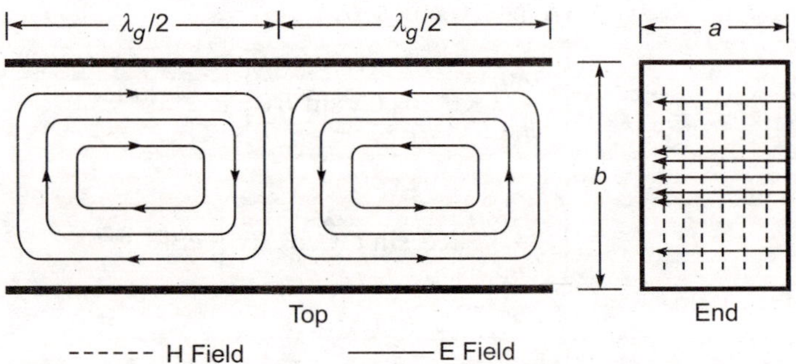

Fig. 6.14 Field patterns of TE_{10} mode

6.39 DEGENERATE MODES

They are defined as higher order modes which have the same cut-off frequency, TE_{mn} and TM_{mn} for $m \neq 0$, $n \neq 0$ are degenerate modes.

6.40 TRANSVERSE MAGNETIC WAVES (*TM* WAVES)

6.40.1 Definition

TM waves are electromagnetic waves for which there is no component of magnetic field in the direction of propagation. That is, $H_z = 0$.

6.40.2 Field Equations of *TM* Waves

The field equations are obtained from the wave equations, Maxwell's equations and boundary conditions.

TM$_{mn}$ wave field components

$$E_z = (K \sin Cx \sin By)\, e^{(-\gamma_g z + j\omega t)}$$

$$E_x = \left(-\frac{\gamma_g K}{h_g^2} C \cos Cx \sin By \right) e^{(-\gamma_g z + j\omega t)}$$

$$E_y = \left(-\frac{\gamma_g K}{h_g^2} B \sin Cx \cos By \right) e^{(-\gamma_g z + j\omega t)}$$

$$H_x = \left(\frac{j\omega \in K}{h_g^2} B \sin Cx \cos By \right) e^{(-\gamma_g z + j\omega t)}$$

$$H_y = \left(\frac{-j\omega \in K}{h_g^2} B \sin Cx \cos By \right) e^{(-\gamma_g z + j\omega t)}$$

$$C = \frac{m\pi}{a} \text{ and } B = \frac{n\pi}{b}$$

Field variations of *TM$_{11}$* mode/wave are shown in fig. 6.15.

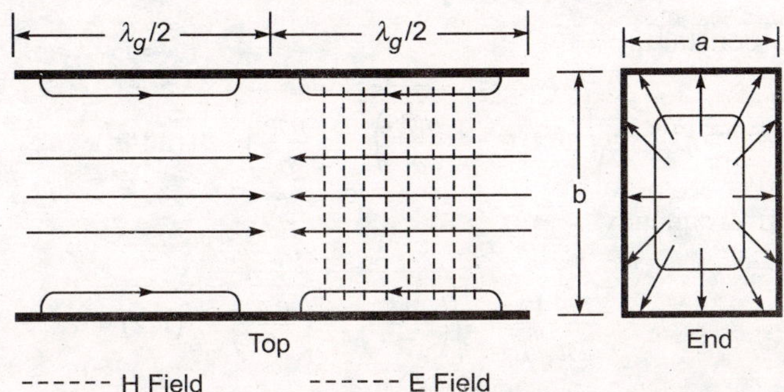

Fig. 6.15   Field variations of *TM$_{11}$* wave in hollow rectangular waveguide

6.41 TRANSVERSE ELECTROMAGNETIC WAVES (*TEM* WAVES)

6.41.1 Definition

TEM wave is defined as an electromagnetic wave in which both *E* and *H* fields are entirely transverse to the direction of propagation.

That is, $E_z = 0$, $H_z = 0$ for *TEM* wave. *TEM* wave is called Principal wave.

6.41.2 Characteristics of *TEM* Waves

1. $TEM = TM_{00}$
2. $E_z = 0$, $H_z = 0$
3. Cut-off frequency, $f_c = 0$.
4. It exists only for two conductors.
5. Transmission lines of free space
6. It does not exist in hollow waveguide.
7. $\lambda_g = \lambda$
8. $\alpha_g = 0$
9. $\eta_g = \eta_0$
10. $\lambda_C = \infty$

6.41.3 Summary of Propagation Parameters of *TE* and *TM* Waves

(1) Propagation constant,

$$\gamma_g = \sqrt{\left(\frac{m\pi}{a}\right)^2 + \left(\frac{n\pi}{b}\right)^2 - \omega^2\mu\in}, \; \left(\frac{1}{m}\right)$$

$$= \alpha_g \; \text{if} \; \left(\frac{m\pi}{a}\right)^2 + \left(\frac{n\pi}{b}\right)^2 > \omega^2\mu\in$$

(2) Phase constant,

$$\beta_g = \sqrt{\omega^2\mu\in - \left(\frac{m\pi}{a}\right)^2 - \left(\frac{n\pi}{b}\right)^2}, \; \text{(rad/m)}$$

(3) Cut-off frequency,

$$f_c = \frac{1}{2\pi\sqrt{\mu\in}} \sqrt{\left(\frac{m\pi}{a}\right)^2 + \left(\frac{n\pi}{b}\right)^2}, \; \text{(Hz)}$$

(4) Cut-off wavelength,

$$\lambda_c = \frac{2}{\sqrt{\left(\frac{m}{a}\right)^2 + \left(\frac{n}{b}\right)^2}}, \; \text{(m)}$$

(5) Phase velocity,

$$v_p = \frac{\omega}{\sqrt{\omega^2\mu\in - \left(\frac{m\pi}{a}\right)^2 - \left(\frac{n\pi}{b}\right)^2}}, \; \text{(m/s)}$$

(6) Guide wavelength,

$$\lambda_g = \cfrac{2\pi}{\sqrt{\omega^2 \mu \in -\left(\dfrac{m\pi}{a}\right)^2 - \left(\dfrac{n\pi}{b}\right)^2}}, \ (m)$$

or

$$\lambda_g = \cfrac{\lambda}{\left[1 - \left(\dfrac{\lambda}{\lambda_c}\right)^2\right]^{1/2}}, \ (m)$$

6.42 CIRCULAR WAVEGUIDES

6.42.1 Definition

A circular waveguide is a waveguide which has a circular cross-section.

6.42.2 Structure

A typical circular waveguide is shown in fig. 6.16.

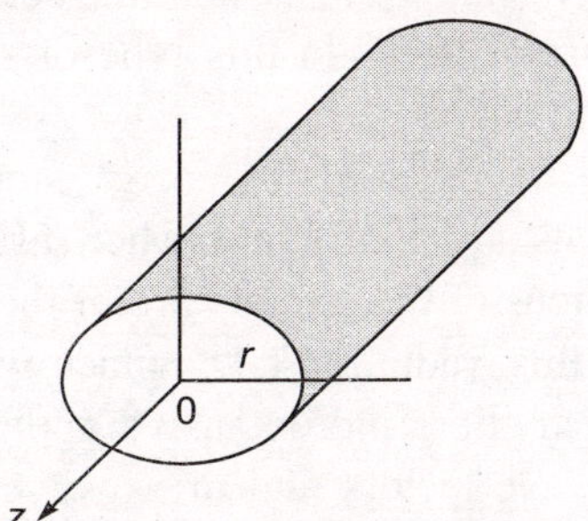

Fig. 6.16 Circular waveguide

6.42.3 Salient Features of a Circular Waveguide

- It is used as transmission line.
- It is used as a radiator.
- It produces circular polarization.
- It is easy to manufacture.
- They are used in rotational coupling.
- There exists rotation of polarization and this can be overcome by rotating modes symmetrically.
- TM_{01} mode is preferred to TE_{01} as it requires a smaller diameter for the same cut-off wavelength.
- TE_{01} does not have practical application.

- For $f > 10$ GHz, TE_{01} has the lowest attenuation per unit length of the waveguide.
- The main disadvantage is that its cross-section is larger than that of rectangular waveguide for carrying the same signal.
- The space occupied by circular waveguides is more than that of rectangular waveguides.
- The determination of fields here consists of differential equations of certain type. Their solutions involve Bessel functions.
- Here also *TE* and *TM* modes exists.

6.43 WAVEGUIDE RESONATORS

A waveguide resonator is made up of a rectangular waveguide with its open ends closed by shorts (fig. 6.17).

It is used for energy storage. As there is no propagation through the shorted ports, standing waves exist inside the cavity. These resonators are often used for various applications particularly, in klystrons and wave meters.

6.43.1 Salient Features of Resonators

1. A rectangular cavity (fig. 6.17) is nothing but a rectangular waveguide whose open ends are shorted. In this type of structure, standing waves, *TE* and *TM* waves exist.
2. The resonators are mainly used for energy storage. At high frequencies *RLC* circuit elements are inefficient when used as resonators. This is because the dimensions of the circuits are of the order of operating wavelength. Because of this, radiation takes place which is undesirable.
3. The *EM* resonator cavities find extensive applications in klystron tubes, band pass filters, wave meters and microwave ovens, etc.

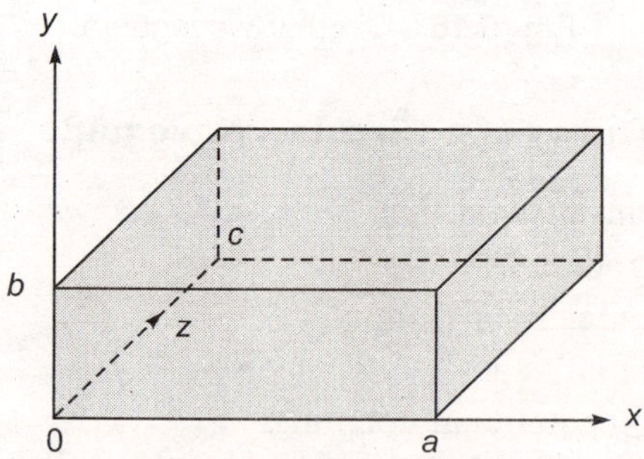

Fig. 6.17   Rectangular cavity

TM Mode $(H_z = 0)$

If $\qquad E_z = (x, y, z) = X(x)\, Y(y)\, Z(z)$, we can write,

$$X(x) = C_1 \cos Ax + C_2 \sin Ax$$
$$Y(y) = C_3 \cos By + C_4 \sin By$$
$$Z(z) = C_5 \cos Cz + C_6 \sin Cz$$

Here,
$$K_r^2 = A^2 + B^2 + C^2 = \omega^2 \mu \in = \beta_r^2$$

and
$$A = \left(\frac{m\pi}{a}\right), \; B = \left(\frac{n\pi}{b}\right), \; C = \left(\frac{l\pi}{c}\right)$$

$$m = 1, 2, 3, \ldots\ldots, \; n = 1, 2, 3, \ldots\ldots, \; l = 0, 1, 2, 3, \ldots.$$

Here, we have three boundary conditions to solve the constants, C_1, C_2, ... etc.

$$E_z = 0 \text{ at } x = 0 \text{ and at } x = a$$
$$E_z = 0 \text{ at } y = 0 \text{ and at } y = b$$
$$E_x = 0, \; E_y = 0 \text{ at } z = 0 \text{ and at } z = c$$

After simplification, E_z becomes

$$E_z = E_m \sin\left(\frac{m\pi x}{a}\right) \sin\left(\frac{n\pi y}{b}\right) \cos\left(\frac{l\pi z}{c}\right)$$

Here, $E_m = C_2 C_4 C_5$

The resonant frequency is given by

$$f_r = \frac{\beta_r}{2\pi\sqrt{\mu \in}} = \frac{\beta_r v}{2\pi}$$

or

$$f_r = \frac{v}{2} \sqrt{\left(\frac{m}{a}\right)^2 + \left(\frac{n}{b}\right)^2 + \left(\frac{1}{c}\right)^2}$$

and

$$\lambda_r = \frac{v}{f_r} = \frac{2}{\sqrt{\left(\frac{m}{a}\right)^2 + \left(\frac{n}{b}\right)^2 + \left(\frac{1}{c}\right)^2}}$$

TE Mode ($E_z = 0$)

As in the case of *TM* mode, H_z is expressed

$$H_z(x, y, z) = X(x) \, Y(y) \, Z(z)$$

Here, $X(x) = p_1 \cos Bx + p_2 \sin Bx$
$$Y(y) = p_3 \cos Ay + p_4 \sin Ay$$
$$Z(z) = p_5 \cos Cz + p_6 \sin Cz$$

The set of boundary conditions are

$$H_z = 0 \text{ at } z = 0 \text{ and at } z = c$$

$$\frac{\partial H_z}{\partial x} = 0 \text{ at } x = 0 \text{ and at } x = a$$

$$\frac{\partial H_z}{\partial y} = 0 \text{ at } y = 0 \text{ and at } y = b$$

Using the boundary conditions, and simplifying, we get

$$H_z = H_m \cos\left(\frac{m\pi x}{a}\right) \cos\left(\frac{l\pi z}{c}\right) \sin\left(\frac{l\pi z}{c}\right)$$

where
$$m = 0, 1, 2, 3, \ldots.$$
$$n = 0, 1, 2, 3, \ldots.$$
$$l = 1, 2, 3, \ldots.$$

6.43.2   Dominant Mode

The dominant mode is defined as the mode which has the lowest resonant frequency for a given cavity size (a, b, c).

The waves are represented by TE_{mnl}, TM_{mnl}

6.43.3   Degenerate Mode

The modes having the same resonant frequency, are called degenerate modes. Ideally, the walls of resonant cavity have infinite conductivity. But practical cavity walls have finite conductivity. As a result, some stored energy is lost.

6.43.4   Quality Factor, Q

Quality factor is defined as

$$Q \equiv 2\pi \frac{\text{average stored energy}}{\text{loss of energy in a cycle}}$$

i.e.
$$Q = 2\pi \frac{W_{av}}{W_L}$$

Here, $\omega = 2\pi f$,
$$W_L = \text{average power loss in a cycle}$$
$$W_{av} = \text{average stored energy}$$

Quality factor, for dominant mode, TE_{101} is

$$Q = \frac{(a^2 + c^2)\,abc}{\delta[2b(a^3 + c^3) + ac(a^2 + c^2)]}$$

Here, δ = depth of penetration in cavity walls

i.e.,

$$\delta = \frac{1}{\sqrt{\pi f_{101}\mu_0\sigma_c}} \ \text{(m)}$$

6.43.5 Salient Features of Cavity Resonators

1. A completely closed metallic structure forms a cavity and it is called cavity resonator.
2. It stores energy.
3. *TE* and *TM* modes exists in the cavity.
4. In *TE* mode, $E_z = 0$ (z is propagation direction) and E_x, E_y, H_x, H_y and H_z are present.
5. In *TM* mode, $H_z = 0$ and H_x, H_y, E_x, E_y and E_z are present.
6. In cavities, the electric and magnetic fields do not propagate along z-axis but they oscillate with time at a specified location.
7. The lowest order of TM_{mnl} mode is TM_{110}.
8. The resonant frequency of the lowest order *TM* mode is

$$f_{110} = \frac{v}{2}\sqrt{\frac{1}{a^2}+\frac{1}{b^2}}$$

9. The lowest order for TE_{mnl} is TE_{101}.
10. The resonant frequency of the lowest order *TE* mode is

$$f_{101} = \frac{v}{2}\sqrt{\frac{1}{a^2}+\frac{1}{c^2}}$$

PROBLEM 6.21   A copper made rectangular cavity resonator is structured by $3 \times 1 \times 4$ cm. Find out its resonant frequency for TM_{110} mode.

Solution   The dimensions of resonator are

$$a = 3 \ \text{cm} = 0.03 \ \text{m}$$
$$b = 1 \ \text{cm} = 0.01 \ \text{m}$$
$$c = 4 \ \text{cm} = 0.04 \ \text{m}$$

For TM_{110}, the resonant frequency is

$$f_r = \frac{v}{2}\sqrt{\frac{1}{a^2}+\frac{1}{b^2}+\frac{1}{c^2}}$$

$$f_r = \frac{3\times10^8}{2}\sqrt{\frac{1}{(0.03)^2}+\frac{1}{(0.01)^2}+\frac{1}{(0.04)^2}}$$

$$f_r = \frac{3 \times 10^8}{2} \sqrt{\frac{10000}{9} + 10000 + \frac{10000}{16}}$$

$$f_r = \frac{3 \times 10^8}{2} \times 100 \sqrt{\frac{169}{144}}$$

$$f_r = \frac{3 \times 10^{10}}{2} \times \frac{13}{12}$$

$$f_r = 1.625 \times 10^{10} \text{ Hz}$$

$$\boxed{f_r = 16.25 \text{ GHz}}$$

PROBLEM 6.22 A copper made rectangular cavity resonator is structured by $3 \times 1 \times 4$ cm and operates at the dominant modes of *TE* and *TM*. Find out its resonant frequency and quality factor. The conductivity of copper is 5.8×10^7 mho/m. Inside the cavity there is air.

Solution For *TM* mode, the dominant mode is TM_{110}. Its resonant frequency is

$$f_r = \frac{v}{2} \sqrt{\frac{1}{a^2} + \frac{1}{b^2}}$$

Here, $v = \dfrac{1}{\sqrt{\mu_0 \in_0}} = 3 \times 10^8$ m/s.

$$a = 3 \text{ cm} = 0.03 \text{ m}$$
$$b = 1 \text{ cm} = 0.01 \text{ m}$$

$$\therefore \quad f_r = \frac{3 \times 10^8}{2} \sqrt{\left(\frac{1}{0.03}\right)^2 + \left(\frac{1}{0.01}\right)^2}$$

$$= \frac{3 \times 10^8}{2} \sqrt{1111.1 + 10000}$$

$$= 1.5 \times 10^8 \times 105.4$$

$$\therefore \quad \boxed{f_r = 15.8 \text{ GHz}}$$

For *TE* mode, the resonant frequency of the dominant, TE_{101} is

$$f_r = \frac{v}{2} \sqrt{\left(\frac{1}{a}\right)^2 + \left(\frac{1}{\ell}\right)^2}$$

$$= 1.5 \times 10^8 \sqrt{1111.1 + \left(\frac{1}{0.04}\right)^2}$$

$$= 1.5 \times 10^8 \sqrt{1111.1 + 625.0}$$

$\therefore$ $\boxed{f_r = 6.249 \text{ GHz}}$

The quality factor, Q for TE_{101}

$$= \frac{(a^2 + c^2)\, abc}{\delta[2b(a^3 + c^3) + ac(a^2 + c^2)]}$$

Here, $\delta = \dfrac{1}{\sqrt{\pi f_{101} \mu_0 \sigma_c}}$

$$= \frac{1}{\sqrt{\pi \times 6.249 \times 10^9 \times 4\pi \times 10^{-7} \times 5.8 \times 10^7}}$$

$$= 8.3598 \times 10^{-7} \text{ m}$$

$\therefore$ $Q = \dfrac{(3^2 + 4^2) \times 3 \times 1 \times 4 \times 10^{-2}}{8.3598 \times 10^{-7} [2 \times 1(3^3 + 4^3) + 3 \times 4(3^2 + 4^2)]}$

$$= \frac{25 \times 12 \times 10^{-2}}{8.3598 \times 10^{-7} [2 \times 91] + 12 \times 25}$$

$$= \frac{300 \times 10^{-2}}{8.3598 \times 10^{-7} [182 + 300]}$$

$$= \frac{300 \times 10^{-2}}{482} \times \frac{1}{8.3598} \times 10^7$$

$\therefore$ $\boxed{Q = 7445}$

PROBLEM 6.23   A copper made resonant cavity is dielectric ($\epsilon_r = 4$) filled and its dimensions are $5 \times 4 \times 10$ cm. Determine the resonant frequency of TE_{101} and its quality factor.

Solution   The conductivity of copper,

$$\sigma_c = 5.8 \times 10^7 \text{ mho/m.}$$
$$a = 5 \text{ cm}$$
$$b = 4 \text{ cm}$$
$$c = 10 \text{ cm}$$

The resonant frequency of TE_{101} is

$$f_r = \frac{v}{2} \sqrt{\left(\frac{1}{a}\right)^2 + \left(\frac{1}{c}\right)^2}$$

$$= \frac{v_0}{2\sqrt{\epsilon_r}} \sqrt{\left(\frac{1}{a}\right)^2 + \left(\frac{1}{c}\right)^2}$$

$$= 0.75 \times 10^8 \sqrt{\left(\frac{1}{a}\right)^2 + \left(\frac{1}{c}\right)^2}$$

$$= 0.75 \times 10^8 \sqrt{\left(\frac{1}{0.05}\right)^2 + \left(\frac{1}{0.1}\right)^2}$$

$$= 0.75 \times 10^8 \times \sqrt{400 + 100}$$

$$\therefore \qquad \boxed{f_r = 1.677 \text{ GHz}}$$

The quality factor, Q for this mode is

$$Q = \frac{(a^2 + c^2)abc}{\delta[2b(a^3 + c^3) + ac(a^2 + c^2)]}$$

$$\delta = \frac{1}{\sqrt{\pi f_{101} \mu_0 \sigma_c}}$$

$$= \frac{1}{\sqrt{\pi \times 1.677 \times 10^9 \times 4\pi \times 10^{-7} \times 5.8 \times 10^7}}$$

$$= 0.032275 \times 10^{-4}$$

$$= 32.275 \times 10^{-7} \text{ (m)}$$

and

$$= \frac{(a^2 + c^2)abc}{2b(a^3 + c^2) + ac(a^2 + c^2)}$$

$$= \frac{(25 + 100)5 \times 4 \times 10}{2 \times 4(125 + 1000) + 5 \times 10(25 + 100)}$$

$$= \frac{125 \times 200}{9000 + 6250} = \frac{25000}{15250}$$

$$= 1.6393 \text{ cm}$$

$$= 1.6393 \times 10^{-2} \text{ m}$$

$$\therefore \qquad Q = \frac{1.6393 \times 10^{-2}}{32.275 \times 10^{-7}}$$

$$= 0.05079 \times 10^5$$

$$\therefore \qquad \boxed{Q = 5079}$$

6.44 STRIPLINES

Stripline is a three-conductor transmission line. It supports *TEM* wave for propagation. The striplines are nothing but planar transmission lines which are used over a frequency range of 100 MHz to 100 GHz.

6.44.1 Construction

The stripline consists of a thin metal strip made between two insulators laid on two metal strips. Its structure is shown in fig. 6.18.

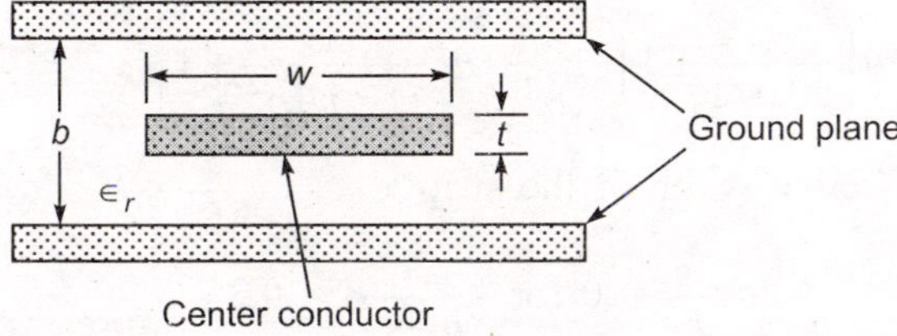

Fig. 6.18 Geometry of stripline transmission line

The structure consists of a thin conducting strip inside a low loss dielectric substrate between two ground plates. It is basically a three-conductor transmission line which supports *TEM* wave. The field distribution in the stripline is shown in fig. 6.19.

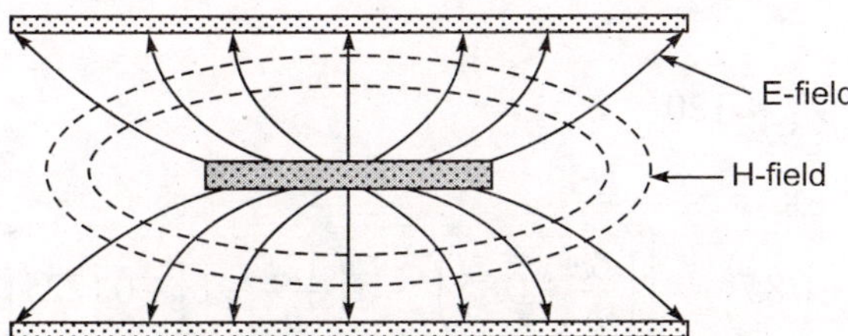

Fig. 6.19 Field distribution in stripline

In this transmission line, there are no fringing fields away from the edges of the central conductor.

6.44.2 Design Considerations

- ➢ The width of strip is greater than its thickness.
- ➢ The substrate is made of a low loss dielectric.
- ➢ The ground plates and the strip have the same thickness.
- ➢ The width of the greater plates is greater than their spacing by five times.

6.44.3 Characteristic Impedance, Z_0

It is given by

$$Z_0 = \frac{30\pi}{\sqrt{\epsilon_r}}\left[\frac{(1 - t/b)}{w_e\,(b + c_f)\,\pi}\right]$$

Here, ϵ_r = relative permittivity of substrate

b = spacing between conducting plates.

c_f = fringing capacitance.

$$= 2\ell n\left(\frac{1}{1 - t/s} + 1\right) + \frac{1}{b}\,\ell n\left[\frac{1}{(1 - t/b)^2} + 1\right]$$

w_e = effective width of the strip

i.e.
$$\frac{w_e}{b} = \frac{w}{b} = 0 \ \text{for} \ \frac{w}{(b - t)} > 0.35$$

$$= \frac{\left(0.35 - \dfrac{w}{b}\right)^2}{1 + \dfrac{12t}{b}} \ \text{for} \ w(b - t) < 0.35$$

The ratio $\dfrac{w}{b}$ can be found if Z_0 and $\dfrac{t}{b}$ are known.

i.e. $\dfrac{w}{b} = x$ for $Z_1 < 120$

$$\frac{w}{b} = \left(\frac{6t}{b} + 0.85\right) - \left[\left(\frac{6t}{b} + 0.85\right)^2 - x\left(\frac{12t}{b} + 1\right) - 0.1225\right]^{\frac{1}{2}} \ \text{for} \ Z_1 > 120$$

Here, $x = \dfrac{30\pi\left(1 - \dfrac{t}{b}\right)}{Z_0\sqrt{\epsilon_r}} - \dfrac{c_f}{\pi}$

and $Z_1 = Z_0\sqrt{\epsilon_r}\left(1 + \dfrac{2.3t}{b}\right)$

6.44.4 Attenuation in Stripline

The attenuation takes place due to conductor loss and due to dielectric loss. The attenuation due to conductor loss is given by

$$\alpha_c = \frac{0.0231\, R_0 \in_r Z_0}{30\pi\,(b-t)}\,(x+y)$$

Here, $\quad x = 1 + \dfrac{2w}{b-t} + \dfrac{1}{\pi}\dfrac{b+t}{b-t}\,\ell n\!\left(\dfrac{2b-t}{t}\right)$

$\qquad y = 0$ for $Z_b < 120$

$$y = \frac{\left(0.35 - \dfrac{w}{b}\right)}{(b-t)\left(1+\dfrac{12t}{b}\right)} \times \left[\frac{t}{b}(17.45b + 35w) - 9w + 5.85 - 32.4\frac{t^2}{b}\right]$$

$\qquad$ for $Z_b > 120$

The attenuation due to dielectric loss is given by

$$\alpha_d = \frac{8.68\,\pi f \sqrt{\in_r}}{v_0}\ \tan\delta\ \text{dB/m}$$

The velocity of propagation for the stripline transmission line is given by

$$v = \frac{v_0}{\sqrt{\in_r}}\ \text{m/s}$$

The wavelength of electromagnetic signal on the stripline is given by

$$\lambda = \frac{v_0}{f\sqrt{\in_r}}$$

6.45 MICROSTRIPLINE

Microstripline is an unsymmetrical stripline which resembles a parallel plate transmission line. It has a dielectric substrate, a metallic ground and a thin conducting strip.

6.45.1 Salient Features of Striplines

- It is a three-conductor line.
- It supports *TEM* wave.
- It is preferred to waveguide due to its small miniaturization.
- It is used in receiver front ends.
- Low power stages of transmission lines, and low power microwave circuits.
- The field in the stripline around the strip conductor and decay rapidly with distance away from the strip in the lateral direction.

6.45.2 Geometry of Microstripline

Its geometry is shown in fig. 6.20.

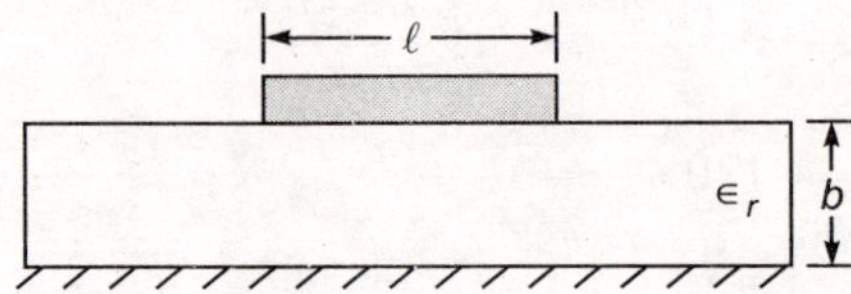

Fig. 6.20 Geometry of a microstripline

The field distribution is shown in fig. 6.21.

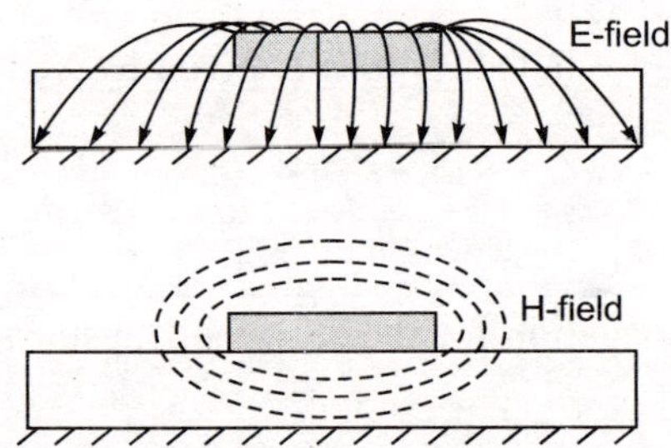

Fig. 6.21 Fields in a microstrip line

The effective dielectric constant is given by

$$\epsilon_e = \frac{\epsilon_r + 1}{2} + \frac{\epsilon_r - 1}{2} \frac{1}{\sqrt{1 + \frac{12h}{w}}} + 0.04\left(1 - \frac{w}{h}\right)^2 \quad \text{for} \quad \frac{w}{h} < 1$$

$$\epsilon_e = \frac{\epsilon_r + 1}{2} + \frac{\epsilon_r - 1}{2} \frac{1}{\sqrt{1 + \frac{12h}{w}}} \quad \text{for} \quad \frac{w}{h} > 1$$

Here, h = thickness of substrate

ϵ_r = permittivity of substrate

w = the line width

6.45.3 Characteristic Impedance, Z_0

$$Z_0 = \frac{377}{2\pi\sqrt{\epsilon_r}} \, \ell n\left(\frac{8h}{w} + \frac{w}{4h}\right) \Omega \text{ for } \frac{w}{h} < 1.$$

$$Z_0 = \frac{377}{\sqrt{\epsilon_r}}\left[\frac{1}{\frac{w}{h} + 1.393 + 0.667 \, \ell n\left(\frac{w}{h} + 1.444\right)}\right]$$

If Z_0 is specified, $\dfrac{w}{h}$ can be found. That is,

$$\frac{w}{h} = \frac{4}{\dfrac{1}{2}e^{x} - e^{-x}} \quad \text{for } \frac{w}{h} < 2$$

$$\frac{w}{h} = \frac{\epsilon_r - 1}{\pi\,\epsilon_r}\left[\ln(y-1) + 0.39 - \frac{0.61}{\epsilon_r} \right] + \frac{2}{\pi}\left[y - 1 - \ln(2y-1) \right] \quad \text{for } \frac{w}{h} > 2.$$

Here, $\quad x = \pi\sqrt{2(\epsilon_r - 1)}\,\dfrac{Z_0}{377} + \dfrac{\epsilon_r - 1}{\epsilon_r + 1} \times \left(0.23 + \dfrac{0.11}{\epsilon_r} \right)$

$$y = \frac{377\pi}{2\sqrt{\epsilon_r}\,Z_0}$$

6.45.4 Attenuation in Microstripline

It has attenuation due to conductor loss and dielectric loss.

The attenuation due to conductor loss is given by

$$\alpha_c = \frac{8.68\,R_S}{WZ_0}\ \text{dB/m}$$

The attenuation due to dielectric loss is given by

$$\alpha_e = 27.3\,\frac{\epsilon_r - 1}{\epsilon_r + 1}\,\frac{\epsilon_r}{\sqrt{\epsilon_r}}\,\frac{f}{v_0}\ \tan\delta\ \text{dB/m}$$

6.45.5 Advantages

- ➤ It has better interconnection features.
- ➤ Easy fabrication.
- ➤ Due to its planar structure, packaged and unpackaged semiconductor chips can be easily attached to it.
- ➤ Fabrication cost is less.
- ➤ It is easy to mount passive or active discrete devices.
- ➤ It is easy to make minor adjustments even after the fabrication of the circuit is completed.
- ➤ It allows access for probing and measurements.

6.45.6 Disadvantages

- ➤ It exhibits high radiation losses due to open structure.

➢ Interference due to nearby conductor is high.

➢ It exhibits discontinuity in the generation of electric and magnetic fields.

➢ The modes in the microstripline is only quasi-*TEM*.

6.45.7 Salient Features of Microstripline

➢ It is an unsymmetrical stripline.

➢ It is similar to parallel plate transmission line.

➢ It supports quasi-*TEM* wave.

➢ It exhibits high radiation losses due to open strip structure.

➢ Its radiation losses can be reduced by the use of thin, high dielectric materials.

➢ Its fabrication cost is less.

➢ It is easy to attach discrete components to the strip.

➢ Minor adjustments possible after the circuit fabrication is completed.

➢ Probing and measurements are easy

6.46 OPTICAL FIBRES

6.46.1 Definition

An optical fibre is a thin, transparent transmission line and it is made of glass or plastic or a combination of both. It allows light to travel through it.

6.46.2 Characteristics of Optical Fibres

➢ It confines the light energy inside its walls.

➢ Light travels in the fibre by internal reflection or by refraction.

➢ For visible light, the refractive index of the fibre is much higher than that of air.

➢ The fibre bundles are used to transfer a number of light beams.

➢ A single light beam is modulated simultaneously by large number of signals.

➢ The fibres are made of pure glass or plastic or by both of them.

➢ In optical fibres, the propagation takes place by single or by multimode. Mode means path.

6.46.3 Types of Fibres

There are three types of optical fibres.

1. Single mode step-index fibre.
2. Multimode step-index fibre
3. Multimode graded-index fibre.

6.46.4 Characteristics of Single Mode Step-Index Fibre

- It has a small central core.
- There exists only one light path.
- Its outside cladding can be of air or glass.
- The refractive index of the glass core is about 1.5 times that of air cladding.
- The air cladded fibre is weak and hence it has limited applications.
- The glass cladded fibre is stronger than air cladded fibre.
- The refractive index of cladding is uniform.
- The refractive index of the glass cladding is slightly less than that of central core.
- It transits single mode for all wavelengths.

- The cut-off wavelength is in the fibre is given by $\lambda_c = \dfrac{2\pi r n_1 \sqrt{2k}}{2.405}$ (µm)

  Here, λ_c = cut-off wavelength (mm)
  n_1 = refractive index of core
  n_2 = refractive index of cladding
  r = radius of core (µm)

  $$k = \frac{n_1 - n_2}{n_1}$$

- The dispersion is minimum.
- Bandwidth is large.
- The coupling of light into and out of the fibre is difficult, as central core is small.
- Highly directive light source is required for this fibre.
- This fibre is expensive.
- Its manufacturing is difficult.

6.46.5 Characteristics of Multistep-Index Fibre

- It is similar to single step-index fibre.
- Its central core is much larger.
- It allows more light to enter into the cable as it has a large light to fibre aperture end.
- The light travels in zig-zag fashion.
- There exists many path rays.
- It is a low cost fibre.
- It is easy to manufacture.
- Coupling of light to fibre is easy.
- Light ray is distorted when it propagates.
- Bandwidth is small.

6.46.6 Characteristics of Multimode Graded-Index Fibre

- ➤ The refractive index of its central core is not uniform.
- ➤ n is maximum at the center and it is minimum at the edges.
- ➤ Light propagates by refraction.
- ➤ Bending of light rays takes place.
- ➤ Light enters the fibre at different angles.
- ➤ Light coupling is easy.
- ➤ Distortion is more.
- ➤ It is easy to manufacture.
- ➤ It exhibits intermodulation properties compared to the other two.

6.47 POINTS TO REMEMBER

- ➤ A transmission line is a device which carries power from transmitter to receiver.
- ➤ A transmission line is an impedance transformer.
- ➤ Free space is a transmission channel.
- ➤ At microwave frequency, the power is delivered through electric and magnetic fields.
- ➤ Striplines and microstriplines are useful transmission lines at high frequencies.
- ➤ The equivalent circuit of a transmission line is a distributed network.
- ➤ The primary constant of a transmission line are R, L, C, G.
- ➤ R, L, C and G are expressed as Ω/km, H/km, F/km and mho/km.
- ➤ The secondary constants of a transmission line are γ, Z_0.
- ➤ Z_0 of parallel wire transmission line is $\sqrt{Z/Y}$.
- ➤ The propagation constant, γ_ℓ of the line is $\sqrt{ZY}$.
- ➤ The condition for lossless line is $R = 0, G = 0$.
- ➤ The condition for distrotionless line is $\dfrac{R}{L} = \dfrac{C}{G}$.
- ➤ The reflection coefficient of a line, $\rho = \dfrac{(Z_L - Z_0)}{(Z_L + Z_0)}$.
- ➤ VSWR, $S = \dfrac{1 + |\rho|}{1 - |\rho|}$.
- ➤ The reflection coefficient, $\rho = \dfrac{S - 1}{S + 1}$.

➤ The equivalent circuit of an open $\lambda/2$ line is a tank circuit.

➤ The equivalent circuit of a shorted $\lambda/2$ line is a series-resonant circuit.

➤ The open wire transmission lines are susceptible to external noise.

➤ Open wires radiates.

➤ Z_0 is defined as V_f/I_f.

➤ For *RF* lines, $\omega L \gg R$, $\omega C \gg G$.

➤ For lossless lines, $Z_0 = \sqrt{L/C}$.

➤ For lossless lines, $\alpha_\ell = 0$, $\beta_\ell = \omega\sqrt{LC}$.

➤ $Z_i = V_S/I_S$.

➤ Reflection coefficient is the ratio of reflected voltage and incident voltage.

➤ VSWR is the ratio of V_{maximum} and V_{minimum}.

➤ $S = \dfrac{Z_0}{R_L}$ if $Z_0 \gg R_L$

➤ $S = \dfrac{R_L}{Z_0}$ if $R_L \gg Z_0$.

➤ A stub is a piece of transmission line.

➤ $Z_i = Z_0\left[\dfrac{Z_L + jZ_0 \tan \beta\ell}{Z_0 + jZ_L \tan \beta\ell}\right]$.

➤ Twisted pair of lines exhibit high frequency losses.

➤ Shielded pair lines have low noise.

➤ Coaxial lines support *TEM* wave.

➤ Hollow rectangular waveguide does not support *TEM*.

➤ Waveguide is a high pass-filter.

➤ The cut-off frequency is zero for *TEM* wave.

➤ The guide wavelength is greater than free space wavelength.

➤ The dominant mode has the lowest cut-off frequency.

➤ *TEM* wave is a uniform plane wave.

➤ Circular waveguide radiates circularly polarized fields.

➤ The stripline is a planar transmission line.

➤ Striplines are used in the frequency range of 100 MHz to 100 GHz.

➤ Stripline is a 3-conductor transmission line.

➤ The characteristic impedance of the stripline depends on strip width, thickness, dielectric constant of the substrate and fringing capacitance.

➤ The attenuation of stripline is due to conductor and dielectric losses.

➤ The attenuation of stripline depends on Z_0, ϵ_r, R_S, b, t, W, f and v_0.

➤ The velocity of propagation for stripline is $v_0/\sqrt{\epsilon_r}$.

- ➤ The wavelength of EM wave on the stripline $v_0/f\sqrt{\in_r}$.
- ➤ Microstripline is a two-conductor transmission line.
- ➤ Microstripline is an asymmetrical stripline.
- ➤ Stripline is a symmetrical transmission line.
- ➤ The characteristic impedance of microstripline depends on width and height of the strip and $\in_r$ of the substrate.
- ➤ The attenuation of microstripline is due to conductor and dielectric losses.
- ➤ The attenuation of microstripline depends on R_S, W, Z_0 and $\in_r$, f, v_0 and losses in conductor and dielectrics.
- ➤ Smith chart is an impedance/admittance chart.
- ➤ Smith chart contains two sets of lines.
- ➤ The two sets of lines in the Smith chart represents resistive and reactive part of impedance or conductive and susceptance of the admittance.
- ➤ The reactive lines in the Smith chart are all arcs of circles.
- ➤ The resistive lines in the Smith chart are all circles.
- ➤ The Smith chart is known as 'Transmission line calculator.
- ➤ Smith chart is useful to find VSWR and design of stubs.
- ➤ The outer perimeter of Smith chart represents a length of $\lambda/2$.

6.48 MULTIPLE CHOICE QUESTIONS

1. Two-wire lines are useful at
    (a) Microwave frequencies
    (b) millimeter wave frequencies
    (c) submillimeter wave frequencies
    (d) lower end of microwave frequencies
2. The advantage of twisted pair wires is
    (a) flexibility   (b) low cost
    (c) useful at all frequencies   (d) useful for applications
3. The equivalent circuit of a parallel wire transmission line consists
    (a) R and L (b) R, L, C and G
    (c) L and C (d) R and G
4. The propagation velocity in transmission line is
    (a) $\dfrac{\text{travel distance}}{\sqrt{LC}}$ (b) $\dfrac{\sqrt{LC}}{\text{travel distance}}$
    (c) $\sqrt{LC} \times (\text{travel distance})$ (d) $\sqrt{LC}$
5. The wavelength in open air is
    (a) f/v_0 (b) v_0/f
    (c) $v_0 f$ (d) $f v_0$

6. The characteristic impedance of a transmission line is

   (a) $\sqrt{L/C}$ (b) $\sqrt{LC}$

   (c) $\sqrt{C/L}$ (d) $\dfrac{1}{\sqrt{LC}}$

7. The wavelength of a microwave in free space, if its frequency is 1 GHz, is

   (a) 3 m   (b) 30 m
   (c) 0.3 m   (d) 0.03 m

8. If the impedance of a transmission line is 1.119 µH/m, and the capacitance is 12.3 pF/m, the time required for the wave to travel 1.0 m length of the line is

   (a) 0.371 nm   (b) 0.0371 ms
   (c) 3.71 ns   (d) 3.71 ms

9. Z_0 of a transmission line when $L = 1.119$ µH/m and $C = 12.3$ pF/m is

   (a) 302 Ω   (b) 30.2 Ω
   (c) 30.2 KΩ   (d) 3.02 Ω

10. VSWR is

   (a) $\dfrac{V_{max}}{V_{min}}$ (b) $\dfrac{V_{min}}{V_{max}}$

   (c) $\sqrt{\dfrac{I_{max}}{I_{min}}}$ (d) $\dfrac{V_{reflected}}{V_{incident}}$

11. The reflection coefficient in a transmission line is

   (a) $\dfrac{Z_L + Z_0}{Z_L - Z_0}$ (b) $\dfrac{Z_L - Z_0}{Z_L + Z_0}$

   (c) $2\dfrac{Z_L}{Z_0}$ (d) $\dfrac{Z_0}{2Z_L}$

12. If $R_L = 105$, $Z_0 = 75$, the reflection coefficient is

   (a) 1.67   (b) 0.167
   (c) 0.0167   (d) 16.7

13. If V_{max} is 45, V_{min} is 22.5, VSWR is

   (a) 2   (b) 0.5
   (c) −2.0   (d) −0.5

14. If VSWR is 1.67, VSWR in dB is

   (a) 44.4   (b) 4.44
   (c) 0.444   (d) 444

15. If the reflection coefficient in a transmission line is 0.376, the VSWR is
    (a) 2.21
    (b) 22.1
    (c) 0.221
    (d) −2.21

16. If VSWR = 1, the reflection coefficient is
    (a) zero
    (b) infinite
    (c) −1
    (d) 1.0

17. If VSWR is infinite, the reflection coefficient is
    (a) −1
    (b) 1
    (c) zero
    (d) infinite

18. If VSWR is −∞, the reflection coefficient is
    (a) −1
    (b) zero
    (c) 1
    (d) infinite

19. When the wavelength of a microwave is 3 cm, the phase constant is
    (a) 2.09 rad/m
    (b) 0.209 rad/m
    (c) 20.9 rad/m
    (d) 2.09 rad

20. If $Z_{in} = 75\ \Omega$, $Z_{output} = 50\ \Omega$, Z_0 is
    (a) $61.2\ \Omega$
    (b) $61.2\ K\Omega$
    (c) $6.12\ \Omega$
    (d) $6.12\ K\Omega$

21. Skin depth, δ is equal to

    (a) attenuation constant, α
    (b) $\dfrac{\alpha}{\beta}$

    (c) $\dfrac{1}{\alpha}$
    (d) $\dfrac{\beta}{\alpha}$

22. The characteristic impedance, Z_0 of a parallel line transmission line is

    (a) $\sqrt{\dfrac{Z}{Y}}$
    (b) $\sqrt{\dfrac{Y}{Z}}$

    (c) $\dfrac{Z}{Y}$
    (d) $\sqrt{ZY}$

23. The condition for a transmission line to be distortionless is

    (a) $\dfrac{R}{C} = \dfrac{G}{L}$
    (b) $\dfrac{R}{L} = \dfrac{G}{C}$

    (c) $RG = LC$
    (d) $GC = RL$

24. The primary constants of a transmission line are
    (a) R, L, C, G
    (b) α, β
    (c) γ, Z_0
    (d) $\lambda, \in$

25. The condition for a transmission line to be lossless is

    (a) $R = G$
    (b) $\dfrac{R}{G} = 1$

    (c) $R = 0,\ G = 0$
    (d) $R = 0$

26. The VSWR of a transmission line is

    (a) $\dfrac{1+|\rho|}{1-|\rho|}$

    (b) $\dfrac{1-|\rho|}{1+|\rho|}$

    (c) $\dfrac{1+\rho}{1-\rho}$

    (d) $\dfrac{1-\rho}{1+\rho}$

27. The equivalent circuit of a shorted line of length $\ell < \dfrac{\lambda}{4}$ is

    (a) a capacitor
    (b) an inductor
    (c) a resistor
    (d) a series $L\text{-}C$

28. The free space characteristic impedance is
    (a) 377 Ω
    (b) 120 Ω
    (c) 377 $\pi\Omega$
    (d) 292 Ω

29. Conduction current in free space is
    (a) zero
    (b) infinite
    (c) moderate
    (d) greater than displacement current

30. The velocity of propagation of microwave in free space is
    (a) 3×10^{10} m/s
    (b) 3×10^{8} m/s
    (c) 3×10^{8} cm/s
    (d) 3×10^{6} m/s

31. The propagation constant in a transmission line is

    (a) $\sqrt{Z/Y}$

    (b) ZY

    (c) $\sqrt{ZY}$

    (d) $\sqrt{\dfrac{Y}{Z}}$

32. The phase constant of RF line is

    (a) $\sqrt{LC}$

    (b) $\sqrt{L/C}$

    (c) $\omega\sqrt{LC}$

    (d) $\omega\sqrt{\dfrac{L}{C}}$

33. For RF lines,
    (a) $\omega L \ll R$
    (b) $\omega C \gg G$
    (c) $\omega C \ll G$
    (d) $\omega L \gg R$ and $\omega C \gg G$

34. The velocity of propagation in a transmission line is

    (a) $\dfrac{1}{\sqrt{LC}}$

    (b) $\sqrt{LC}$

    (c) $\sqrt{L/C}$

    (d) $\sqrt{C/L}$

35. The characteristic impedance of a transmission line is

    (a) $\sqrt{(Z_i)_{SC}\,(Z_i)_{OC}}$

    (b) $\sqrt{\dfrac{(Z_i)_{SC}}{(Z_i)_{OC}}}$

(c) $\sqrt{\dfrac{(Z_i)_{OC}}{(Z_i)_{SC}}}$ (d) $(Z_i)_{SC}\,(Z_{i_{OC}})$

36. For open-circuited transmission line, VSWR is
    (a) 1   (b) zero
    (c) ∞ (d) -1

37. For a short circuited transmission line, VSWR is
    (a) ∞ (b) 1
    (c) -1 (d) zero

38. Z_0 of coaxial cable with its outside conductor of 24 mm and diameter of inside conductor is
    (a) 65.84 Ω (b) 6.584 Ω
    (c) 65.85 KΩ (d) 6.584 K

39. The inductance of coaxial cable of $Z_0 = 75\ \Omega$, $C = 70$ pF/m, is
    (a) 3.937 μH/m   (b) 0.3937 μH/m
    (c) 3.937 mH/m   (d) 0.3937 mH/m

40. Parallel plate transmission line supports
    (a) only *TE*   (b) only *TM*
    (c) *TEM*   (d) only *TE* and *TM* only

41. A rectangular hollow waveguide supports
    (a) *TEM*   (b) *TE* only
    (c) *TM* only   (d) *TE* and *TM*

42. The cut-off frequency of *TEM* wave is
    (a) finite   (b) zero
    (c) 6.8 GHz   (d) 9.375 GHz

43. The cut-off wavelength of the dominant mode is
    (a) $2 \times$ broadwall dimension   (b) $2 \times$ narrow wall dimension
    (c) $2\,\lambda$ (d) λ_g

44. Stub is
    (a) a transmission line   (b) an antenna
    (c) a waveguide   (d) a parallel plate

45. A waveguide is
    (a) low-pass filter   (b) high-pass filter
    (c) non-resonant   (d) band-pass filter

46. If $Z_0 = 75\ \Omega$, VSWR is 2.0, the maximum impedance is
    (a) 150 Ω (b) 37.5 Ω
    (c) 300 Ω (d) 75 Ω

47. The maximum value of reflection coefficient in a transmission line is
    (a) ∞ (b) 1
    (c) 10   (d) 0

48. Power loss caused by a mismatched line is

    (a) $10 \log_{10} \dfrac{p_i}{p_i + p_r}$ (b) $10 \log_{10} \dfrac{p_i}{p_r}$

    (c) $\log_{10} \dfrac{p_i}{p_i - p_r}$ (d) $20 \log_{10} \dfrac{p_i}{p_i - p_r}$

49. If $Z_L = 100 + j\,150$, and $Z_0 = 50\ \Omega$ in a transmission line, the normalized impedance is
    (a) $2 + j\,3$ (b) $2 - j\,3$
    (c) $500 + j750\ \Omega$ (d) $2 + j\,3\ \Omega$

50. If D is spacing between two lines, d is the diameter of the conductor, Z_0 of the line is

    (a) $276 \log_{10} \dfrac{2D}{d}$ (b) $276 \log_{10} \dfrac{d}{2D}$

    (c) $276\ \ell n \dfrac{2D}{d}$ (d) $276\ \ell n \dfrac{d}{2D}$

51. Skin depth is
    (a) proportional to frequency
    (b) inversely proportional to square root of frequency
    (c) proportional to square root of frequency
    (d) proportional to attenuation

6.49 ANSWERS

1. d	2. a	3. b
4. a	5. b	6. a
7. c	8. c	9. a
10. a	11. b	12. b
13. a	14. b	15. a
16. a	17. b	18. a
19. a	20. a	21. c
22. a	23. b	24. a
25. c	26. a	27. b
28. a	29. a	30. b
31. c	32. c	33. d
34. a	35. a	36. c
37. a	38. a	39. b
40. c	41. d	42. b
43. a	44. a	45. b
46. a	47. b	48. a
49. a	50. a	51. b

6.50 EXERCISE PROBLEMS

1. The diameter of outer conductor of a coaxial cable is 11 mm, the diameter of inner conductor is 4.9 mm. The dielectric constant of dielectric material between the conductor is $\epsilon_r = 2.3$. Find L, C and Z_0.

2. A typical air filled coaxial line has $L = 0.051\ \mu H/m$, $C = 20.4\ pF/m$. Find the phase velocity, phase constant, λ, Z_0 and travel time over a length of 10 m.

3. Find the time required to travel 50 meters length of the line if $L = 2.0\ \mu H/m$, $C = 10\ pF/m$.

4. If $L = 1.0\ \mu H$ and $C = 12\ pF/m$ for a transmission line, find the velocity of the wave traveling along the line. Also find out percentage of free space velocity the wave.

5. A transmission line is connected to load impedance of $Z_0 = 50\ \Omega$. It is given to 100 V dc. The source resistance, $R_S = 50\ \Omega$. What is Input voltage to the line and the current in the load.

6. Find the power dissipated in the load of the transmission line when it is terminated in a short. The voltage is 100 V, $Z_0 = 50\ \Omega$, $R_S = 50$.

7. If the incident voltage is 50 V and the reflected voltage is 30 V, find the reflection coefficient. Also find the percentage of reflected voltage and reflected power.

8. If the load impedance of a transmission line 100 Ω, $Z_0 = 75$, find the reflection coefficient.

9. If a transmission line is terminated with a load resistance of 34 and its $Z_0 = 50\ \Omega$. Find VSWR in dB.

10. A transmission line has $Z_0 = 50\ \Omega$. Identify the values of (a) $R = 50\ \Omega$ (b) $R_2 = 75\ \Omega$ (c) $G_1 = 0.03$ and (d) $G_2 = 0.04$ siemens.

11. VSWR is 2 in a transmission line terminated in a normalized impedance of Z_n and the voltage minimum is at a distance of $\lambda/4$ from load. Find Z_n.

12. A lossless transmission line is terminated in $Z_L = 150 - j\,50\ \Omega$. Its characteristic impedance is 50 Ω. Determine VSWR, reflection coefficient and maximum and minimum impedances.

13. A rectangular waveguide has the dimensions of 3.4×7.22 cm. Determine λ_g, phase constant, phase velocity and group velocity, if $f = 3.0$ GHz.

Chapter 7

Microwave Integrated Circuits

The microwave integrated circuit is popularly known as MIC.

7.1 INTRODUCTION

In MICs, both passive and active devices are grown on semiconductor substrate. These circuits are useful for linear and digital applications. The BJTs or MOSFETs are used for resistors, capacitors, rectifiers and amplifiers. MICs are made either in monolithic or hybrid form.

Monolithic technology has some disadvantages, namely, complex and difficult processing, poor performance and low yields and hence it is not preferred for MICs.

On the other hand, MICs with hybrid technology is preferred. In this technology, the active devices are mounted and the passive devices are deposited on the substrates. In this, high quality ceramic, glass and ferrites are used as substrates.

7.2 SALIENT FEATURES OF MICs

1. MIC represents Microwave Integrated Circuits.
2. It consists of both passive and active microwave elements.
3. It is a multifunction circuit.
4. The active elements in MIC are usually silicon planar chips.
5. The passive elements in MIC are thin or thick film components.
6. The solid-state devices are more suitable for MIC fabrication compared to waveguides and coaxial lines.
7. Miniaturization, mass production, high performance, low cost, less weight, reliability are the merits of MICs.
8. MIC fabrication consists of thin film, thick film, monolithic and hybrid technologies.
9. The resistors are fabricated with thin film and thick film technologies.
10. Capacitors are fabricated with thick film technology.
11. The thickness of thick film is much higher than that of thin film. It is more than a few thousand angstroms.

12. The conducting and non-conducting materials are deposited on passive insulated substrates.
13. The common substrates are silicon dioxide, glass, ceramic etc.
14. The materials are deposited by vacuum deposition.
15. Monolithic Microwave Integrated Technology (MMIT) is used for fabricating single function devices.
16. Hydrid technology is used for fabricating multifunction devices.
17. MMIC represents Monolithic Microwave Integrated Circuit.
18. In MMIC, the substrates like silicon, or gallium arsenide are usually used.
19. The passive components are grown on the substrates in MMIC.
20. The active devices are grown on situ material in MMIC.
21. In hybrid circuits, passive components are associated with ferrite substrates and active components are attached with quartz, glass and ceramic etc.
22. In Hybrid Integrated Circuits (HIC), two or more IC types like monolithic, thick film, thin film and one IC type along with the discrete elements are used.
23. In Greek, *mono* means single and *litho* means stone.
24. Monolithic means single crystal.

7.3 TYPES OF ELECTRONIC CIRCUITS

The classification of electronic circuits is given below

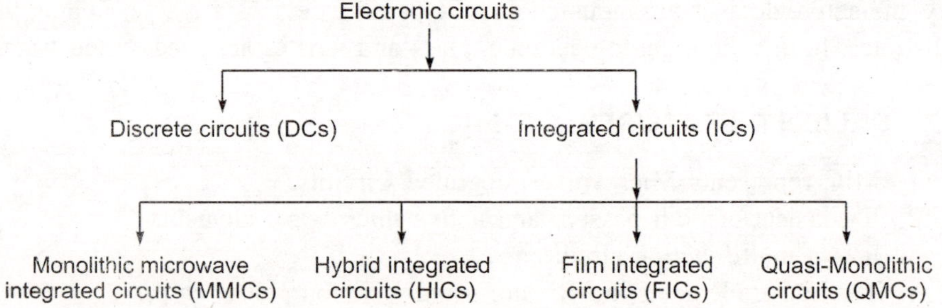

7.4 DISCRETE CIRCUIT (DC)

- Discrete circuits consist of discrete components.
- The components are made separately.
- The components are connected by wires.
- It consists of passive and active components and interconnections.

7.5 INTEGRATED CIRCUIT (IC)

- Integrated means combined.
- It consists of integrated version of electronic circuits.

- The components are not made separately.
- It consists of a single chip made of semiconductor.
- It consists of passive, active and interconnecting elements.
- Packaging density is high in ICs.

7.6 MONOLITHIC MICROWAVE INTEGRATED CIRCUIT (MMIC)

- *Mono* means single and *litho* means stone.
- It consists of an IC on a single crystal useful at microwave frequency.
- Its fabrication involves different processes like oxidation masked impurity growth and oxide etching.
- It consists of active, passive and interconnecting elements.
- Packing density is small in MMIC.
- It consists of two or more ICs.

7.7 HYBRID INTEGRATED CIRCUIT (HIC)

- It consists of two or more ICs.
- It consists of one MMIC, one IC, DCs and a film IC.

7.8 FILM INTEGRATED CIRCUIT (FIC)

- It is an MMIC in which the components are made on an insulating substrate.
- The substrate is usually glass or ceramic.

7.9 QUASI-MONOLITHIC INTEGRATED CIRCUIT (QMIC)

- In this, heteroepitaxy techniques are used for growing layers of semiconductors on substrates.
- The substrates are passive.
- The substrates are sapphire and spinel etc.
- The semiconductors are silicon and gallium arsenide.
- Typically silicon is grown on sapphire and GaAs is grown on spinel.
- The substrates used in this have low resistivity and low loss.
- Active devices are formed in semiconductors.
- Passive circuits are deposited in semiconductor removed areas.

7.10 MERITS OF MIC

- Light weight
- Small size
- High reliability

- Compactness
- High performance
- Reproducibility
- Economical
- Fast construction
- Low cost
- Withstand shock
- Withstand vibration
- Compatibility
- Uneffected of temperature variations

7.11 APPLICATIONS OF MMICs

They are useful in

- Satellites, aircraft, submarines
- Military applications
- Applications where compactness and lightweight are required.

7.12 MMIC MATERIALS

The MMICs consists of following materials:

- Substrate materials
- Metals
- Resistive films
- Dielectric film

7.13 EXAMPLES OF SUBSTRATE MATERIALS

- Alumina
- Sapphire
- Beryllia
- Rutile
- Gallium Arsenide
- Ferrite/garnet
- Glass
- High resistivity silicon
- Ceramic – filled resigns

7.14 CHARACTERISTICS OF IDEAL SUBSTRATES

- Low loss tangent
- High resistivity

- High dielectric constant
- High dielectric strength
- High purity
- Uniform thickness
- Uniform dielectric constant at wide frequency range
- Uniform dielectric constant over wide range of temperature.
- High surface smoothers
- High thermal conductivity

7.15 APPLICATIONS AND PROPERTIES OF SUBSTRATE MATERIALS AT 10 GHz

1. Alumina is used in microstrip and suspended substrates.

   Its $\epsilon_r = 10$

   Loss tangent $(D) = 1 \times 10^4$ to 2×10^4

   Thermal conductivity, $K = 0.3$ W/cm^2 °C

2. Sapphire is used in microstrip and lumped element substrates.

   Its $\epsilon_r = 9.3 - 11.7$

   $D = 1.0 \times 10^4$

   $K = 0.4$ W/cm^2 °C

3. Beryllia is used as compound substrate.

   Its $\epsilon_r = 6.0$

   $D = 1.0 \times 10^{+4}$

   $K = 2.5$ W/cm^2 °C

4. Rutile is used in microstrip substrates

   Its $\epsilon_r = 100$

   $D = 4.0 \times 10^{+4}$

   $K = 0.02$ W/cm^2 °C

5. GaAs is used for high frequency applications

   Its $\epsilon_r = 13.0$

   $D = 16 \times 10^{+4}$

   $K = 0.03$ W/cm^2 °C

6. Ferrite/garnet is used as compound substrates in microstrip and coplanar circuits.

   Its $\epsilon_r = 13 - 16$

   $D = 2 \times 10^{+4}$

   $K = 0.03$ W/cm^2 °C

7. Glass is used in lumped element circuits.

   Its $\epsilon_r = 4.0$

   $D = 5 \times 10^{+4}$

   $K = 0.01$ W/cm^2 °C

8. Silicon is used in high resistivity circuits

   Its $\epsilon_r = 12$

   $D = 10 - 100 \times 10^{+4}$

   $K = 0.9$ W/cm^2 °C

9. Quartz is used in very low dissipation circuits

   Its $\epsilon_r = 3.8$

   $D = 1 \times 10^{+4}$

10. Ceramic filled resin is used in the circuits where some dissipation can be tolerated.

    Its $\epsilon_r = 1.7 - 2.5$

    $D = 5 - 50 \times 10^{+4}$

7.16 EXAMPLES OF METALS (CONDUCTOR MATERIALS)

The common metals used in MICs are Ag, Cu, Au, Al, W, Mo, Cr, Ta, Ti, Pt, Pd etc.

7.17 IDEAL CHARACTERISTICS OF METALS

These are

- Low temperature coefficient of resistance.
- High conductivity
- Good adhesiveness to the substrate
- Good solderability and etchability
- Ease of deposition and electroplating.

7.18 APPLICATIONS OF METALS

The metals are commonly used for MIC transmission lines and lumped elements etc.

7.19 PROPERTIES OF COMMON METALS

The properties of conductors are described in terms of skin depth (δ), surface resistivity (ρ_s), coefficient of thermal expansion (α_t), method of deposition and adherence to dielectric film of substrate.

1. Silver (Ag)

   The adherence property of Ag is poor. It is deposited by evaporation and screening.

   δ at $f_{\mathrm{GHz}} = 1.4$ μm

   $\rho_s = 2.5 \times 10^{-7} \sqrt{f}$ Ohm/sq

   $\alpha_t = 21 \times 10^6$ /°C

2. Copper (Cu)

   The adherence property of Cu is very poor. It is deposited by evaporation and plating.

   δ at f_{GHz} = 1.5 μm

   $$\rho_s = 2.6 \times 10^{-7}\sqrt{f} \ \ \text{Ohm/sq}$$

   $$\alpha_t = 18 \times 10^6 \ /°C$$

3. Gold (Au)

   The adherence property of Au is very poor and it is deposited by evaporation and plating.

   δ at f_{GHz} = 1.7 μm

   $$\rho_s = 3.0 \times 10^{-7}\sqrt{f} \ \ \text{Ohm/sq}$$

   $$\alpha_t = 15 \times 10^6 \ /°C$$

4. Aluminium (Aℓ)

   The adherence property of Aℓ is very good and it is deposited by evaporation

   δ at f_{GHz} = 1.9 μm

   $$\rho_s = 3.3 \times 10^{-7}\sqrt{f} \ \ \text{Ohm/sq}$$

   $$\alpha_t = 26 \times 10^6 \ /°C$$

5. Tungsten (W)

   The adherence property of W is good and it is deposited by sputtering and evaporation.

   δ at f_{MHz} = 2.6 μm

   $$\rho_s = 4.7 \times 10^{-7}\sqrt{f} \ \ \text{Ohm/sq}$$

   $$\alpha_t = 4.6 \times 10^6 \ /°C.$$

6. Molybdenum (Mo)

   The adherence property of Mo is good and it is deposited by electron beam evaporation.

   δ at f_{GHz} = 2.7 μm

   $$\rho_s = 4.7 \times 10^{-7}\sqrt{f} \ \ \text{Ohm/sq}$$

   $$\alpha_t = 6.0 \times 10^6 \ /°C$$

7. Tantalum (Ta)

   The adherence property of Ta is very good. It is deposited by electron beam sputtering.

   δ at f_{GHz} = 4.0 μm

   $$\rho_s = 7.2 \times 10^{-7}\sqrt{f} \ \ \text{Ohm/sq}$$

   $$\alpha_t = 6.6 \times 10^6 \ /°C$$

8. Cromium (Cr)

The adherence property of Cr is good and it is deposited by evaporation

δ at $f_{GHz} = 2.7$ μm

$$\rho_s = 4.7 \times 10^{-7} \sqrt{f} \ \ \text{Ohm/sq}$$

$$\alpha_t = 9.0 \times 10^6 \ /°C$$

7.20 EXAMPLES OF DIELECTRIC MATERIALS USED IN MICs

SiO, SiO_2, Si_3N_4, Al_2O_3 and Ta_2O_5 etc.

7.21 APPLICATIONS

The dielectric materials are used in MMICs for fabricating capacitors, blockers and couple line structures.

The properties of dielectric materials are

- breakdown value (> 200 V)
- Reproducible
- Low radio frequency dielectric loss
- Capable of undergoing processes without developing pin holes
- Compatibility

The dielectric materials are deposited by the following methods.

- Evaporation
- Deposition
- Vapor phase
- Sputtering
- Anodization
- Evaporation and
- Sputtering

7.22 EXAMPLES OF RESISTIVE MATERIALS USED IN MICs

Cr, NiCr, Ta, Cr-SiO and Ti etc. They are deposited either by evaporation or sputtering.

Applications

They are used MMICs for bias networks, attenuators and terminations etc.

7.23 PROPERTIES

The required properties are

- High stability
- Good dissipation capability

- Low temperature coefficient of resistance
- Sheet resistivity in the range of 10-1000 Ω/sq.

7.24 METHODS OF MMIC FABRICATION

The Classification of Fabrication Techniques of MMICs

- Diffusion ion implementation
- Oxidation deposition
- Epitaxial growth
- Lithography
- Etching photoresist
- Deposition

Diffusion

This is used to control the amount of dopants in semiconductors. The diffusion of impurities in pure semiconductor alters the properties. It is possible to form n or p type semiconductors by this method. This method is used in both discrete and integrated devices fabrication.

Ion Implementation

Like diffusion, this is also used to control the amount of dopants. In this, the substrate crystal is doped with high energy ion impurities.

It is used to form n or p type semiconductors. The doping by ion implementation is in wide use as it has advantages like precise control of doping materials, better reproductivity with reduced processing temperature.

Oxidation

It is possible to fabricate integrated circuits for making different devices, using the methods of oxidation and deposition. The commonly used thin films are:

- Thermal oxides
- Metal film
- Dielectric film
- Polycrystalline silicon

Epitaxial Growth

In Greek, *epi* means on and *taxi* means arrangement. In epitaxial growth, doping is well controlled. This method consists of 3 types. They are vapor-phase liquid phase and molecular-beam epitaxis.

The first one vapor phase epitaxy is useful for both silicon and GaH devices. The second type, molecular beam epitaxy is useful to control doping as well as chemical composition.

The third type, liquid-phase epitaxy, is useful to grow thin layers and multilayered structures. It controls doping as well chemical composition.

Deposition

The deposition is made by vacuum, electron-beam evaporation techniques. The vacuum deposition is also called as d.c. sputtering.

Lithography

The different types of lithography are electron beam, ion beam, optical and *x*-ray lithographies.

In lithography, the patterns of geometric shapes are transferred to a resist. The resist is a radiation sensitive material. It represents the replicas of a circuit.

Photo Etching

The etching is done to remove specified parts of SiO_2 to create openings. The impurities are diffused through the created openings.

7.25 STEPS INVOLVED IN FABRICATION

- Deposition
- Photographic mask
- Etching

In the first step of fabrication, the deposition of an oxide layer is made on the semiconductor material followed by the deposition of a photo resist layer. The photo resist layer covers the oxide on the semiconductor.

In the second step, photographic masking to the photo resist is done for shining by the application of ultraviolet light.

In the third step, hydraulic acid is applied to etch out undesired oxide regions. Finally, an organic solvent is used to dissolve photo resist in the oxide to keep the required openings.

7.26 TRANSMISSION LINES IN MICs

The transmission lines are of the following types:

- Microstrip line
- Slot line
- Coplanar waveguide
- Suspended substrate transmission line

The microstrip line is shown in fig. 7.1.

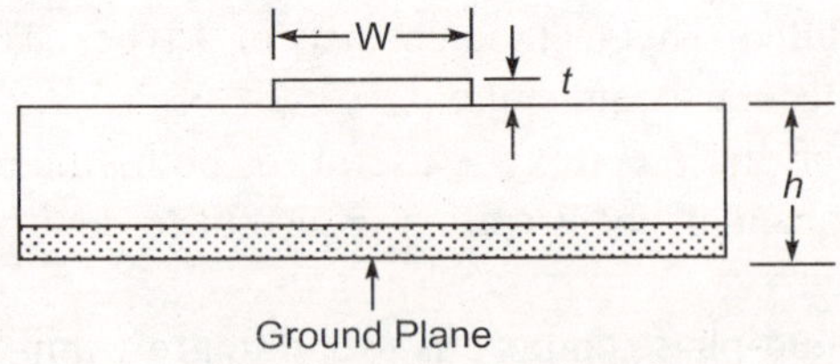

Fig. 7.1 Microstrip line

The slot line is shown in fig. 7.2.

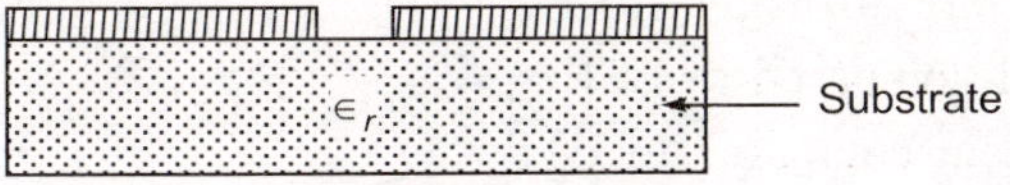

Fig. 7.2   Slot line

The glass-reinforced Teflon, ceramic filled resins, 99 percent alumina, quartz, 99 percent Beryllia are used as substrates in microstrip lines.

The characteristic impedance of a microstrip line is given by

$$Z_0 = \sqrt{\frac{\mu_0}{\epsilon_0\,\epsilon_r}}\,\frac{h}{w}$$

or

$$Z_0 = \frac{377}{\sqrt{\epsilon_r}}\,\frac{h}{w}\ \text{ohm}$$

7.27   FABRICATION OF MOSFET IN MMICs

The methods of fabrication has the following steps.

- Oxidation
- Diffusion
- Photo etching
- Oxidation
- Deposition
- Aluminum metallization
- Etching

The method of oxidation is used to create silicon dioxide layer on a substrate (say p type).

The diffusion is used to create n^+ - layer through windows. The windows are formed by photo resist method. Photo etching is used to remove some portion of oxide layer. Now oxidation is used again to cover the top surface by silicon dioxide.

The deposition technique is again used to deposit phosphorous glass to cover the oxide layer. Photo etching is reused to create two windows on the two n^+ - type regions.

Aluminum metallization is made on the surface of the device. Now etching is made to remove unwanted metal.

Finally, metal contacts are made to form source, drain and gate.

7.28 ADVANTAGES OF MOSFET IN MMICs

1. It requires one diffusion process.
2. Economical compared to BJT.
3. More efficient compared to BJT.
4. It is the best suited for VLSI's to be used in microprocessors and memories.

7.29 DISADVANTAGES OF BJT IN MMICs

- Costlier than MOSFET fabrication.
- Less efficient.
- Requires 3 diffusion process.

7.30 FABRICATION OF CMOS

CMOS stands for Complementary Metal Oxide Semiconductor Field Effect Transistor.

CMOS is made of NMOS or PMOS.

NMOS means n-channel MOS

PMOS means p-channel MOS

CMOS is characterized by high speed and low power.

NMOS is preferred to PMOS as the mobility of electrons is higher than that of holes.

7.31 STEPS INVOLVED IN THE FABRICATION OF CMOS

1. Epitaxy
2. Deposition
3. Implantation
4. Oxidation
5. Deposition
6. Implantation
7. Deposition
8. Metallization

In the 1st step, a semiconductor is lightly doped over a heavily doped n^+ substrate material.

In step 2, a layer of an oxide and nitride is formed. The silicon is then exposed over an n-region. The phosphorous is implanted and masked by nitride layer.

In the third step, the wafers are oxidized over n regions and the nitride is stripped. Now boron is implanted for p-tub.

In the fourth step, namely, oxidation, the boron is allowed to enter the silicon and it is masked by SiO_2 layer. By stripping the oxides, the tubs are made.

In the 5th step, n^+ polysilicon is deposited to create drain and source.

In the 6th step, phosphorous is implanted into source and drain with high concentration.

In the 7th step, a phosphorous layer is deposited.

7.32 NMOS FABRICATION

This consists of the following steps.

1. Deposition of silicon nitride layer Si_3N_4 on oxide surface is made p-type substrate is lightly doped and SiO_2 layer is grown in it.
2. Etching of field areas. This is done by either plasma etching or reactive ion etching.
3. Implantation of boron ion is done subsequently, an n-channel is implanted to create depletion device.
4. The deposition of polysilicon is done.
5. Metallization of films is done by evaporation.
6. Etching is carried out to form electrode contacts.

7.33 APPLICATIONS OF MMICs

They are used in

- Satellite systems
- Receivers and transmitters for communications
- For making phased array antennas
- Varactor diode
- Schottky diode
- PIN diode
- HBT
- MESFET
- HEMT
- PHEMT

7.34 FABRICATION OF PASSIVE COMPONENTS

The fabrication of passive components, resistors, inductors and capacitors are fabricated using this technology.

7.35 DESIGN OF PLANAR RESISTORS

It involves the following factors for the design of the components:

- Bandwidth
- Resistivity
- Thermal resistance of the load
- Temperature coefficient

The planar resistor films are made of aluminum, copper, nicrome, titanium, tantalum etc. The resistive film is deposited on a substrate which is basically an insulator.

7.36 APPLICATION OF PLANAR RESISTORS

They are basically used as terminations. They are used in

- Hybrid amplifiers
- Power diodes
- Bias circuits
- Current controllers

7.37 RESISTANCE OF PLANAR RESISTORS

It is given by

$$R_p = \frac{\ell \rho_s}{wd} = \frac{\ell}{wd\sigma_s}$$

Here, R_p = Resistance of planar resistor
ℓ = resistive film length (m)
ρ_s = resistivity of film (Ω-m)
w = film width (m)
d = film thickness (m)
σ_s = conductivity, (mho/m)

PROBLEM 7.1   Find the resistance of the aluminum planar resistor which has the following specifications

$$\sigma_s = 4.10 \times 10^7 \text{ (mho/m)}$$
$$\ell = 10 \text{ mm}$$
$$w = 5 \text{ mm}$$
$$d = 0.2 \text{ }\mu\text{m}$$

Solution
$$R_p = \frac{l}{Wd\,\sigma s}$$

$$= \frac{10 \times 10^{-3}}{5 \times 10^{-3} \times 0.2 \times 10^{-6} \times 3.82 \times 10^7}$$

$\therefore$ $$\boxed{R_p = 0.261 \text{ }\Omega}$$

7.38 DESIGN OF PLANAR INDUCTOR

Different types of geometrics of planar inductors are given by

- Rounded-wire inductors
- Flat or ribbon inductors
- Circular spiral inductor
- Circular-loop inductor
- Square spiral inductor

7.39 EXPRESSIONS FOR INDUCTANCE OF DIFFERENT GEOMETRIES

Wire Inductor

Its inductance is given by

$$L_W = 5.08\,\ell\left[\ell n\left(\frac{\ell}{d}\right) + 0.386\right] \text{ pH/mil}$$

Here, ℓ = length of inductor wire (mils)

d = diameter of wire (mils)

Flat Inductor

These are also called ribbon inductors. The expression for its inductance is

$$L_F = 5.08 \times \ell\left[\ell n\left(\frac{\ell}{n}\right) + 0.022\left(\frac{h}{\ell}\right) + 1.19\right] \text{ pH/mil}$$

Here, $h = w + t$

W = width of ribbon (mils)

t = thickness of ribbon (mils)

ℓ = length of ribbon (mils)

Circular Loop Inductor

The expression for the inductance of circular loop inductor is

$$L_{CL} = 5.08 \times \ell\left[\ell n\left(\frac{t}{w + t}\right) - 1.76\right] \text{ PH/mil}$$

Here, $h = w + t$

w = width (mils)

t = thickness (mils)

ℓ = length (mils)

Square Spiral Inductor

The expression for the inductance of spiral inductor of square shape is

$$L_{ss} = 8.5\sqrt{A_s}\ \ N^{5/3} \times 10^3\ \text{PH}$$

Here, A_s = surface area, cm^2

N = Number of turns

Circular Spiral Inductor

The expression for the inductance of a spiral inductor of circular shape is

$$L_{cs} = 31.25\ N^2 D\ \text{PH/mil}$$

Here, N = No. of turns

$D = 5d$

$= 2.5\ N\ (w + s)$

w = width of film, mils

s = separation of turns, mils

PROBLEM 7.2   Round-wire inductor has the following dimensions

$$\ell = 100\ \text{mils}$$
$$d = 10\ \text{mils}$$

Find its inductance.

Solution

$$L_w = 5.08\ell\left[\ell n\left(\frac{\ell}{d}\right) + 0.386\right]\ \text{PH/mil}$$

$$= 5.08 \times 100\left[\ell n\left(\frac{100}{10}\right) + 0.386\right]$$

$$= 5.08 \times 100\,[2.302 + 0.386]$$

$$\doteq 13.66 \times 100\ \text{PH/mil}$$

$$\therefore \quad \boxed{L = 1.366\ \text{nH/mil}}$$

PROBLEM 7.3   The dimensions and parameters of an aluminum planar resistor are:

$$\text{Length} = 11\ \text{mm}$$
$$\text{Thickness} = 0.2\ \mu\text{m}$$
$$\text{Width} = 8\ \text{mm}$$
$$\sigma_s = 3.82 \times 10^7\ \text{mho/m}$$

Determine its resistance.

Solution

$$R_p = \frac{\ell}{wd\,\sigma_s}$$

$$\ell = 11\ \text{mm} = 11 \times 10^{-3}\ \text{m}$$

$$d = 0.2 \ \mu m = 0.2 \times 10^{-6} \ m$$
$$w = 8 \ mm = 8 \times 10^{-3}$$
$$\sigma_s = 3.82 \times 10^7 \ mho/m$$

$$\therefore \quad R_p = \frac{11 \times 10^{-3}}{0.2 \times 10^{-6} \times 8 \times 10^{-3} \times 3.82 \times 10^7}$$

$$= \frac{11}{6.112 \times 10}$$

$$\therefore \quad \boxed{R_p = 0.18 \ \Omega}$$

PROBLEM 7.4   The dimensions of parameters of a gold planar resistor are

$$\ell = 11 \ mm$$
$$d = 0.2 \ \mu m$$
$$w = 8 \ mm$$
$$\sigma_s = 4.10 \times 10^7 \ mho/m$$

Find its resistance.

Solution
$$R_p = \frac{\ell}{wd \, \sigma_s}$$

$$= \frac{11 \times 10^{-3}}{8 \times 10^{-3} \times 0.2 \times 10^{-6} \times 4.10 \times 10^7}$$

$$= \frac{11}{65.6}$$

$$\therefore \quad \boxed{R_p = 0.168 \ \Omega}$$

PROBLEM 7.5   A silver planar resistor has the following dimensions and parameters

$$\ell = 11 \ mm$$
$$d = 0.2 \ \mu m$$
$$w = 8 \ mm$$
$$\sigma_s = 6.17 \times 10^7 \ mho/m$$

Find its resistance.

Solution
$$R_p = \frac{\ell}{wd \, \sigma_s}$$

$$= \frac{11}{8 \times 10^{-3} \times 0.2 \times 10^{-6} \times 6.17 \times 10^7}$$

$$= \frac{11}{98.72}$$

$$\therefore \quad \boxed{R_p = 0.111 \ \Omega}$$

PROBLEM 7.6   A square spiral planar inductor has the dimensions.

$$\text{Surface area} = 0.4 \ \text{cm}^2$$
$$\text{No. of turns} = 4$$

Find its inductance.

Solution

$$A = 0.04 \ \text{cm}^2$$
$$N = 4$$

$$\therefore \quad L_{ss} = 8.5 \times A_s^{\frac{1}{2}} N^{\frac{1}{2}} \times 10^3 \ \text{PH}$$

$$= 8.5 \times (0.04)^{\frac{1}{2}} \times 10^3 \times 4^{\frac{5}{3}}$$
$$= 8.5 \times 0.2 \times 10^3 \times 10.07$$

$$\therefore \quad \boxed{L_{ss} = 17.12 \ \text{nH}}$$

PROBLEM 7.7   A ribbon planar inductor has the following dimensions.

$$\text{Length} = 10 \ \text{mils}$$
$$\text{Thickness} = 0.2 \ \text{mils}$$
$$\text{Width} = 8 \ \text{mils}$$

Find its inductance.

Solution

$$\ell = 10 \ \text{mils}$$
$$t = 0.2 \ \text{mils}$$
$$w = 8 \ \text{mils}$$

$$L_r = 5.08 \times \ell \left[\ell n \left(\frac{\ell}{w+t} \right) + 0.222 \left(\frac{w+t}{\ell} \right) + 1.19 \right]$$

$$\left(\frac{\ell}{w+t} \right) = \frac{10}{8 + 0.2 \times 10^{-3}} = 1.25$$

$$\ell n \left(\frac{\ell}{w+t} \right) = 0.223$$

$$L_r = 5.08 \times 10[0.223 + 1.19 + 0.176]$$

$$\therefore \quad \boxed{L_r = 80.7 \ \text{PH/mil}}$$

7.40   PLANAR CAPACITORS

These capacitors are of two types:

- Metal-oxide-metal capacitor
- Interdigitated capacitor

7.40.1   Metal-Oxide-Metal Capacitor

In this a substrate is sandwiched between two metal layers. The substrate is made of dielectric material. The structure is shown in fig. 7.3.

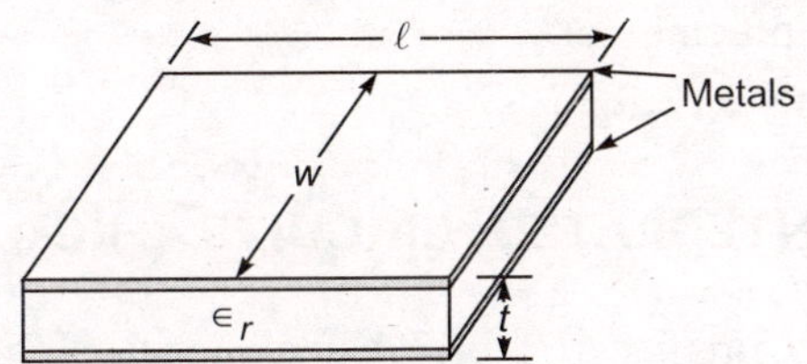

Fig. 7.3   Structure of metal-oxide-metal capacitor

The expression for its capacitance is give by

$$C_{\mathrm{MOM}} = \frac{\in \ell w}{t}, \text{ Farads}$$

Here, $\in = \in_0 \in_r$

$\in_r$ = relative permittivity of substrate

$\in_0 = 8.854 \times 10^{-12}$ F/m

w = width of metal layer

ℓ = length of metal layer.

t = height of substrate

7.40.2   Interdigitated Capacitor

It is shown in fig. 7.4.

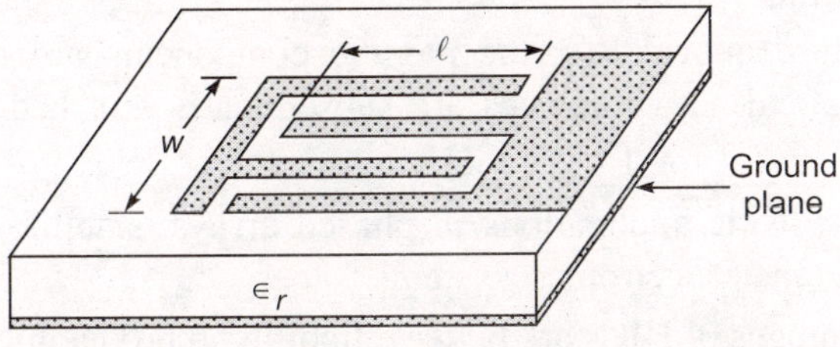

Fig. 7.4   Structure of interdigitated capacitor

It consists of a ground plane and substrate. The expression for its capacitance is given by

$$C_{IS} = \left(\frac{\epsilon_r + 1}{w}\right) \ell[(N - 3)k_1 + k_2]$$

Here, ϵ_r = relative permittivity of substrate

N = Number of projections

w = projection base width, cm

ℓ = projection length, cm

k_1 = 0.089 pF/cm   for $h > w$

k_2 = 0.10 pF/cm   for $h < w$

7.41   HYBRID INTEGRATED CIRCUITS (HICs)

The hybrid integrated circuits are fabricated in integrated form by forming active and passive devices and elements on a substrate. The substrate is made of dielectric.

7.41.1   Salient Features of HICs

➤ In this passive and active devices are integrated.

➤ The passive devices include lumped and distributed elements.

➤ HICs are two types. Normal hybrid IC and miniature hybrid IC.

➤ In normal hybrid IC, a single layer metallization technique is used to create components on the substrates.

➤ The metallization technique is again divided in two techniques, namely, plate-through and etchback.

➤ In plate-through technique, a substrate is coated with a thin evaporated metal layer. When this photoresist pattern is formed subsequently, second metal layer is created. This layer has a specified thickness. At the end of this process, the photoresist layer is removed with the etching of metal.

➤ In etchback technique, thin metal layer is used. In this, an unwanted metal is etched out.

➤ In miniature hybrid ICs, the passive components are deposited on the substrates and active devices are subsequently attached to the chip.

➤ HICs are useful for 1 – 20 GHz.

➤ HICs have wide applications in phased arrays, satellite communications and in defence electronics systems.

➤ The advantages of HICs are better reliability, reproducibility, small in size, economical, better performance etc.

7.41.2 Applications of HICs

They are used in

> - phased array radars
> - satellite communications
> - defence electronic systems
> - ECM
> - ECCM etc.

7.41.3 Advantages of HICs

> - Small size
> - Low cost
> - Small spaces
> - Better performance
> - High reliability
> - Greater reproductibility

PROBLEM 7.8   Determine the resistance of planar resistor when its aluminium film is characterized by

$$\text{Length} = 8 \text{ mm}$$
$$\text{Width} = 8 \text{ mm}$$
$$\text{Thickness} = 0.1 \ \mu\text{m}$$
$$\text{Resistivity} = 0.262 \times 10^{-7} \ \Omega - \text{m}$$

Solution   The resistance of planar resistor is given by

$$R_p = \frac{\ell}{wt\sigma_s}$$

Here, $\ell = 8 \text{ mm} = 8 \times 10^{-3} \text{ m}$

$w = 8 \text{ mm} = 8 \times 10^{-3} \text{ m}$

$t = 0.1 \times 10^{-6} \text{ m}$

$\sigma_s = 0.26 \times 10^{-7} \text{ mho/m}$

$\therefore \qquad \sigma_s = 3.82 \times 10^{-7} \text{ mho/m}$

$$\therefore \qquad R_p = \frac{8 \times 10^{-3}}{8 \times 10^{-3} \times 0.1 \text{ kt } 10^{-6} \times 3.82 \times 10^{7}}$$

$$= 2.62 \times 10^{-1}$$

$$\boxed{R_p = 0.262 \ \Omega}$$

PROBLEM 7.9   The planar resistor is made of silver film. The film has the following dimensions.

$$\text{Length} = 15 \text{ mm}$$
$$\text{Width} = 15 \text{ mm}$$
$$\text{Thickness} = 0.1 \ \mu\text{m}$$
$$\text{The conductivity of silver} = 6.17 \times 10^7 \text{ mho/m}$$

Determine its resistance per square.

Solution   The resistance of planar resistor is given by

$$R_p = \frac{\ell}{wt\sigma_s}$$

Here, $\ell = 8 \text{ mm} = 8 \times 10^{-3} \text{ m}$
$w = 8 \text{ mm} = 8 \times 10^{-3} \text{ m}$
$t = 0.1 \times 10^{-6} \text{ m}$
$\sigma_s = 0.26 \times 10^{-7} \text{ mho/m}$

$$R_p = \frac{15 \times 10^{-3}}{15 \times 10^{-3} \times 0.1 \times 10^{-6} \times 6.17 \times 10^7}$$

$$\therefore \quad \boxed{R_p = 0.62 \ \Omega/\text{square}}$$

PROBLEM 7.10   A planar resistor made of gold film is in rectangular shape with length × width dimensions of 12×10 mm and with a thickness of $0.12 \ \mu$m. Find its planar resistance.

Solution   Resistance of planar resistor,

$$R_p = \frac{\ell}{wt\sigma_s}$$
$$\ell = 12 \text{ mm} = 12 \times 10^{-3}$$
$$w = 10 \text{ mm} = 10 \times 10^{-3}$$
$$t = 0.12 \ \mu\text{m} = 0.12 \times 10^{-6} \text{ m}$$
$$\sigma_s \text{ of gold} = 4.10 \times 10^7 \text{ mho/m}$$

$$\therefore \quad R_p = \frac{12 \times 10^{-3}}{12 \times 10^{-3} \times 0.12 \times 10^{-6} \times 4.10 \times 10^7}$$

$$\therefore \quad \boxed{R_p = 0.2439 \ \Omega}$$

PROBLEM 7.11   A planar resistor is made of copper film. It has the following dimensions.
$$\ell = 20 \text{ mm}$$
$$w = 10 \text{ mm}$$

$$t = 15 \ \mu m$$
$$\sigma_s = 5.8 \times 10^7 \ \text{mho/m}$$

Solution
$$R_p = \frac{\ell}{wt\sigma_s}$$

$$= \frac{20 \times 10^{-3}}{10 \times 10^{-3} \times 15 \times 10^{-6} \times 5.8 \times 10^7}$$

$$= \frac{20}{100 \times 15 \times 5.8}$$

$$\therefore \qquad \boxed{R_p = 0.23 \ \Omega}$$

PROBLEM 7.12   Design a planar resistor made of gold film to have a resistance of 0.30 Ω.

Solution   The resistor is made of gold.

Its $\sigma_s = 4.7 \times 10^7$ mho/m

The required resistance = 0.30 Ω

Assume its thickness, $t = 0.1 \ \mu m$

Its length, $\ell = 30$ mm

$$\therefore \qquad R_p = \frac{\ell}{wt\sigma_s}$$

$$0.3 = \frac{30 \times 10^{-3}}{w \times 0.1 \times 10^{-6} \times 4.1 \times 10^7}$$

or
$$w = \frac{30 \times 10^{-3}}{0.3 \times 0.1 \times 10^{-6} \times 4.1 \times 10^7}$$

$$= \frac{10^{-2}}{0.41} = 24 \ \text{mm}$$

$$\therefore \qquad \boxed{w = 24 \ \text{mm}}$$

The designed parameters of planar resistor are

$$\boxed{\begin{array}{c} \ell = 30 \ \text{mm} \\ w = 24 \ \text{mm} \\ t = 0.1 \ \mu m \end{array}}$$

PROBLEM 7.13   Design a planar resistor made of silver film to have a resistance of 0.15 Ω.

Solution
$$R_p = 0.15 \ \Omega$$

$$\text{For silver, } \sigma_s = 6.17 \times 10^7$$
$$\text{Assume } w = 10 \text{ mm}$$
$$t = 0.08 \ \mu\text{m}$$

$$R_p = \frac{\ell}{wt\sigma_s}$$

$$\therefore \quad \ell = R_p \times w \times t \times \sigma_s$$
$$= 0.15 \times 10 \times 10^{-3} \times 0.08 \times 10^{-6} \times 6.17 \times 10^7$$
$$= 15 \times 0.05 \times 6.17 \times 10^{-3}$$

$$\therefore \quad \boxed{\ell = 7.404 \text{ mm}}$$

$\therefore$ The designed parameters are

$$\boxed{\begin{aligned} \ell &= 7.404 \text{ mm} \\ w &= 10 \text{ mm} \\ t &= 0.08 \ \mu\text{m} \end{aligned}}$$

PROBLEM 7.14   Determine the inductance of a circular spiral inductor if it has

$$\text{Number of turns} = 10$$
$$\text{Separation} = 50 \text{ mils}$$
$$\text{Film width} = 20 \text{ mils}$$

Solution

$$L = 31.25 \ N^2 d \text{ pH/mil}$$
$$d = 2.5 \ N \ (w + s) \text{ mil}$$
$$= 2.5 \times 10 \times (50 + 20)$$
$$= 1750$$

$$\therefore \quad L = 31.25 \times 100 \times 1750$$
$$= 551 \times 10^4 \times 10^{-12}$$

$$\therefore \quad \boxed{L = 5510 \text{ nH/mil}}$$

7.42   POINTS TO REMEMBER

- MIC contains both passive and active elements.
- Active elements are silicon planar chips.
- The passive elements are made thin and thick films.
- Some substrates are ceramic, glass and silicon dioxide.
- Thick film means a film having several thousand angstroms thick.
- VLSI circuit has more than 1 million components.
- Types of electronic circuits are discrete circuits, ICs, MMICs, HICs.
- Hybrid integrated circuits consists of ICs and discrete elements.
- Advantages of MMICs are low cost, small size, light weight, high reliability, reproductivity with high performance.

- ➢ MMIC materials are substrates, conductors, dielectrics and resistive materials.
- ➢ Substrates have high dielectric constants.
- ➢ Conductors have high conductivity.
- ➢ Dielectrics withstand high voltage.
- ➢ Resistive materials have low temperature coefficient of resistivity.
- ➢ MMIC fabrication consists of oxidation, film deposition, epitaxial growth, lithography and deposition.
- ➢ In MMICs, MOSFET formation, NMOS and CMOS are developed.
- ➢ In planar resistors thin resistive film is deposited on insulating substrates.
- ➢ Planar resistors are useful for terminations for hybrid computers, bias-circuits.
- ➢ Design of planar resistors takes care of sheet resistivity, thermal stability, thermal resistance, bandwidth etc.
- ➢ The resistance of planar resistor is $R = \dfrac{\ell}{wt\sigma_s}$.
- ➢ Planar inductor films have shapes like circular spiral, round wire, flat, square spiral etc.
- ➢ Planar capacitors are two types, namely, metal-oxide-metal capacitors and interdigitated capacitors.
- ➢ Metal-oxide-metal capacitor has a capacitance given by $C = \dfrac{\in \ell W}{h}$.
- ➢ HICs are used over a frequency range of 1 – 20 GHz.
- ➢ HICs are used in satellite communication and phased arrays.
- ➢ HICs are two types, namely, hybrid ICs and miniature hybrid ICs.
- ➢ HIC fabrication consists of etchback and plate through techniques.

7.43 MULTIPLE CHOICE QUESTIONS

1. MMIC means
    (a) miniature microwave integrated circuits
    (b) millimeter miniature integrated circuit
    (c) Monolithic microwave integrated circuit
    (d) millimeter integrated circuit
2. VLSI circuits have more number of components compared to
    (a) MSI only   (b) SSI only
    (c) LSI only   (d) MSI, SSI, LSI
3. A film integrated circuit is
    (a) an MMIC   (b) HIC
    (c) not useful at microwave frequencies   (d) discrete circuit

4. Substrate material in MMIC is
   (a) glass
   (b) Cu
   (c) gold
   (d) SiO

5. Conductor material in MMIC is
   (a) Aluminia
   (b) Ag
   (c) GaAs
   (d) SiO

6. The property of dielectric material in MMIC is
   (a) reproductivity
   (b) low resistivity
   (c) low temperature coefficient
   (d) high *RF* dielectric loss

7. LPE means
   (a) low profile epitaxy
   (b) liquid phase epitaxy
   (c) light profile epitaxy
   (d) light phase epitaxy

8. Capacitance of metal-oxide-metal capacitor

   (a) $\in \dfrac{\ell W}{h}$
   (b) $\dfrac{\in W}{\ell h}$

   (c) $\in_0 \dfrac{\ell w}{h}$
   (d) $\dfrac{\in_0 hw}{\ell}$

9. The resistance of a planar resistor is

   (a) $\dfrac{wt}{\ell \rho_s}$
   (b) $\dfrac{\ell \rho_s}{wt}$

   (c) $\dfrac{\ell w}{\rho_s t}$
   (d) $\dfrac{\ell w}{\rho_s t}$

10. DC sputtering is nothing but
    (a) vacuum deposition
    (b) evaporation
    (c) vacuum evaporation
    (d) etching

11. Lithography is
    (a) the process of deposition
    (b) the process of evaporation
    (c) the process of transferring patterns of geometric shapes
    (d) the process of etching

12. Diffusion is
    (a) adding dopants
    (b) the same as evaporation
    (c) epitaxial growth
    (d) a method of lithography

7.44   ANSWERS

1. c	2. d	3. a
4. a	5. b	6. a

| 7. b | 8. a | 9. b |
| 10. a | 11. c | 12. a |

7.45 EXERCISE PROBLEMS

1. Design a circular spiral inductor to have an inductance of 5 μH/mil.
2. Design a square spiral inductor to give an inductance of 10 μH/mil.
3. Design the inductance of ribbon inductor to give an inductance of 15 μH/mil.
4. Design a round wire inductor to produce an inductance of 8 μH/mil.
5. Find the capacitance of planar resistor made of GaAs substrate if it has the following

   the height of substrate = 3 mm

   finger length = 0.03 mm

   finger width = 0.6 mm

   ϵ_r = 13.10

   number of finger = 10
6. Design a planar capacitor to exhibit a capacitance of 0.5 pF/cm.

Chapter 8

Microwave Antennas

Antenna is an essential device in all wireless communications and radars.

8.1 INTRODUCTION

Antenna is a source and a sensor of microwaves. The size of the antenna and frequency of operation are inversely proportional to each other.

Antennas are essential devices in all types of communication and radar systems. Without an antenna of one type or the other, there is no communication and there is no radar system. These devices are essential both in the transmitting and receiving sides.

Antenna is a source as well as a sensor of microwaves. An antenna is basically a transducer. It converts *RF* electrical current into an electromagnetic wave of the same frequency. It also can be used as a temperature sensor. It is characterized by the following properties.

1. Its impedance is the same when used for transmitting and receiving purposes. This property is called equality of impedances.
2. Its directional characteristics / patterns are the same when it is used for transmitting and receiving purposes. This property is called equality of directional patterns.
3. Its effective length is the same when it is used for transmitting and receiving purposes. This property is called equality of effective lengths.

The microwave communication and radar systems do not exist without one type of antenna or the other.

The size of the antenna mainly depends on the frequency of operation. If the frequency is low, the size of the antenna becomes large and vice-versa.

The antennas at microwave frequencies are small in size and they exhibit better radiation characteristics. In the present-day miniaturized world small is smart. The smaller the antennas in communication, more and more is the attraction.

The antennas are used in one or more frequency bands. Different popular and important microwave antennas and their characteristics are presented in this chapter.

The microwave region extends from 1 GHz to 100 GHz. The transmitting and receiving antennas in microwave frequencies are directive with high gain and narrow beam width in both vertical and horizontal planes.

The antennas are described by the following parameters.

8.2 ANTENNA PARAMETERS

- Directional Characteristics
- Antenna impedance
- Effective length
- Radiation intensity
- Directive gain
- Directivity
- Power gain
- Efficiency
- Effective area
- Admittance characteristics
- Bandwidth
- Front-to-back ratio
- Polarization
- Main lobe of radiation pattern
- Sidelobes of the radiation pattern

8.3 DIRECTIONAL CHARACTERISTICS

These are known as radiation patterns. An antenna radiation pattern is a three dimensional variation of radiation field. It is a pattern drawn as a function of θ and ϕ. The pattern consists of a main lobe and a number of side lobes. The patterns are presented in two forms, namely, field strength and power patterns. Field pattern is the variation of absolute value of field strength as a function of θ. Power strength pattern is the variation of radiated power with θ. A typical 3-dimensional radiation pattern is shown in fig. 8.1.

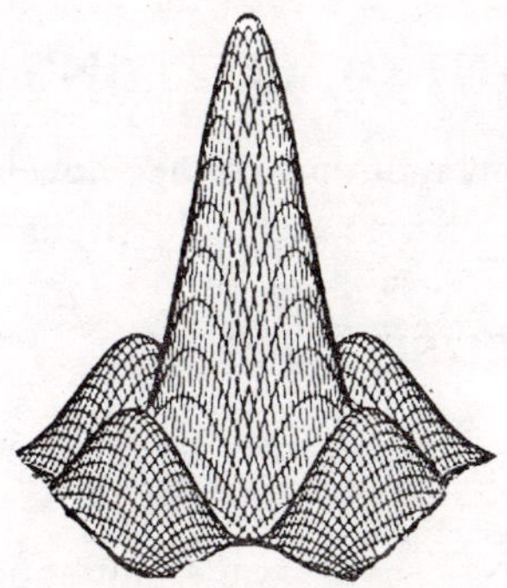

Fig. 8.1 Typical 3-D radiation patterns

8.4 ANTENNA IMPEDANCE, Z_a

It is defined as the ratio of input voltage to input current i.e.

$$Z_a \equiv \frac{V_i}{I_i}\ \Omega$$

Z_a is a complex quantity and it is written as

$$Z_a = R_a + j\, X_a$$

Here, the reactive part X_a results from fields surrounding the antenna. The resistive part, R_a is given by

$$R_a = R_\ell + R_r$$

Here, R_ℓ represents losses in the antenna. R_r is called radiation resistance.

8.5 RADIATION RESISTANCE, R_r

R_r is defined as the hypothetical resistance that would dissipate an amount of power equal to the radiated power. It is also defined as the ratio of radiated power to the square of the RMS current in the antenna.

8.6 EFFECTIVE LENGTH OF ANTENNA (L_{eff})

It is used to indicate the effectiveness of the antenna as a radiator or receiver of *EM* energy.

8.7 EFFECTIVE LENGTH OF TRANSMITTING ANTENNA

It is equal to the length of an equivalent linear antenna which radiates the same field strength as the actual antenna and the current is constant throughout the length of the linear antenna. Effective length of an antenna is always less than the actual length. That is, $L_{\text{eff}} < L$.

8.8 EFFECTIVE LENGTH OF RECEIVING ANTENNA

It is defined as the ratio of open circuit voltage developed at the terminals of the antenna under the influence of received field strength, E.

8.9 RADIATION INTENSITY (ψ)

It is defined as radiated power in a given direction per unit solid angle. That is,

$$\psi = r^2 P = \frac{r^2 E^2}{\eta_0}\ \text{watts/unit solid angle}$$

Here, η_0 = Intrinsic impedance of the medium (Ω)

r = Radius of the sphere, (m)

P = Power radiated – instantaneous

E = Electric field strength, (V/m)

$\psi = \psi\,(\theta,\,\phi)$ is a function of θ and ϕ.

The unit of a solid angle is steradian (sr)

8.10 DIRECTIVE GAIN g_d

It is defined as the ratio of radiation intensity in a specified direction to the average radiation intensity. That is,

$$g_d \equiv \frac{\psi}{\psi_{av}} = \frac{\psi}{w_r/4\pi}$$

or

$$g_d = \frac{4\pi\,\psi}{w_r}$$

w_r = Radiated power.

8.11 DIRECTIVITY, D

It is defined as the ratio of the maximum radiation intensity to the average radiation intensity. That is,

$$D \equiv (g_d)_{max}$$
$$D \text{ in dB} = 10\,\log_{10}\,(g_d)_{max}$$

Directivity is also defined as maximum directive gain.

8.12 POWER GAIN, g_p

It is defined as the ratio of radiated power to the total input power. That is,

$$g_P \equiv \frac{4\pi\,\psi}{w_t}$$

Here $w_t = w_r + w_\ell$

w_ℓ = Ohmic losses in the antenna

8.13 ANTENNA EFFICIENCY (η)

It is defined as the ratio of radiated power to the input power. That is,

$$\eta_a \equiv \frac{w_r}{w_t} = \frac{w_r}{w_r + w_\ell} = \frac{g_p}{g_d}$$

8.14 EFFECTIVE AREA

It is defined as

$$A_e \equiv \frac{\lambda^2}{4\pi} g_d \ (\text{m}^2)$$

or

$$A_e \equiv \frac{w_R}{P} \ (\text{m}^2)$$

Maximum effective area is given by

$$A_{e(\max)} = \frac{\lambda^2}{4\pi} D \ (\text{m}^2)$$

Here, w_R = Received power (watt)

P = Power flow per square meter (watts/m^2) for the incident wave.

8.15 ADMITTANCE CHARACTERISTICS

Typical antenna admittance characteristics are shown in fig. 8.2.

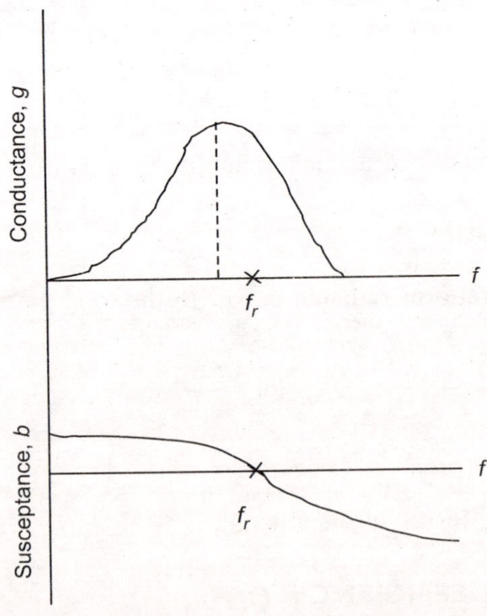

Fig. 8.2 Variation of conductance and susceptance with frequency

In this, the peak of the conductance curve shifts slightly away from resonant frequency.

8.16 ANTENNA BANDWIDTH

It is defined as the range of frequencies over which the antenna maintains its characteristics and parameters like gain, front-to-back ratio, standing wave ratio, radiation pattern, polarization, impedance and directivity etc. without considerable change.

8.17 FRONT-TO-BACK RATIO (FBR)

FBR is defined as the ratio of radiated power in the desired direction to the radiated power in the opposite direction.

i.e.
$$FBR = \frac{\text{Radiated power in desired direction}}{\text{Radiated power in opposite direction}}$$

8.18 POLARIZATION

It is defined as the direction of electric vector of electromagnetic wave produced by an antenna.

It is three types

(i) Linear polarization

(ii) Circular polarization

(iii) Elliptical polarization

Linear polarization is again three types, namely, horizontal, vertical and theta polarizations. On the other hand, these circular and elliptical polarizations can also be described in terms of their sense of rotation. The sense of rotation can be right handed or left handed. Accordingly, they are called as right handed or left handed circular and elliptical polarizations.

8.19 MAIN LOBE OF RADIATION PATTERN

The antenna pattern consists of one main lobe and a few side lobes. The height of the main lobe and its width are important for a given application.

8.20 SIDE LOBES OF THE RADIATION PATTERN

The side lobes are the lobes around the main beam. For an ideal antenna, the side lobes are absent. The presence of side lobes is undesirable as they create electromagnetic interference in the receiving system.

8.21 BEAM WIDTH OF MAIN LOBE

Beam width is described in two forms. One is half-power beam width and null-to-null beam width. It represents the directivity. The null-to-null beam width is approximately twice that of half-power beam width.

8.22 CLASSIFICATION OF ANTENNAS

The microwave antennas are basically classified into

> - Primary antennas
> - Secondary antennas

Some of the important primary antennas and secondary antennas are described in detail in this chapter.

8.23 PRIMARY ANTENNAS

These are

- half-wave dipole
- Quarter wave monopole
- uniform linear arrays
- non-uniform arrays
- slot antennas
- horn antennas
- microstrip antennas

8.24 HALF-WAVE DIPOLE

> It is basically a centre-fed wire antenna of half-wave length long (fig. 8.3).

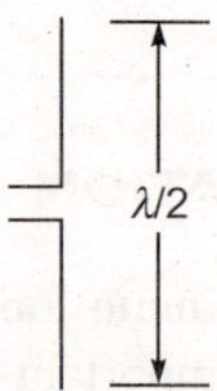

Fig. 8.3 Half wave dipole

8.24.1 Salient Features of Half-Wave Dipole

- Bi-directional radiation pattern.
- The current distribution in the half-wave dipole is sinusoidal.

- The radiation pattern of a dipole of length $2L$ is given by:

$$E(\theta) = \left[\frac{\cos(\beta L\cos\theta) - \cos\beta L}{\sin\theta}\right]$$

- Its radiated power is $73.0\ I_{\text{eff.}}^2$ watts.
- Its radiation resistance is 73 Ω.

- The radiation pattern of a vertical dipole is 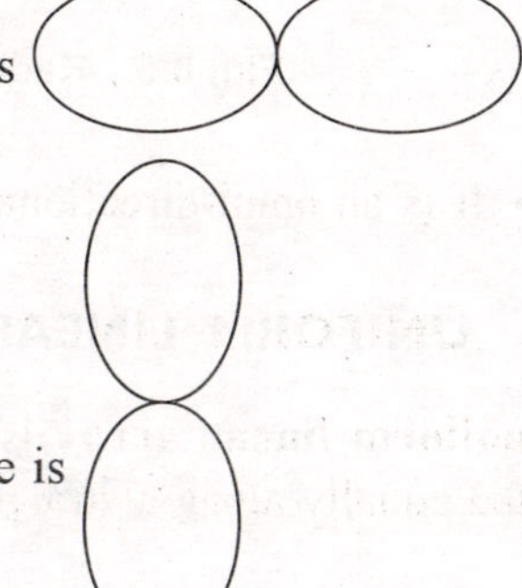

- The radiation pattern of a horizontal dipole is 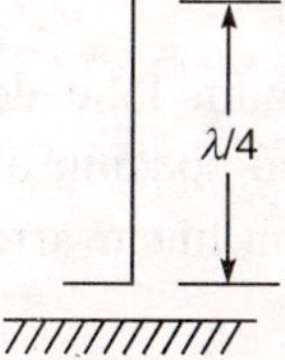

- It is an omni-directional antenna
- It is used as feed element for reflector antennas.

8.25 QUARTER-WAVE MONOPOLE

It is a wire antenna fed at one end with respect to ground plane.

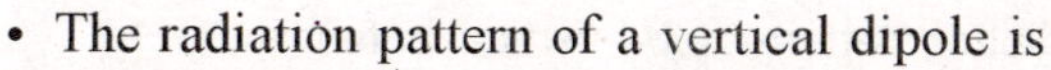

Fig. 8.4 Quarter wave monopole

8.25.1 Salient Features of Quarter-Wave Monopole

- Bi-directional radiation pattern.
- The current distribution is sinusoidal.
- The radiation pattern of a monopole of length L is

$$E(\theta) = \left[\frac{\cos(\beta L\cos\theta) - \cos\beta L}{\sin\theta}\right]$$

- Its radiation power is $36.5\ I_{\text{eff.}}^2$ watts.
- Its radiation resistance is 36.5 Ω.
- Its radiation pattern is given by

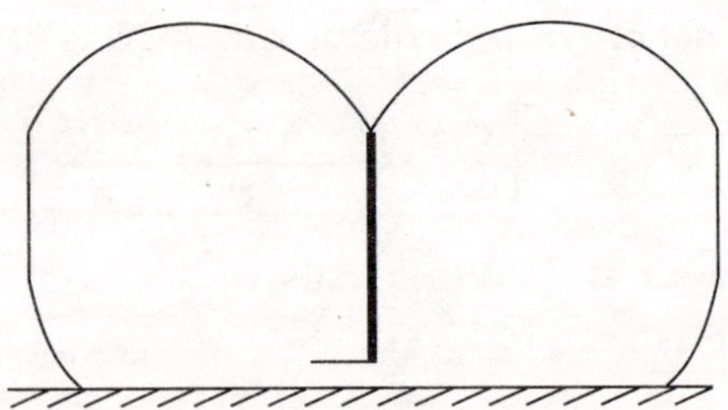

Fig. 8.5   Radiation pattern of λ/4 vertical antenna

- It is an omni-directional antenna.

8.26 UNIFORM LINEAR ARRAYS

An uniform linear array is an array in which the elements are spaced and excited equally along a straight line.

8.26.1   Salient Features of Uniform Linear Arrays

- First side lobe level of − 13.5 dB.
- Radiation pattern with narrow beam width.
- They are used as feed elements in reflector antennas.
- They are useful for point-to-point communication and high angular resolution radars.
- Their radiation pattern consists of one main lobe and a number of side lobes.
- The beam width of the main lobe depends on a number of elements, frequency, array length and spacing of the elements.
- A typical pattern of uniform linear array is given by

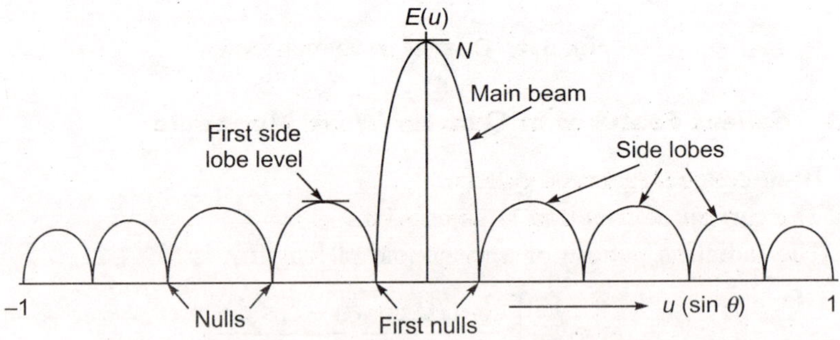

Fig. 8.6   Radiation pattern of uniform linear array

- They are used to increase the gain and directivity.

- The radiation patterns of small uniform array are obtained either analytically or by the method of multiplication of patterns.
- The side lobe levels are reduced with tapered amplitude distributions.
- Beam width is reduced by increasing array length.
- The normalized height of the main beam is N where, N is number of elements in the array.
- The normalized field strength of a uniform linear array is

$$E = \left| \frac{\sin \dfrac{N\psi}{2}}{\sin \dfrac{\psi}{2}} \right|$$

Here, N is number of elements,

$\psi = \beta d \cos \phi + \alpha_e$

β = Wave number = $2\pi/\lambda$

d = Spacing between the elements

ϕ = angle between the axis of the array and line of observer

α_e = Excitation phase (progressive phase shift)

8.27 NON-UNIFORM ARRAYS

The non-uniform arrays are characterized by

- Non-uniform amplitude distribution.
- Non-uniform spacing between the elements

Some of the non-uniform amplitude distributions are

- Uniform
- Circular
- Parabolic
- Cosinusoidal
- Raised cosine
- Triangular
- The characteristics of the above amplitude distribution are given in table 8.1.

Table 8.1   Different types of amplitude distributions and pattern characteristics

Amplitude distribution	First Side lobe level (dB)	Null-to-null B.W
Uniform	−13.5	0.9
Circular	−17.5	1.3
Parabolic	−22.0	1.2
Cosine	−23.5	1.5
Raised cosine	−32.0	2.0
Triangular	−26.8	1.3

Apart from the above standard distributions it is possible to find out the required distribution functions for specified radiation patterns. Some of the methods are:

- Schelkunoff polymomial method
- Fourier transform method
- Woodward Lawson method
- Dolph-Chebycheve method
- Taylor's method etc.

8.28 SLOT ANTENNAS

Slot is an antenna formed by a slot in a metallic surface. It is an opening cut in a conducting sheet or in one of the walls of waveguide. It is excited suitably either by a co-axial cable or through the waveguide.

8.28.1 Salient Features of Slot Antennas

1. A slot can be rectangular, circular and can be of any shape.

2. A resonant slot has a length of $\dfrac{\lambda}{2}$ and its width is much less than $\dfrac{\lambda}{2}$.

3. Slots are excited by a coaxial cable at a distance of about $0.05\ \lambda$ from one end of the slot to get reasonable impedance properties.

4. The slot radiates from both sides. A horizontal slot with such an excitation produces vertical polarization and vice versa.

5. A slot is used to radiate into free space.

6. A dipole and a slot of $\dfrac{\lambda}{2}$ lengths have similar gain and radiation characteristics with a difference in polarization.

7. The array of slots are used to increase gain and directivity.

8. The slots are omni-directional radiators.

9. The electric field in the slot is proximately sinusoidal.

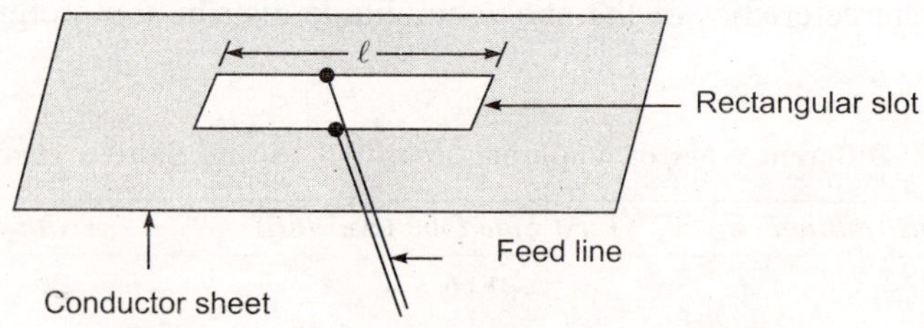

Fig. 8.7.   Slot antenna excited by two-wire line

A dipole antenna is shown in fig. 8.8.

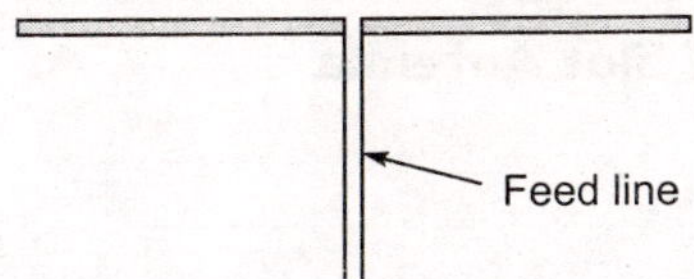

Fig. 8.8. Dipole antenna

A dipole and a slot in a conducting sheet have similar shapes except, the metal region and free space are interchanged as shown in fig. 8.8.

The slot and dipole impedances are related by

$$Z_s\, Z_d = \frac{\eta_0^2}{4}$$

Here, η_0 is free space
    Intrinsic impedance = 377 Ω

If the width of the slot is very small, i.e. $w << \lambda$
Then

$$Z_d = 73 + j\,42.5 \ \Omega$$

$\therefore$ Slot impedance, Z_s

$$Z_s = \frac{\lambda_0^2}{4\,Z_d} = \frac{(377)^2}{4\,(73 + j\,42.5)}$$

i.e. $\qquad\qquad Z_s = 363 - j\,211$ ohms

Centre-fed slot is not suitable for matching with conventional coaxial cable as the impedance of centre-fed slot in large plane is characterized by high impedance ($\approx$ 500 Ω). But, off set centre fed slot is suitable to match 50 Ω cable by choosing suitable feed point.

The rectangular slot in a conducting sheet and its complimentary dipole are shown in fig. 8.9.

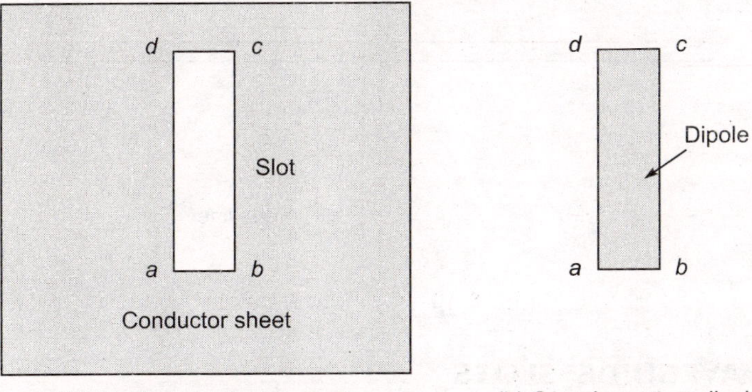

(a) Slot in a conductor sheet   (b) Complementary dipole

Fig. 8.9. Slot and complementary dipole

8.28.2 Impedance of Slot Antenna

The slot impedance is

$$
Z_s = \frac{\eta_0^2}{4 Z_d}
$$

$$
= \frac{\eta_0^2}{4 (R_d^2 + X_d^2)} (R_d - j X_d),\ \Omega
$$

Here, $\eta_0 = 120\ \pi\ \Omega$

Z_d = Impedance of complementary dipole, $\Omega = R_d + j X_d,\ \Omega$
R_d = Real part of Z_d, Ω
X_d = Reactive part of Z_d, Ω

8.28.3 Impedance of a Few Typical Dipoles

1. If the length of dipole, $\ell = \dfrac{\lambda}{2}$, diameter of dipole, $d = 0$

$$
Z_d = 73 + j\ 42.5\ \Omega
$$

2. If $\ell \leq \dfrac{\lambda}{2},\ \dfrac{\ell}{d} = 100$,

$$
Z_d \approx R_d = 67\ \Omega
$$

3. If $\ell \approx \lambda,\ \dfrac{\ell}{d} = 28$,

$$
Z_d = 710\ \Omega
$$

The length, diameter and impedance of cylindrical dipole are shown in fig. 8.10.

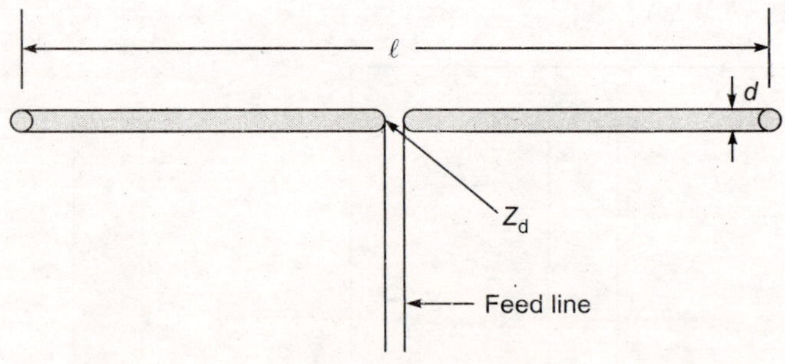

Fig. 8.10   Cylindrical dipole

8.29   WAVEGUIDE SLOTS

The slots can be cut in one of the walls of a rectangular waveguide. Different slot configurations are shown in fig. 8.11.

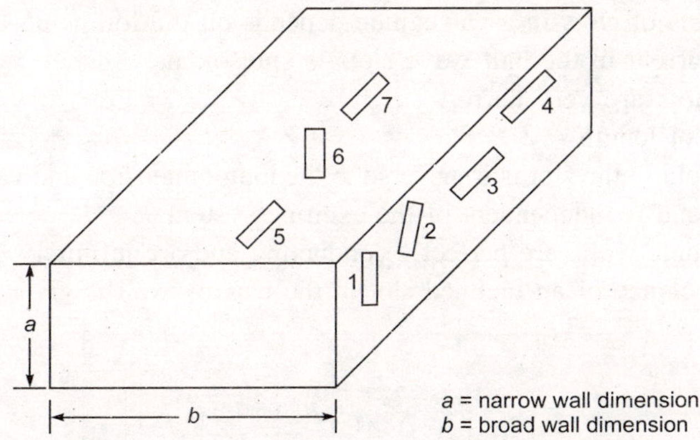

Fig. 8.11   Slot configurations in a waveguide

Slots 2, 3, 4, 5, 6 are radiating slots. The slots 1 and 7 do not radiate and hence they are not used as radiators. But they can be used for the field or power measurements inside the guide.

Slots 2, 3, 4, 5 and 6 are used as antennas. Their equivalent circuits are given fig. 8.12.

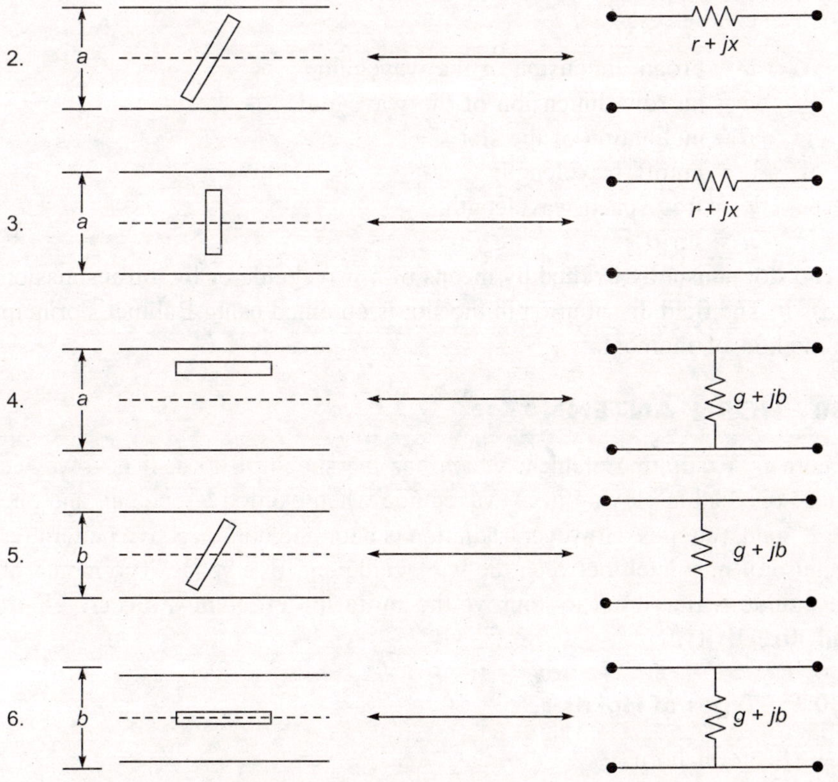

Fig. 8.12   Equivalent circuits of slots

The analysis of slots in a waveguide depends on the following assumptions:
1. The current in the half-wave slots is sinusoidal.
2. The slots are very narrow.
3. The slot length $\approx \lambda/2$.
4. The field in the slot is transverse to the long dimension and varies sinusoidally and is independent of the exciting system.
5. The guide walls are perfectly conducting and are infinitely thin.

The conductance of an inclined slot in the narrow wall is given by

$$g = \frac{30}{73\pi}\left(\frac{\lambda_g}{\lambda}\right)\frac{\lambda^4}{a^3 b}\left[\frac{u\cos\left(\dfrac{\pi\lambda}{2\lambda_g}u\right)}{1-\left(\dfrac{\lambda}{\lambda_g}\right)^2 u^2}\right]^2$$

Here, λ_g = waveguide length

$$= \frac{\lambda}{\sqrt{1-\left(\dfrac{\lambda}{\lambda_c}\right)^2}}$$

$\quad a$ = broad dimension of the waveguide
$\quad b$ = narrow dimension of the waveguide
$\quad \theta$ = inclination of the slot
$\quad \lambda_c$ = cutoff wavelength
$\quad \lambda$ = free space wavelength
$\quad u$ = $\sin\theta$

The slot is usually excited by means of a waveguide or by a transmission line across it. The field distribution in the slot is obtained using Babinet's principle or the method of moments.

8.30 HORN ANTENNA

A horn is a radiating element which has the shape of horn. It is a waveguide whose one end is blown out. A waveguide when excited at one end and open at the 2nd end, radiates. However, radiation is poor and non-directive pattern results because of mismatch between the waveguide and free space. The mouth of the waveguide is flared out to improve the **radiation efficiency, directive pattern and directivity.**

8.30.1 Types of Horns

1. Sectoral horn
2. Pyramidal horn
3. Conical horn

Sectoral horn is of two types.
* Sectoral H – plane horn
* Sectoral E – plane horn

These are shown in fig. 8.13 and fig. 8.14.

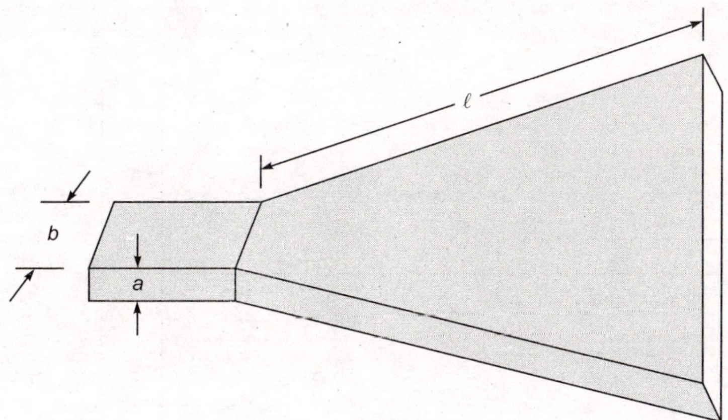

Fig. 8.13 *H*-plane horn

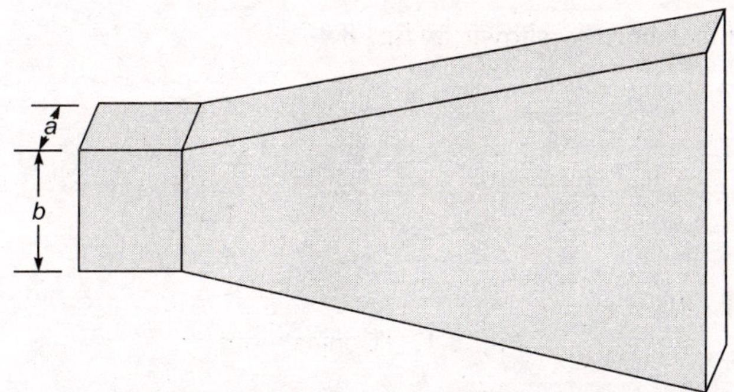

Fig. 8.14 Sectoral *E*-plane horn

A typical pyramidal horn is shown in fig. 8.15.

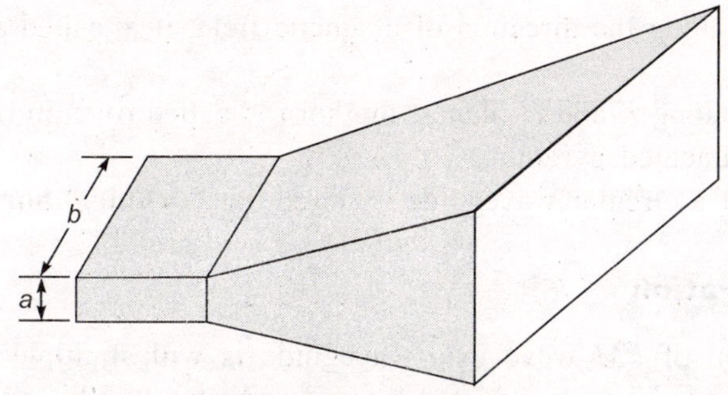

Fig. 8.15 Pyramidal horn

The horn parameters are described in fig. 8.16.

Δ = path difference,

ℓ = axial length,

D_a = aperture dimension,

α = flare angle

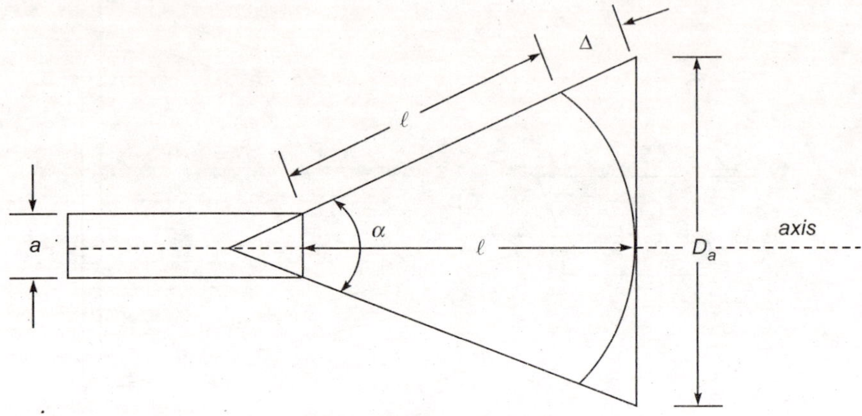

Fig. 8.16   Horn parameters

Typical conical horn is shown in fig. 8.17.

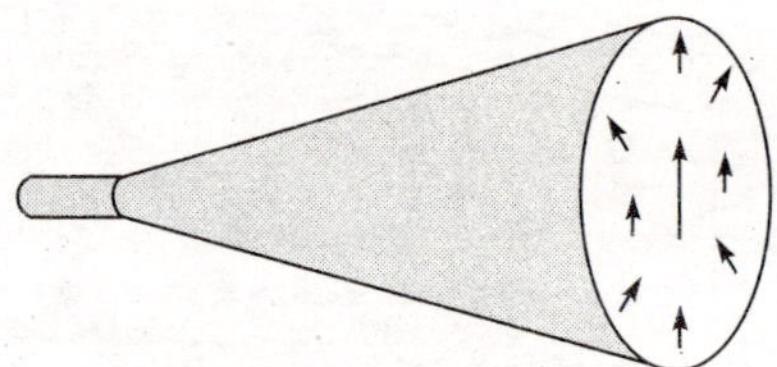

Fig. 8.17   Conical horn

Sectoral horn is a horn which has flaring in only one direction. Flaring provides impedance matching.

If flaring is along the direction of electric field, it is called as **sectoral *E*-plane** horn.

If flaring is along the direction of magnetic field, it is called as **sectoral *H*-plane** horn.

If flaring is along E and H planes, the horn is called **pyramidal horn**. It has the shape of truncated pyramid.

If one end of a circular waveguide is flared out, a **conical horn** is produced.

8.30.2 Operation

The propagation of *EM* wave in a waveguide is with multiple reflections. In waveguide, propagation is restricted by its conducting walls and waves do not spread. At the mouth of the waveguide, waves spread laterally and wave front becomes spherical. At the mouth of the waveguide, there exists near field region where the wave front is a complicated one. This is considered to be the transition region where change of propagation from waveguide to free space takes place. As the waveguide impedance and free space impedance do not match, walls of the waveguide are flared out. This flaring provides impedance matching, more directivity and narrow beamwidth. Horn produces a uniform plane wave front with a larger aperture in comparison to a waveguide. As the aperture is large, directivity is high.

The design equations of horn antenna are

$$\alpha = 2\,\tan^{-1}\left(\frac{D_a}{2\ell}\right)$$

$$= 2\,\cos^{-1}\left(\frac{\ell}{1+\Delta}\right)$$

and

$$\ell = \frac{D_a^2}{8\,\Delta}$$

Half power beam width of optimum flared horns are

$$\phi_E = \frac{56\lambda}{d_E}\ \text{degrees}$$

$$\phi_H = \frac{67\lambda}{d_H}\ \text{degrees}$$

Here, ϕ_E = HPBW in E-plane

ϕ_H = HPBW in H-plane

d_E = aperture in E-plane in free space wavelength

d_H = aperture in H-plane in free space wavelength

The directivity of horn is

$$D_i = \frac{7.5\,A_a}{\lambda^2}$$

Power gain,

$$g_p = \frac{4.5\,A_a}{\lambda^2}$$

Here, $A_a = d_E\,d_H$ = aperture physical area

8.30.3 Applications of Horns

1. They are used as feed elements for reflectors.
2. Horns are useful to obtain moderate gains.
3. They are often used in laboratory measurements.

8.30.4 Salient Features of Horns

1. It is used as feed antenna for paraboloid.
2. If $\alpha = 15°$, when $\ell/\lambda = 50$, the B.W. $= 23°$ and directivity is 120.
3. Horn becomes small if the flare angle is small. Its radiation pattern is directive, wave front is spherical, mouth area is small, and its directivity is small.
4. Directivity of pyramidal horn is more as the flare is more than in one direction.
5. Flare angle and axial length are dependent on each other.
6. The gain of the conical antenna is optimum for a given slant length of flare,

$$\ell \text{ and } D_a \approx \left(\frac{3}{\lambda}\right)^{\frac{1}{2}}$$

   Here, D_a = diameter of the aperture
7. It is easy to use with the waveguide.
8. It is used as an antenna.
9. Its directivity is not as high as that of paraboloid.
10. The directivity of a lossless horn antenna is nothing but its gain and it is given by

$$D_i = \frac{4\pi A_e}{\lambda^2}$$

$$= \frac{4\pi \eta_a}{\lambda^2} A_a$$

   Here, A_e is effective aperture, m^2

   A_a is actual area or physical area, m^2

   η_a is aperture efficiency $= \dfrac{A_e}{A_a}$

   λ is operating wavelength, m
11. In rectangular horn $A_a = d_E\, d_H$

    Here, d_E is aperture size in E-plane

    d_H is aperture size in H-plane

    The rectangular horn is nothing but a pyramidal horn.
12. In conical horn $A_a = \pi d$

    Here d is aperture diameter
13. If aperture efficiency is about 0.6,

$$D_i \approx 7.5\ A_a/\lambda^2$$

8.31 CORRUGATED HORNS

Corrugation means groove. A typical corrugated horn is shown in fig. 8.18.

$\alpha_f =$ half of the flare angle
$w =$ width of the groove
$s =$ separation between adjacent grooves
$D_a =$ aperture dimension
$d =$ groove depth

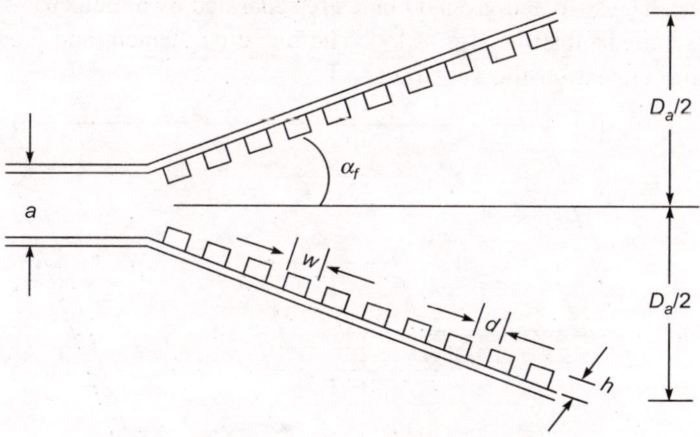

Fig. 8.18 Corrugated horn

8.31.1 Salient Features of Corrugated Horn

1. In corrugated horns, spill-over efficiency and cross-polarization losses are small and aperture efficiency is high. These are used in radio astronomy and satellite communications.

   Spill-over is defined as part of the power radiated by a feed which is not intercepted by the secondary radiator.
2. For horn feeds, the aperture efficiency is about 50 – 60 %.
3. For corrugated horn feed system, the aperture efficiency is 75 – 80 %.
4. The corrugations on the walls of a horn modify the E-field distribution in the E-plane from uniform at the waveguide-horn junction to cosine at the aperture.
5. The corrugated surface reactance is given by

$$x = \frac{d}{d + w}\ \eta_f \tan{(kh)}$$

Here, $\quad \dfrac{d}{d + w} = 1$, if $w \le w/10$, $d < \lambda/10$ and $d \ll D$

$$\eta_f = 120 \ \pi \ \Omega$$
$$k = 2\pi/\lambda$$

8.32 MICROSTRIP OR PATCH ANTENNAS

These antennas are made from patches of conducting material on a dielectric substrate above a ground plane.

8.32.1 Construction Details

It consists of a very thin ($\ell << \lambda$) metallic strip called a patch placed above a ground plane. The strip and ground plane are separated by a dielectric sheet. This dielectric is called substrate (fig. 8.19). The radiating element and feed lines are usually photo etched on the substrate.

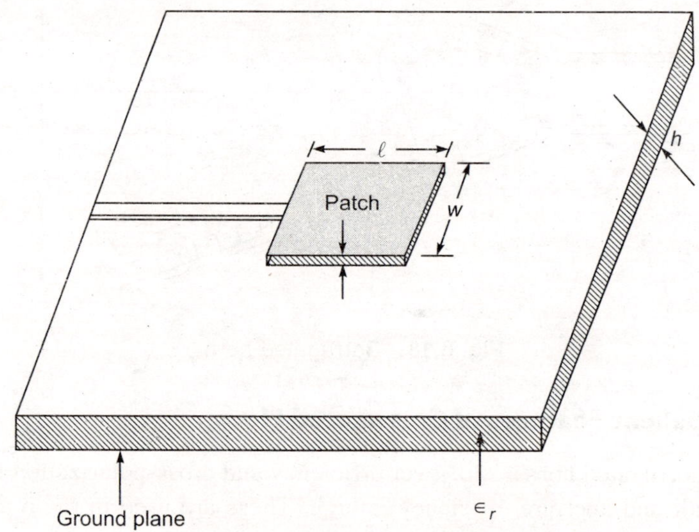

Fig. 8.19   Microstrip antenna

The side view of patch antenna is shown in fig. 8.20.

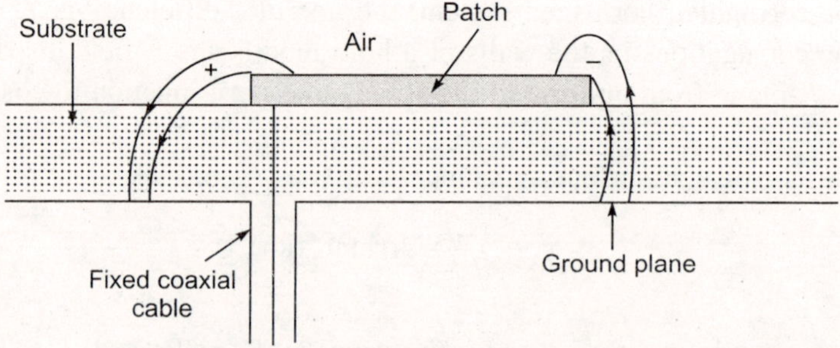

Fig. 8.20   Side view of a patch antenna

Typical patch parameters:

1. Dielectric constant of substrate, $\in_r \approx 2$.
2. The thickness of the patch, $t \approx \lambda/100$.
3. The height of the substrate, $h \ll \lambda$.
4. Width of the patch is less than λ
5. $\ell \leq \lambda/2$.

8.32.2 Applications

They are used in

1. Spacecraft and aircraft.
2. Telemetry, satellite communications and defence radars systems to operate over a frequency range of 1 to 10 GHz.

8.32.3 Advantages

1. Microstrip antennas are small in size and they are low profile antennas.
2. Installation of these antennas is simple and easy.
3. It does not give rise to aerodynamic drag when used in aircraft.
4. Their weight is small.
5. These antennas can be flush mounted on a metallic conductor.
7. They are economical.
8. They do not require space for feed line. The feed line can be placed behind the ground plane.

8.32.4 Disadvantages

1. Efficiency is less.
2. Bandwidth is small and it is typically a few percent.

8.32.5 Shapes of Patch Antennas

1. Rectangular
2. Square
3. Circular
4. Elliptical
5. Triangular
6. Diamond
7. Any other shape

These are shown in fig. 8.21.

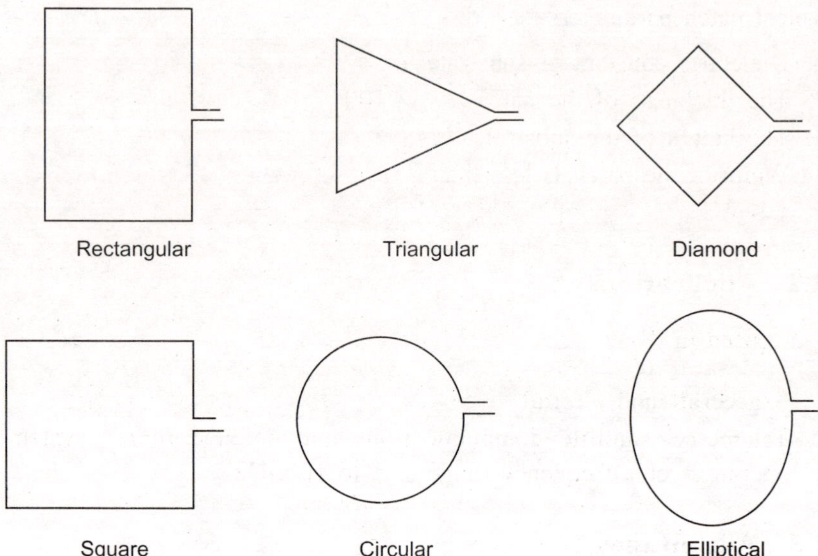

Fig. 8.21　Patch antennas with different shapes

The patch antenna acts as a resonant $\lambda/2$ parallel-plate microstrip transmission line. Its characteristic impedance Z_0 is equal to the reciprocal of the number, n of parallel field cell transmission lines.

Each field-cell transmission line has a characteristic impedance, Z_i of the medium. Here

$$Z_i = \sqrt{\frac{\mu}{\epsilon}} = \eta_0 \sqrt{\frac{\mu_r}{\epsilon_r}} \ \Omega$$

The characteristic impedance of patch antenna

$$Z_P = \frac{Z_0}{n\sqrt{\epsilon_r}}$$

If $n = 10$,

$$Z_P = \frac{120\,\pi}{n\sqrt{\epsilon_r}} = \frac{120\,\pi}{10\sqrt{2}} = 26.63 \ \Omega$$

As $n = w/t$, Z_P becomes

$$Z_P = \frac{Z_0\,t}{w\sqrt{\epsilon_r}}$$

This expression does not take care of fringing field at the edges.
As $w \gg t$, this effect is small on the patch.
For **microstrip line** where w/t is small, the fringing effect is accounted by approximately adding 2 cells. This gives more accurate formula for the impedance of the microstrip line. That is given by

$$Z_P \text{ (for microstrip line)} = \frac{\eta_0}{\sqrt{\epsilon_r}\left[\dfrac{w}{t} + 2\right]}$$

The radiation pattern of patch antenna is broad.
The directivity, D_p of the patch antenna,

$$D_p = \frac{4\pi}{\Omega_A} = \frac{4\pi}{\pi} = 4$$

$$D_p = 10 \log_{10} 4 = 6.021 \text{ dB}$$

The effective height h_e of the antenna is

$$h_e = \sqrt{\frac{2R_r A_e}{\eta_0}}$$

Here, R_r = Radiation resistance, Ω

A_e = Effective aperture, m^2

η_0 = Intrinsic impedance of free space, Ω

R_r = 50 Ω for a typical patch

$$A_e = \frac{D\lambda^2}{4\pi} = \frac{4\lambda^2}{4\pi} = \frac{\lambda^2}{\pi}$$

$$\therefore \quad \boxed{A_e = \frac{\lambda^2}{\pi}}$$

For matching purposes, feed point can be moved from the edge. Arrays of patches provide more directivity.

Monolithic Microwave Integrated Circuit (MMIC) technology is used to manufacture patch antennas and associated circuitry in a compact form. These are used for frequencies ranging from 50 MHz to 100 GHz.

Printing technology is suitable for a variety of antenna elements like dipoles,
The conductance of the patch antennas is given by

$$g \approx \frac{1}{90}\left(\frac{w}{\lambda}\right)^2 \text{ for } w \ll \lambda$$

$$g \approx \frac{1}{120}\left(\frac{w}{\lambda}\right) \text{ for } w \gg \lambda$$

8.32.6 Methods of Bandwidth Control

It is a narrow bandwidth antenna. However, its bandwidth can be increased by

1. Using high dielectric constant ($\in_r$) substance.
2. Adding reactive component to the patch. This reduces VSWR.
3. Increasing the thickness of parallel plate transmission line.
4. Cutting holes or slot in the patch. These holes or slots increase its inductance.

8.33 SECONDARY ANTENNAS

These antennas are always associated with the primary antennas. These are basically of two types.

1. Lens antennas
2. Reflector antennas

8.34 LENS ANTENNAS

A **lens antenna** consists of electromagnetic lens with a feed. In other words, a lens antenna is a three-dimensional electromagnetic device whose refractive index is different from unity. It is similar to a glass lens in optics.

It is usually made of polystrene ($\in_r = 2.56$, $n = 1.6$) or lucite and polyethylene, ($\in_r = 2.25$, $n = 1.5$).

8.34.1 Functions of Lens Antennas

It is used to

1. Generate plane wavefront from spherical wavefront.
2. Form the incoming wavefront at its focus.
3. Generate directional characteristics.
4. Collimate the electromagnetic rays.
5. Control the aperture illumination.

8.34.2 Principle of Operation

The lens antenna in association with a primary feed antenna in transmitting mode is shown in fig. (8.22).

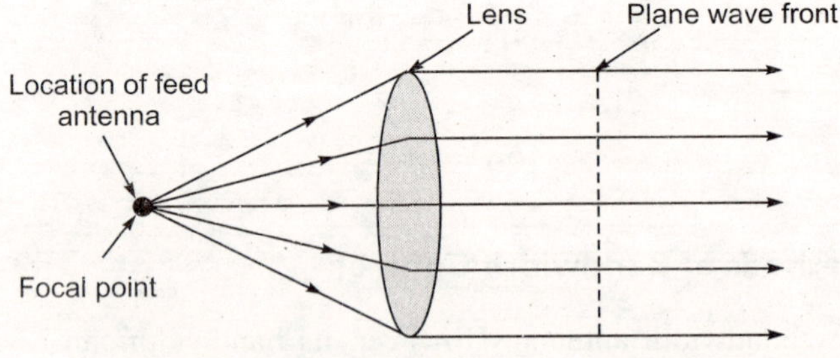

Fig. 8.22 Lens antenna in transmitting mode

When the feed antenna is kept at focal point of lens antenna, the diverging rays are collimated (parallel rays) forming plane wave front after their incidence on the lens and passing through it. Collimation occurs because of refraction mechanism.

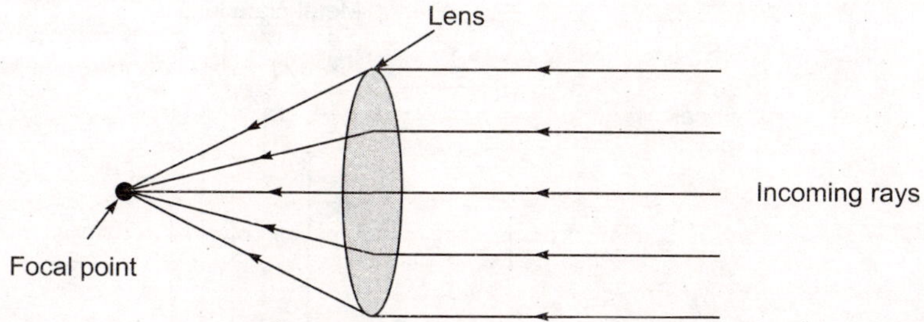

Fig. 8.23   Lens antenna in receiving mode

A **lens antenna** in receiving mode is shown in fig. 8.23.

The parallel rays converge at the focal point after passing through the lens because of refraction mechanism. The collimation is also possible if lens with refractive index is less than unity. Lens antennas are used in association with a point source. But in practice, it is used with horn-like antennas.

8.34.3   Types of Lens Antennas

These are two types

1. Dielectric lens.
2. Metal plate lens.

1.   Dielectric lenses

Dielectric lens are also known as the **delay lenses.** Here, outgoing electromagnetic rays are collimated and are retarded or delayed by the lens material or media. This is illustrated in fig. (8.24).

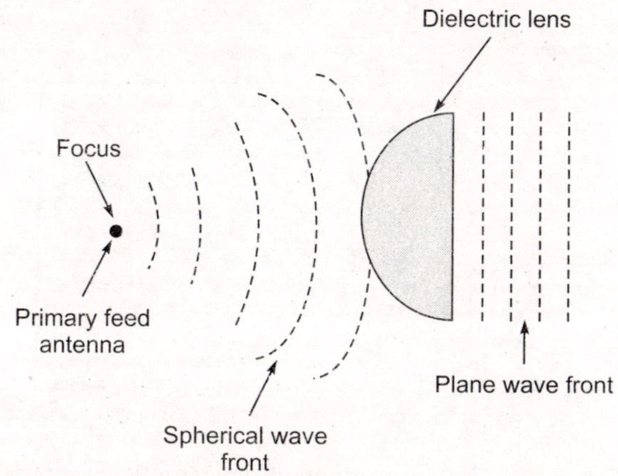

Fig. 8.24   Mechanism of dielectric lens (delay lens)

2. *E*-plane metal plate lens

In this, outgoing wavefront is speeded up by the lens material. This is illustrated in fig. (8.25).

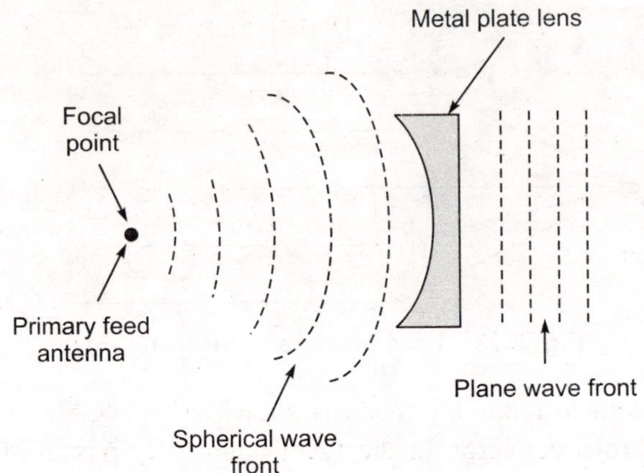

Fig. 8.25 Mechanism of *E*-plane metal plate lens

8.34.4 Salient Features of Dielectric Lens

1. They are extremely useful at high frequencies.
2. They are bulky and heavy for $f < 3$ GHz.
3. They are usually made of polystyrene or lucite and polyethylene.
4. This can be avoided by zoned or stepped dielectric lens as in fig. (8.26).
5. Uniform illumination for lens antennas is better if focal length is long.
6. At $f < 10$ GHz the lens become excessively and undesirably thick.
7. The width w of a stepped lens is

$$w = \frac{\lambda}{n-1}$$

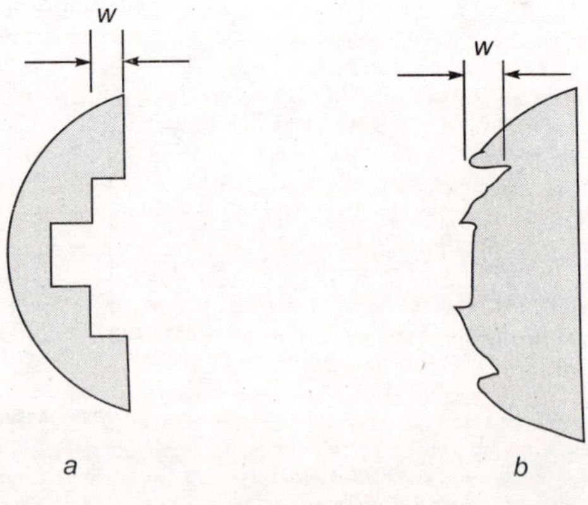

Fig. 8.26 Zoned or stepped dielectric lens antenna

8. Un-stepped dielectric lens antenna is wideband and its shape does not depend on wavelength.

9. The zoned lens of fig. 8.23 (b) is mechanically stable.

10. The main disadvantage of dielectric lens antenna is its frequency sensitivity.

11. The weight of stepped lens is less and the power dissipation is also less.

12. They are also bulky and heavy and design is involved.

13. A typical horn feed is shown in fig. (8.26). To avoid undesirable radiation from sides, horn mouth is extended up to the location of the lens.

14. The main advantage of lens antennas is that the feed antennas do not obstruct the aperture.

15. Feeding it away from the axis is possible and it is suitable to move the beam angularly w.r.t. its axis, if required.

16. The tolerance in the design of lens antenna is more.

17. Lens antennas are costlier for the same gain and bandwidth in comparison with reflector antennas.

18. The bandwidth of zoned lens is

$$\text{B.W.} = \frac{50\,n}{1 + Kn} \quad (K = \text{number of zones})$$

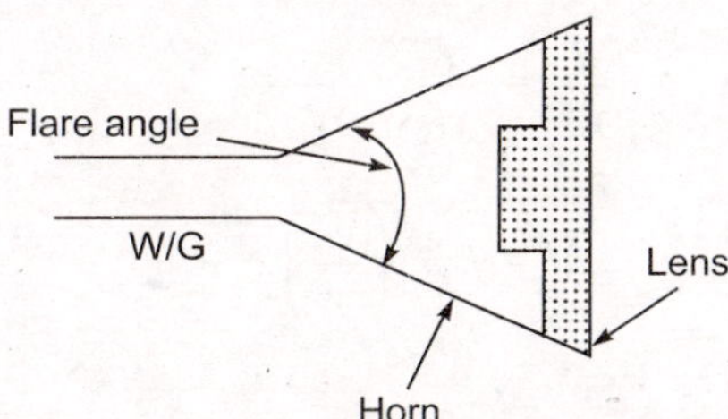

Fig. 8.27   Horn feed of lens antenna

8.34.5   Equation of the Shape of Lens

The equation which represents the shape of the lens is given by

$$d = \frac{\ell\,(n-1)}{(n\cos\phi - 1)}$$

Here d, ℓ, ϕ are shown in fig. (8.25), n is the refractive index of lens.

Proof: Consider fig. (8.28)

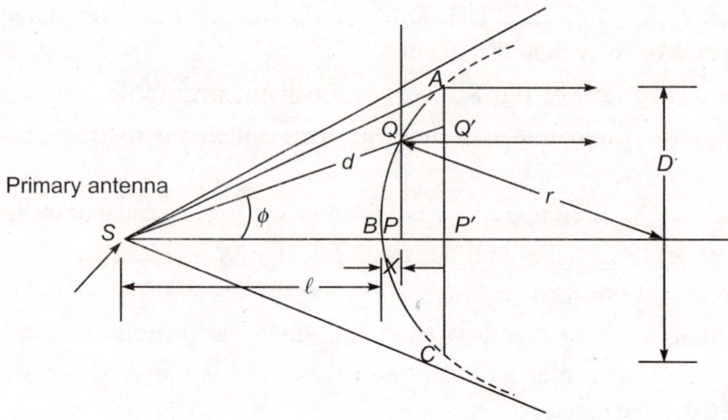

Fig. 8.28 Lens Ray paths

The electrical paths of all rays from S to AB are equal so that the field on the entire plane surface is in phase. The outgoing rays from the source will have constant phase across the aperture, d for suitably shaped lens.

If the velocity of *EM* wave in air is v_0 and in lens medium is v, then due to equality of electrical paths,

$$SQ + QQ' = SB + BP' = SB + BP + PP'$$

i.e.
$$SQ = SB + BP$$

or
$$\frac{d}{v_0} = \frac{\ell}{v_0} + \frac{x}{v}$$

or
$$d = \ell + \left(\frac{v_0}{v}\right)x$$

But the refractive index, $n = \left(\dfrac{v_0}{v}\right)$

and
$$x = SP - SB = d\cos\phi - 1$$

Hence
$$d = \ell + nx$$
$$= \ell + n\,(d\cos\phi - 1)$$

$\therefore$
$$\boxed{d = \frac{\ell(n-1)}{(n\cos\phi - 1)}}$$

This is nothing but the equation of hyperbola of focal length, ℓ and radius of curvature, d

$$d = \ell(n-1) \quad \text{for small } \phi$$

8.35 REFLECTOR ANTENNAS

Types of Reflectors

1. Plane reflector
2. Corner reflector
3. Parabolic reflector

8.36 PLANE REFLECTOR

This reflector is useful to radiate *EM* energy in a desired direction. But it is difficult to collimate the energy in the forward direction. In fact, the polarization of primary antenna and its position with respect to the reflecting surface is used to control pattern characteristics, impedance, power gain and directivity of complete system. Infinitely large reflector is ideal. It is difficult to collimate the energy by plane reflector.

A typical plane reflector is shown in fig. 8.29.

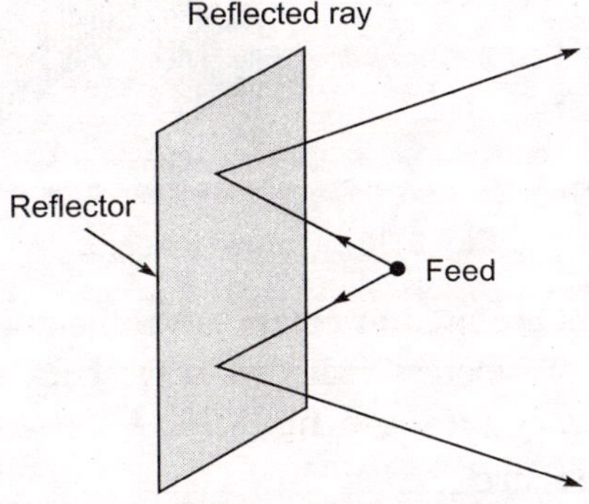

Fig. 8.29 Plane reflector

8.37 CORNER REFLECTOR

A corner reflector is a reflecting object which consists of two or three mutually intersecting conducting flat surfaces.

Dihedral forms of corner reflectors are used in antennas frequently. However, trihedral forms with mutually perpendicular surfaces are used as radar targets.

A typical corner reflector is shown in fig. 8.30.

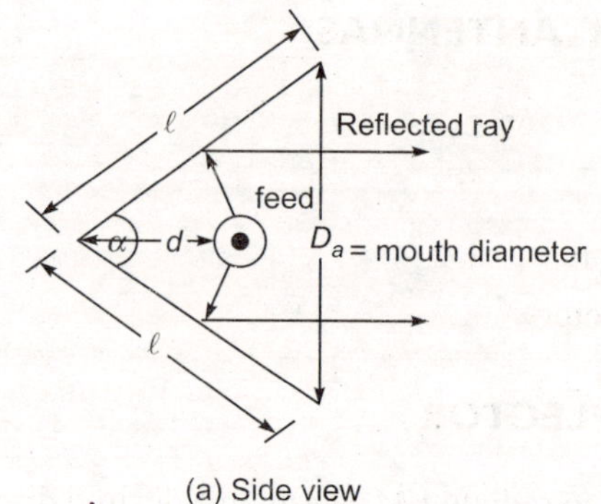

(a) Side view

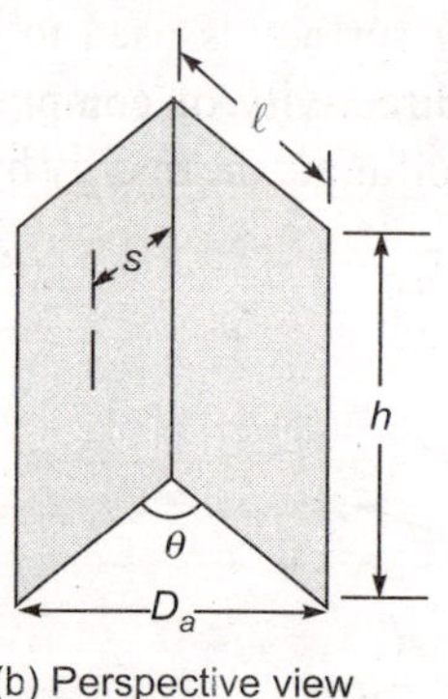

(b) Perspective view

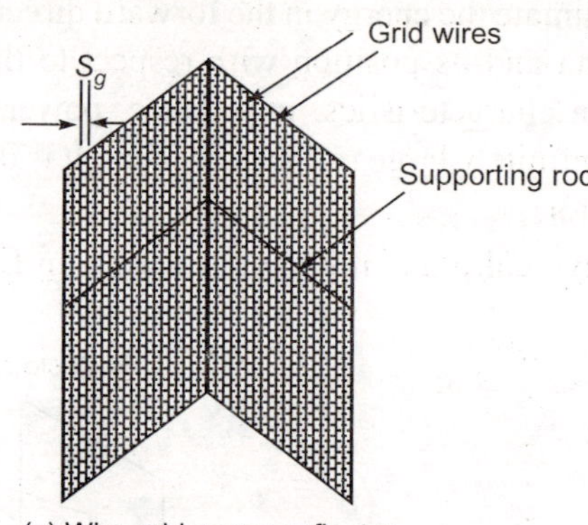

(c) Wire-grid corner reflector

Fig. 8.30   Corner reflector

The corner reflectors are used to obtain the collimation of *EM* energy in the forward direction and to suppress radiation in the back and side directions. For the corner reflector shown in fig. 8.30.

D_a = Aperture size
ℓ = Length
h = Height
d = Spacing between the vertex and feed point location
s_g = Spacing between grid wires
θ = included angle

The ranges of the above parameters for a good corner reflector are

1. $\lambda < D_a < 2\lambda$
2. $\ell \approx 2d$
3. $\dfrac{\lambda}{3} < s < \dfrac{2\lambda}{3}$
4. h is 1.2 to 1.5 times greater than the total length of feed element
5. $s_g \leq \dfrac{\lambda}{10}$

8.37.1 Salient Features of Corner Reflector

1. The feed element can be a dipole or an array of collinear dipoles.
2. If the feed elements are cylindrical or biconical dipoles instead of thin wires, bandwidth and radiation resistance become large. It is a simple reflector and is easy to construct.
3. It is used as a passive target for radar and communication applications.
4. It returns the signal exactly in the same direction by choosing $\alpha = 90°$.
5. Due to this unique feature, most of the defence ships and vehicles are designed with minimum sharp corners to reduce the chances of their detection by enemy radars.
6. It is also used in home television antennas.
7. The intersecting angle, $\theta = 90°$ is preferred.
8. The spacing between the vertex and feed element position is increased if θ is decreased and vice versa, to improve efficiency.
9. When θ is small, gain is increased by increasing the length of the sides of the reflector.
10. When the spacing, s is small, radiation resistance becomes small and hence efficiency falls.
11. If the spacing is very large, the system produces undesirable multiple lobes and it loses its directional characteristics.
12. The main lobe is broad for reflectors with finite sides compared to that of infinite dimensions.
13. The array factor of corner reflector antenna is

    $$E = 2\,[\cos\,(\beta d\,\sin\,\theta\,\cos\,\phi - \cos\,(\beta d\,\sin\,\theta\,\sin\,\phi))]$$

    Here, $\beta = \dfrac{2\pi}{\lambda}$

14. For small θ, the side lengths are long.

8.38 PARABOLOID

Paraboloid: It is a reflector antenna which has the shape of paraboloid and employs the properties of a parabola.

The Parabola: It is a plane curve obtained by the locus of a point which moves so that its distance from another point, called the focus plus its distance from a straight line, called directrix, is uniform.

> **A Paraboloid** is a three-dimensional surface obtained by revolving the parabola about its axis. The paraboloid is called parabolic reflector or dish antenna.

The geometry of a parabolic reflector in transmitting mode is shown in fig. 8.31.

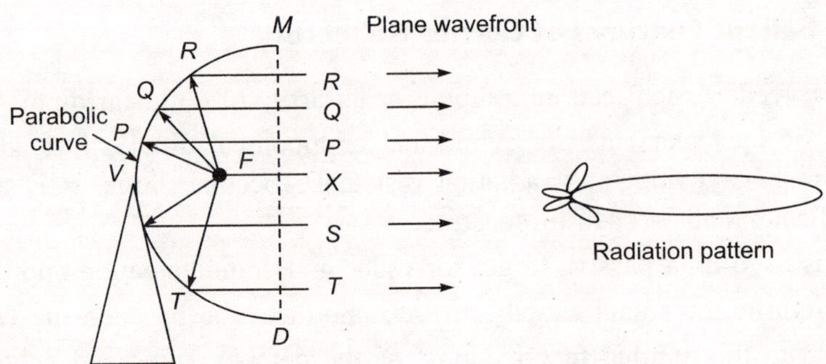

Fig. 8.31 Geometry of parabolic reflector in transmitting mode and radiation pattern

Here VX = axis of the parabola

MD = Mouth diameter, D_a

VF = Focal length = ℓ.

V = Vertex

F = Focus

MVD = Parabola

The line MD = Directrix

AF/MD = Aperture of the parabola or mouth diameter

For a parabola,

$$FP + PP = FQ + QQ = FS + SS = \text{Constant } (k)$$

k changes for different shapes.

The equation of the parabola is represented mathematically

and

$$y^2 = 4\, l_f x$$

the equation of the paraboloid is represented mathematically

$$y^2 + z^2 = 4\, \ell_f x$$

The parabolic reflector in receiving mode is shown in fig. 8.32.

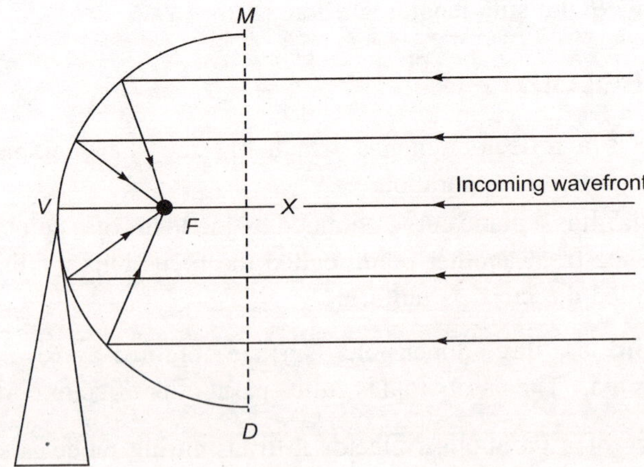

Fig. 8.32 Geometry of parabolic reflector in receiving mode

3.38.1 Operation of Parabolic Reflector

When the feed antenna is placed at the focus, all the waves are incident on the reflector and they are reflected back forming a plane wave front. If the reflected waves reach the directrix, all of them will be in phase irrespective of the point of reflection.

The radiation is concentrated along the axis of the parabola. But the waves will be cancelled in other directions due to path and phase differences.

The parabolic reflector converts a spherical wave into a plane wave.

The difference between the plane wave and spherical wave are shown in fig. 8.33.

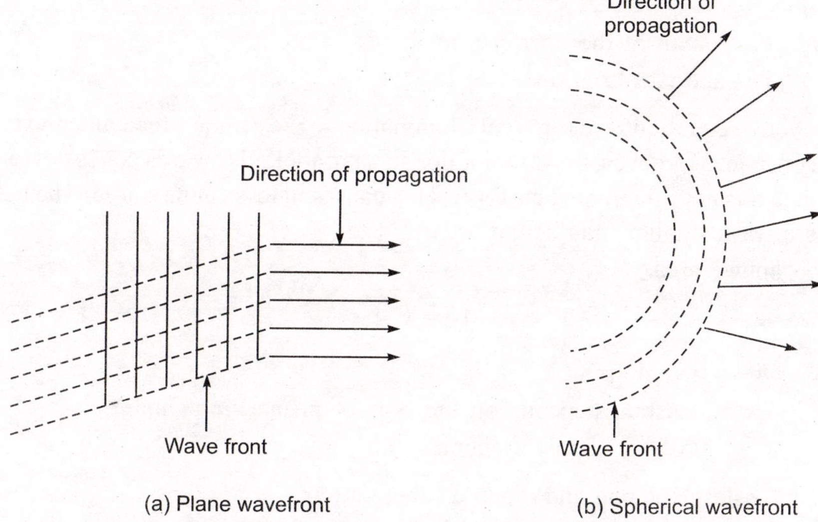

Fig. 8.33 Direction of propagation of plane and spherical waves

When the feed antenna is isotropic, the beam width of the radiation pattern of paraboloid is given by

$$\text{HPBW} = \phi_{HP} = \frac{70\lambda}{D_a}$$

$$\text{NNBW, } \phi_0 = 2\phi_{HP} = \frac{140\lambda}{D_a}$$

$$\text{Directivity, } D = 9.87\left(\frac{D_a}{\lambda}\right)^2$$

Here, ϕ_{HP} = Half power beam width in degrees

ϕ_0 = Beam width between first nulls in degrees

$$\lambda = \text{wave length, m}$$
$$D_a = \text{Mouth diameter, m}$$

For a large uniformly illuminated rectangular aperture,

$$\text{HPBW} = \phi_{HP} = \frac{57.5\lambda}{\ell} \quad \text{(degrees)}$$

$$\text{NNBW} = \phi_0 = \frac{115\lambda}{\ell} \quad \text{(degrees)}$$

$$\text{Directivity, } D = \frac{4\pi A}{\lambda^2}$$

Here, ℓ = length of the aperture, in λ

A = aperture area, m^2

The above equations are for ideal illuminations. The primary feed antenna is not truly isotropic. Moreover, the illumination in a paraboloid towards edges decreases and hence it is not illuminated uniform. This causes in less capture area. The actual area is always smaller than actual area. That is,

The capture area,

$$A_c = k\, A_a$$

Here, A_a = Actual area

k = constant depends on the type of primary antenna

≈ 0.65 for dipole antenna

For a lossless antenna and tapered illumination,

$$\text{Power gain, } g_p = \frac{4\pi}{\lambda^2} A_c$$

$$= \frac{4\pi}{\lambda^2} k A_a$$

For circular apertured paraboloid,

$$A_a = \frac{\pi D_a^2}{4}$$

$$\therefore \quad g_p = \frac{4\pi k}{\lambda^2}\, \frac{\pi D_a^2}{4}$$

$$= \frac{\pi^2 k D_a^2}{\lambda^2}$$

For dipole feed, $k = 0.65$

i.e.
$$g_p = 0.65\,\pi^2\left(\frac{D_a}{\lambda}\right)^2$$

$$= 6.41\left(\frac{D_a}{\lambda}\right)^2$$

$$\boxed{g_p = 6.41\left(\frac{D_a}{\lambda}\right)^2}$$

8.38.2 Salient Features of Paraboloid Reflectors

1. The main beam is narrow with several side lobes around it.

2. If the primary antenna is non-directional, NNBW $= 140\left(\dfrac{\lambda}{D_a}\right)$

3. HPBW $= 70\left(\dfrac{\lambda}{D_a}\right)$.

4. The gain depends on $\left(\dfrac{D_a}{\lambda}\right)$ and type of illumination.

5. For tapered illumination, the power gain, $g_p = 6.41\left(\dfrac{D_a}{\lambda}\right)^2$.

6. ERP is very high even for small input powers.
7. Effective Radiated Power (ERP) = product of input power to the antenna and power gain.
8. Paraboloid provide large gain and narrow beam width reflectors.
9. Paraboloids are not used at low frequencies due to their excessive size.
10. In order to be fully effective and useful, its mouth diameter must be at least 10λ.
11. The performance of paraboloid reflectors is influenced by the characteristics of primary antenna and its size etc.
12. Parabolic reflectors are used in most of the **communications and radars**.
13. The reflector is only a **secondary antenna** and feed antenna is a primary antenna.
14. The radiation pattern of primary antenna placed at focus of the parabolic reflector is called as **primary pattern**. The radiation pattern of entire system consisting of primary and secondary antenna is called **antenna pattern**.
15. A mesh surface is often used to reduce wind effect on the antenna and extra strain on the supports and also to reduce distortion caused by uneven wind force **distribution** over the surface.

8.38.3 Disadvantages of Parabolic Reflectors

1. The presence of side lobes creates electromagnetic interference (EMI) and the effect of EMI is more prominent in low noise receivers due to the imperfections in the reflector.
2. Deviations from a true shape of a paraboloid should not exceed one-sixteenth of λ. Such tolerances may be difficult to achieve in large reflectors whose surface is a network of wires instead of a smooth continuous surface.
3. **Diffraction** is another cause of side lobes and will occur around the edges of the paraboloid, which produces EMI.
4. The bandwidth also depends on the size of primary antenna.
5. As the feed antenna is isotropic and point source, it cannot be located exactly at the focus.
6. Due to the presence of aberrations, the main lobe is broadened and the side lobes are reinforced.
7. Any type of primary antenna introduces distortion. If the dipole feed is used, the radiation in one plane is different from the other and the beam from the reflector becomes broad.
8. Broad flat top beams are eliminated by using horn feed. Here also, paraboloid is not illuminated uniformly and it will taper at the edges. This results into a small capture area.

8.38.4 Types of Parabolic Reflectors

The other types of reflectors are

1. Cut or truncated paraboloid
2. Parabolic cylinder
3. Pill box and cheese antenna
4. Offset paraboloid reflector.
5. Torus antenna

The **advantages of these** are low cost, and small size. The main disadvantage is that the beam is not directional in both azimuth and elevation.

1. **Cut or truncated paraboloid**

   This is shown in fig. 8.34. It is not circular in appearance when viewed from a point on the parabolic axis.

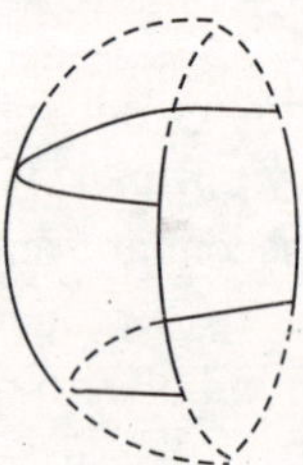

Fig. 8.34 Cut paraboloid

2. Parabolic cylinder

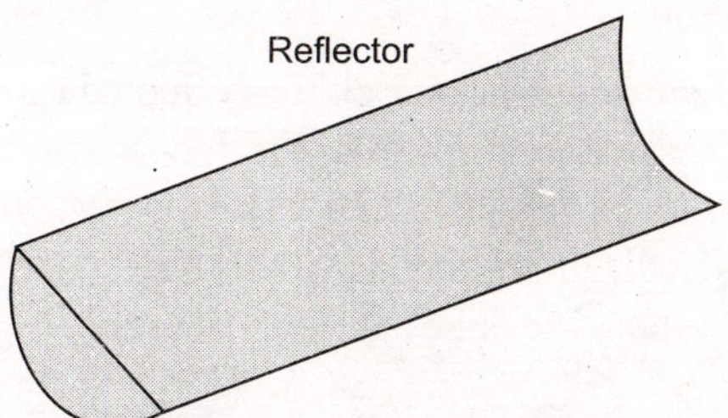

Fig. 8.35 Parabolic cylinder

The parabolic cylinder is created by moving the parabola in sideways. It consists of **a focal line** and **a vertex line.** When a radiating line source is on the focal line, parabolic cylinder is illuminated uniformly. Then the beam results in the vertical plane. Its beam width is slightly less than that of full paraboloid. It gives rise to a wide beam in E-plane and narrow beam in H-plane.

3. Pillbox antenna and Cheese Antenna

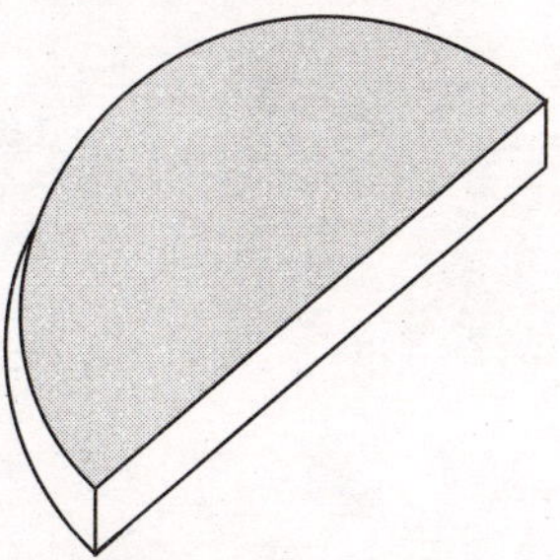

Fig. 8.36 Pillbox antenna

Pillbox antenna is another reflector antenna which has a cylindrical reflector enclosed by two parallel conducting plates perpendicular to the cylinder, spaced less than one wavelength apart. A typical pillbox structure is shown in fig. 8.36. It looks like pillbox and hence the name. It is excited by a probe through a coaxial line. It produces a wide beam in E-plane and narrow beam in H-plane. That is, these reflectors produce shaped beams, namely, fan type of wide beam in one plane and narrow in the other. This is used in **moving radars.**

Cheese antenna is also a reflector antenna which has a cylindrical reflector enclosed by two parallel conducting plates perpendicular to the cylinder but spaced more than one wavelength apart.

Pillbox and cheese antennas are used in marine radars.

4. Offset paraboloid

In this, feed antenna does not block the reflected ray. It is one form of cut-paraboloid in cross-section. In this, the focus is located outside the aperture (fig. 8.37). If the feed antenna is kept at the focus, the reflected and collimated rays will be produced without any interference. This reflector is very useful when the size of feed antenna is comparable with the reflector.

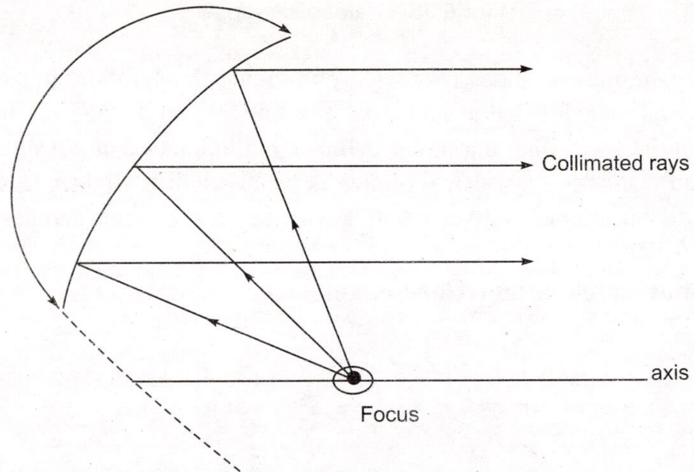

Fig. 8.37 Offset parabolic reflector

5. Torus antenna

It is an improved version of an offset reflector. It is similar to cut paraboloid. It is parabolic along one axis and circular along the other. This antenna is useful to transmit or receive a number of beams simultaneously to or from geostationary satellite orbit. This is possible by placing several feeds at the focus.

8.38.5 Feed Systems for Parabolic Reflectors

The feeding of reflector is done by

- ➢ Half-wave dipole
- ➢ An array of collinear dipoles
- ➢ Yagi – Uda antenna
- ➢ Horn
- ➢ Cassegrain feed

Half-wave dipole feed

It has bidirectional radiation characteristics. Ideally, **unidirectional** antenna is required as feed antenna. Here, the reflected rays and the backward radiated rays interfere and the plane wave front is disturbed due to phase difference.

Yagi-Uda antenna feed

Although this produces unidirectional pattern, the size of the antenna is a common problem. It blocks the reflected rays.

Array of collinear dipoles feed

This is another possible primary feed antenna. But the feeding with dipole array involves changing from unbalanced system to a balanced system.

Horn feed

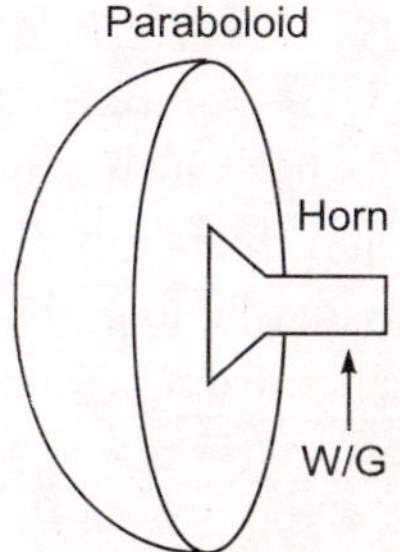

Fig. 8.38   Horn feed

8.38.6   Salient Features of Horn Feed

- Horn radiates directional characteristics towards the reflector.
- There is no direct radiation in the backward direction.
- Horn obstructs the reflected rays when it is placed at the focus. But obstruction is not high. It may be one or two percent of the total reflected energy.

8.38.7   Cassegrain Feed

It is named after an early eighteenth century astronomer. The feed mechanism is shown in fig. 8.38. It consists of

A parabolic reflector,

A hyperbolic reflector and

A feed antenna, horn with waveguide.

One of the foci of hyperbolic reflector coincides with the focus of the paraboloid. When *EM* rays from horn antenna are incident on the hyperboloid reflector, they are reflected back and then incident on the paraboloid. These incident rays are reflected and propagate as a plane wave front.

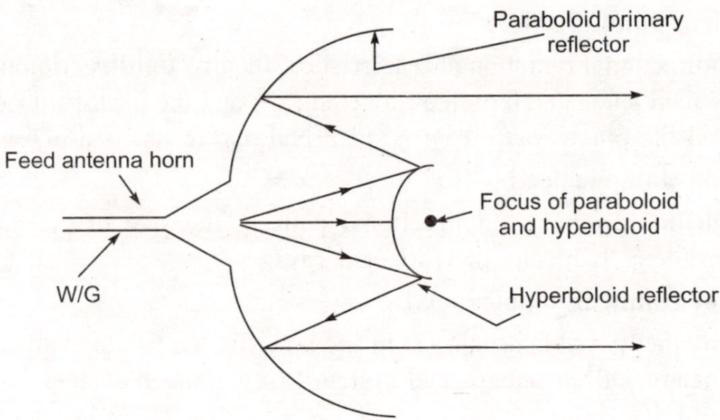

Fig. 8.39 Cassegrain feed in transmission mode

Hyperbolic curve is traced by a point which moves so that its ratio of the distance from a fixed point, the focus to its distance from a fixed straight line, the directrix is a constant and is greater than unity. The hyperboloid is a three dimensional surface obtained by revolving the hyperbola about its axis. The size of the hyperboloid depends on its distance from primary feed antenna, mouth diameter of horn and operating frequency.

8.38.8   Applications of Cassegrain Feed

1. It allows to keep when the primary antenna is in a convenient position.
2. It is useful when a short transmission line or waveguide for connecting the receiver or transmitter to the primary antenna.
3. It is useful for low-noise receiver applications.

   If the active part of the transmitter or receiver is kept at the focus, it is possible to reduce the power loss. But the size of the transmitter or receiver prohibit such a placing. This is the reason why cassegrain feed is best suited for low-noise applications.

8.38.9   Advantages of Cassegrain Feed

1. Spillover and minor lobe radiation is less.
2. Scanning the beam or broadening the beam by moving the reflecting surfaces is possible.
3. Primary antenna can be kept at a convenient location.

Disadvantages:

1. A fraction of the *EM* energy is obstructed by the secondary hyperboloid reflector. This is tolerable when the size of hyperboloid is small compared to paraboloid.

The obstruction is reduced by using large paraboloid reflector and keeping the small distance between the horn feed and hyperboloid reflector. This technique reduces the diameter of the hyperboloid reflector.

2. Large paraboloid is not economical.
3. The vertically polarized waves from the feed antenna are reflected back to the main reflector by hyperboloid.
4. Polarization of the waves is changed from vertical to horizontal when they are reflected by the paraboloid. That is, the reflected waves are horizontally polarized and they propagate through the vertical bars of the hyperboloid.

8.39 SHAPED BEAM ANTENNAS

Narrow radiation beams (pencil beams) are useful

> ➤ in point to point communication system
> ➤ to obtain large gain
> ➤ to obtain precision direction finding and
> ➤ in high resolution of radars.

However, when fast scanning is required, such narrow beams are not preferred as the involved scan time is more. As a result, broad beams are useful in such applications. It is well known that small apertures produce broad. But this is not sufficient. In several applications well shaped beams are required.

The antennas that produce shaped beams are known as **shaped beam antennas.**

Some of the popular shaped beams are

1. Fanned beams
2. Sector beams
3. Cosecant beams

1. Fanned beams

The fan beam is characterized by broadbeam characteristics in one of the principal planes. These beams are generated from the ship antennas. These are used to compensate role and pitch of the ships.

Fanned beams can be generated from the following antennas.

- An array antenna with suitably designed amplitude and phase distributions.
- A shaped parabolic reflector with the primary antenna at its focus.
- A parabolic cylinder with a line source producing a rectangular aperture.

Applications of fanned beams: These are used in

1. groundborne radars for air search.
2. airborne radars for surface search.

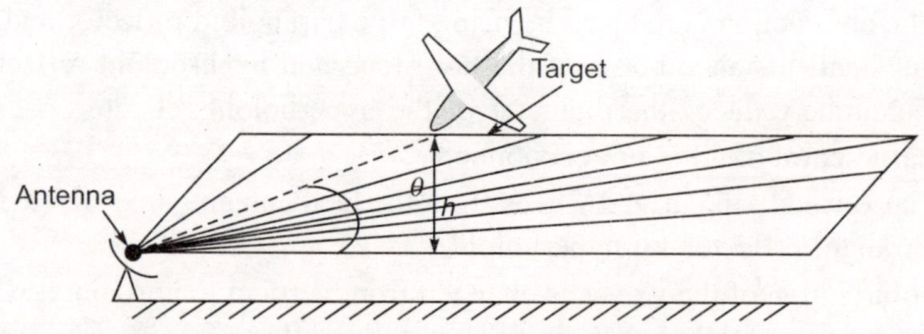

Fig. 8.40 Fanned Beam from Ground based antenna for air search

In fig. 8.40, the beam is from ground based antenna for air search. The beam in elevation provides coverage on aircraft. Azimuth coverage is obtained with scanning. The beam shown in fig. 8.41 is from airborne antenna for surface search.

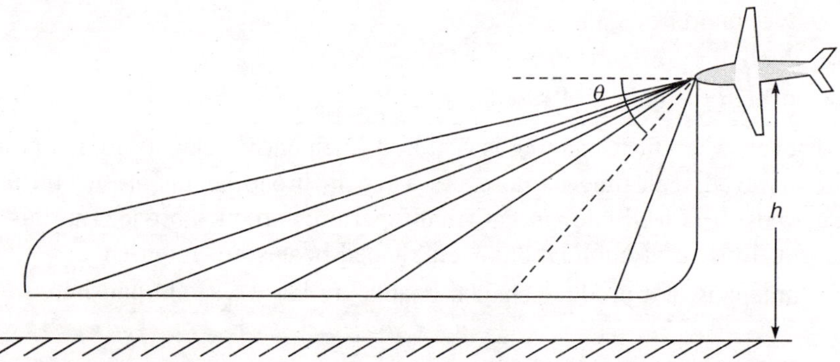

Fig. 8.41 Fanned beam from airborne antenna for surface search

2. Sector beams

This beam is basically flat over a desired angular region as shown in fig. 8.41. This beam is again sharp in azimuth and is broad in elevation to accommodate roll and pitch. This type of beam provides a more constant illumination of the target and is also more conservative.

Typical sector beam is shown in figs. 8.42 and 8.43.

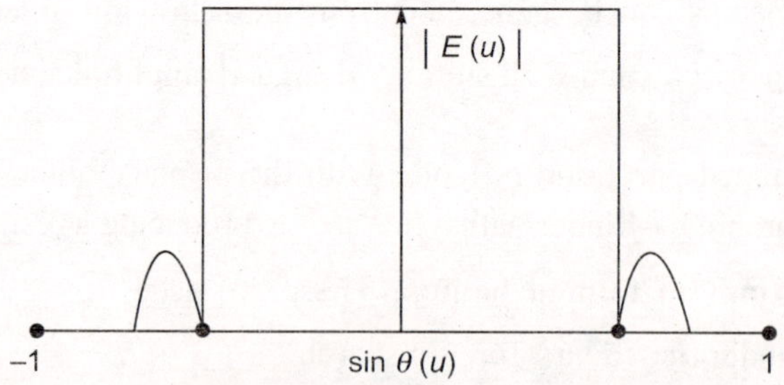

Fig. 8.42 Sector beam

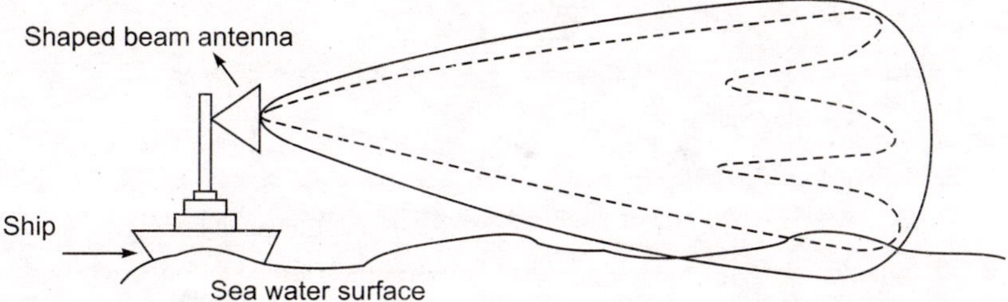

Fig. 8.43 Sector beam for surface search from shipborne antenna

The sector beams are also used for surface search from shipborne antennas and air search from ground based antennas.

3. Cosecant square beams

The cosecant square beams are used for ground mapping, airport surveillance purposes. Typical cosecant square beams is shown in figs. 8.44.

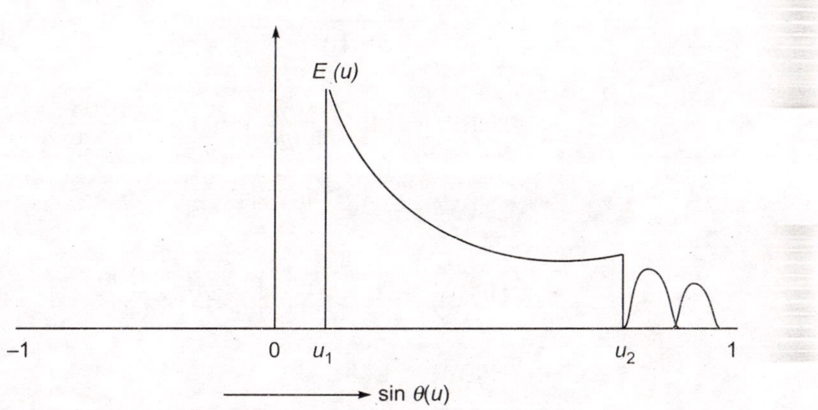

Fig. 8.44 Cosecant square beam

These shaped beams can be precisely produced from array antennas with appropriate amplitude and phase distributions.

When the beam is stationary in azimuth, an aeroplane flying across the beam is illuminated for a certain period. This period is proportional to its distance from the antenna. Low intensity broadened beam is desired to increase time of illumination on moving targets. If a fixed minimum illumination is required at a specified linear distance, a shaped beam called $\mathrm{cosec}^2\ \theta$ is used.

For height finding applications, a ground based or shipborne antenna produces a sharp elevation beam for getting precise elevation information and a rapid elevation scan. A typical beam is shown in fig. 8.45.

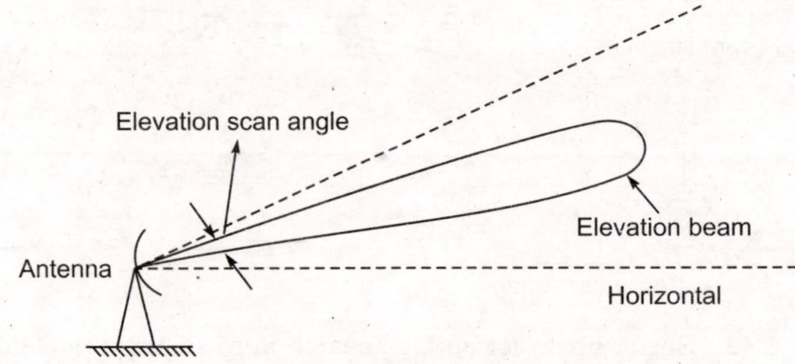

Fig. 8.45 Elevation narrow beam

PROBLEM 8.1 Determine the NNBW of 2.5 m paraboloid reflector used at 5 GHz. Also find HPBW.

Solution

$$D_a = 2.5 \text{ m}$$

Frequency, $\qquad f = 5 \text{ GHz}$

$$\lambda = \frac{3 \times 10^8}{5 \times 10^9} = 0.06 \text{ m}$$

$$\text{NNBW} = 140 \times \left(\frac{\lambda}{D_a}\right) = 140 \times \frac{0.06}{2.5}$$

$$\boxed{\text{NNBW} = 3.36°}$$

$$\text{HPBW} = 70 \times \left(\frac{\lambda}{D_a}\right) = 70 \times \frac{0.06}{2.5}$$

$$\boxed{\text{HPBW} = 1.68°}$$

PROBLEM 8.2 Determine the gain of paraboloid of 2.5 m diameter operating at 5 GHz when half-wave dipole feed is used.

Solution

$$D_a = 2.5 \text{ m}$$

Frequency, $\qquad f = 5 \text{ GHz}$

$$\lambda = \frac{3 \times 10^8}{5 \times 10^9} = 0.06 \text{ m}$$

The gain of the paraboloid is

$$g_p = 6.4\left(\frac{D}{\lambda}\right)^2 = 6.4 \times \left(\frac{2.5}{0.06}\right)^2$$

$\therefore \qquad \boxed{g_p = 11,111 = 40.45 \text{ dB}}$

PROBLEM 8.3   What is the bandwidth between first nulls and half power points of radiation pattern of paraboloid operating at 9 GHz which has a mouth diameter of 0.15 m. Also find out power gain.

Solution $\qquad\qquad\qquad D_a = 0.15$ m

Frequency, $\qquad\qquad\qquad f = 9$ GHz

Mouth diameter, $\qquad\qquad D_a = 0.15$ m

$$\lambda = \frac{3 \times 10^8}{9 \times 10^9} = 0.033 \text{ m}$$

$$\frac{\lambda}{D_a} = 0.222$$

$\therefore \qquad\qquad$ NNBW $= \phi_0 = 140 \times \left(\frac{\lambda}{D_a}\right) = 31.11°$

$$\text{HPBW} = \phi = 70 \times \left(\frac{\lambda}{D_a}\right) = 15.55°$$

Power gain, $\qquad\qquad g_p = 6.4 \left(\frac{D_a}{\lambda}\right)^2 = 129.63 = 21.13 \text{ dB}$

$\therefore \qquad\qquad \boxed{\phi_0 = 31.11°,\ \phi = 15.55°,\ g_p = 129.63 = 21.13 \text{ dB}}$

PROBLEM 8.4   If a paraboloid reflector antenna with 2.0 m diameter is operating at 2 GHz, find out the power gain in dB.

Solution   Frequency, $\qquad\quad f = 2$ GHz

Diameter, $\qquad\qquad\quad D_a = 2.0$ m

$$\therefore \qquad\qquad \lambda = \frac{3 \times 10^8}{2 \times 10^9} = 0.15 \text{ m}$$

The power gain, $\qquad\quad g_p = 6.4 \left(\frac{2.0}{\lambda}\right)^2 = 1137.78$

$$g_p(\text{dB}) = 10 \log_{10}(1137.78) = 30.56 \text{ dB}$$

$\therefore \qquad\qquad \boxed{g_p = 29.64 \text{ dB}}$

PROBLEM 8.5   A paraboloid is operating at 6 GHz has a radiation pattern with null-to-null beam width of 5°. Find out the mouth diameter of paraboloid, HPBW and power gain.

Solution   Frequency, $\qquad\qquad f = 6$ GHz

$$\lambda = \frac{3 \times 10^8}{6 \times 10^9} = 0.05 \text{ m}$$

We have
$$\phi_0 = \text{NNBW} = 140 \times \left(\frac{\lambda}{D_a}\right)$$

or
$$D_a = 140 \times \left(\frac{\lambda}{\text{NNBW}}\right)$$

$$= 140 \times \left(\frac{0.05}{5}\right) = 1.4 \text{ cm}$$

Mouth diameter, $D_a = 1.4$ m

$$\text{HPBW} = \phi = 70 \times \left(\frac{\lambda}{D_a}\right) = 2.5°$$

Power gain, $g_p = 6.4\left(\frac{D_a}{\lambda}\right)^2$

$\therefore$ $g_p = 5017.6$

$\therefore$ $\boxed{D_a = 0.84 \text{ m}, \phi = 5°, g_p = 1254.4}$

PROBLEM 8.6 For a paraboloid reflector of diameter of 5 m, illumination efficiency, $b = 0.65$. The frequency of operation is 9 GHz. Find out its beamwidth, directivity and capture area.

Solution Frequency, $f = 9$ GHz

$\therefore$
$$\lambda = \frac{3 \times 10^8}{9 \times 10^9} = 0.0333 \text{ m}$$

Mouth diameter, $D_a = 5$ m

Actual area, $A = \dfrac{\pi D_a^2}{4} = \pi \times 25/4 = 19.63 \text{ m}^2.$

Capture area, $A_c = 0.65\, A = 12.76 \text{ m}^2$

Directivity, $D = 6.4 \times \left(\dfrac{D_a}{\lambda}\right)^2 = 144288.43 = 51.59$ dB

$$\text{HPBW} = \phi = 70 \times \left(\frac{\lambda}{D_a}\right) = 0.466°$$

$$\text{NNBW} = \phi_0 = 2\phi = 0.9324°$$

$\therefore$ $\boxed{\begin{array}{l}\phi_0 = 0.9324°, \phi = 0.466°, A_c = 12.76 \text{ m}^2, \\ \text{Directivity} = 51.59 \text{ dB}\end{array}}$

PROBLEM 8.7   A paraboloid reflector operates at 5 GHz. Its mouth diameter is 5 m. It is required to measure far-field pattern of the paraboloid. Find out the minimum distance required between the two antennas.

Solution   The minimum distance required

$$r = \frac{2D_a^2}{\lambda}$$

Where
$$D_a = 5.0 \text{ m}$$
$$f = 5 \text{ GHz}$$

$\therefore$
$$\lambda = \frac{3 \times 10^8}{5 \times 10^9} = 0.06 \text{ m}$$

$\therefore$
$$r = \frac{2D_a^2}{\lambda} = 833.33 \text{ m}$$

PROBLEM 8.8   A paraboloid reflector is required to have a power gain of 500 at a frequency of 4 GHz. Determine mouth diameter and beam width of the antenna.

Solution   Frequency,
$$f = 4 \text{ GHz}$$

$\therefore$
$$\lambda = \frac{3 \times 10^8}{4 \times 10^9} = 0.075 \text{ m}$$

Required power gain,
$$g_p = 500$$

We have,
$$g_p = 6.4 \times \left(\frac{D_a}{\lambda}\right)^2 \quad \text{or}$$

Mouth diameter,
$$D_a = \lambda\sqrt{\frac{g_p}{6.4}}$$

$$= 0.075\sqrt{\frac{500}{6.4}} = 0.663 \text{ m}$$

$$\text{HPBW} = 70 \times \left(\frac{\lambda}{D_a}\right) = 7.92°$$

$$\text{NNBW} = 140 \times \left(\frac{\lambda}{D_a}\right) = 15.84°$$

$\therefore$
$$D_a = 0.663 \text{ m}, \ \text{HPBW} = 7.92°, \ \text{BWFN} = 15.84°$$

PROBLEM 8.9   A paraboloid reflector operates at a frequency of 10 GHz and it provides a power gain of g_p = 100 dB. Find out capture area of the paraboloid and beamwidth.

Solution   Frequency, $f = 9$ GHz

$$\therefore \qquad \lambda = \frac{3 \times 10^8}{9 \times 10^9} = 0.0333 \text{ m}$$

Power gain, $$g_p = 6.4 \left(\frac{D_a}{\lambda}\right)^2$$

But, $g_p = 100$ dB

i.e. $g_p = 10 \log_{10} g = 100$

or $\log_{10} g_p = 100/10 = 10$

or $g_p = 10^{10}$

$$\therefore \qquad g_p = 6.4 \left(\frac{D_a}{\lambda}\right)^2 = 10^{10}$$

or $$\left(\frac{D_a}{\lambda}\right)^2 = \frac{10^{10}}{6.4}$$

or $$D_a = 0.0333 \times \sqrt{\frac{10^{10}}{6.4}}$$

$$D_a = 0.0333 \, \frac{10^5}{\sqrt{6.4}}$$

Mouth diameter, $D_a = 1317.48$ m

Actual area, $$A = \frac{\pi D_a^2}{4} = 1363257.65 \text{ m}^2$$

Capture area, $A_c = 0.65 \times A$
$$= 0.65 \times 1363257.65 = 886117.47 \text{ m}^2$$

$$\text{NNBW} = \phi_0 = 140 \times \left(\frac{\lambda}{D_a}\right) = 0.0035°$$

$$\text{HPBW} = \phi = 70 \times \left(\frac{\lambda}{D_a}\right) = 0.00175°$$

$$\therefore \qquad \boxed{A_c = 886117.47 \text{ m}^2, \ \phi_0 = 0.0035°, \ \phi = 0.00175°}$$

PROBLEM 8.10   A parabolic reflector is operated at 10 GHz and it has mouth diameter of 5 m. If it is fed by a non-directional antenna, find out HPBW, BWFN and power gain.

Solution   Frequency, $f = 10$ GHz

$$\therefore \qquad \lambda = \frac{3 \times 10^8}{2 \times 10^9} = 0.03 \text{ m}$$

Mouth diameter, $D_a = 5$ meters

Power gain, $g_p = 6.4 \left(\frac{D_a}{\lambda} \right)^2 = 177777.78$

Or $g_p = 52.499$ dB

$$\text{HPBW} = \phi = 70 \times \left(\frac{\lambda}{D_a} \right) = 0.42°$$

$$\text{NNBW} = \phi_0 = 140 \times \left(\frac{\lambda}{D_a} \right) = 0.84°$$

$$\therefore \qquad \boxed{g_p = 52.499 \text{ dB}, \ \phi = 0.42°, \ \phi_0 = 0.84°}$$

PROBLEM 8.11   A parabolic reflector with a mouth diameter of 12 meters operates at $f = 10$ GHz. It has illumination efficiency of 0.6. Find out the power gain.

Solution   Mouth diameter, $D_a = 12$ m
   Frequency, $f = 10$ GHz

$$\therefore \qquad \lambda = \frac{3 \times 10^8}{5 \times 10^9} = 0.03 \text{ m}$$

Illumination efficiency = 0.6

Power gain, $g_p = \text{illumination efficiency} \times \left(\frac{D_a}{\lambda} \right)^2$

$$= 0.6 \times \left(\frac{12}{0.03} \right)^2 = 96000 \text{ or } 49.82 \text{ dB}$$

$$\therefore \qquad \boxed{\text{Power gain } g_p = 49.82 \text{ dB}}$$

PROBLEM 8.12   For what mouth diameter and capture area of a paraboloid reflector a NNBW of 8° is obtained when it is operated at 4 GHz.

Solution   Frequency, $f = 4$ GHz

$$\therefore \qquad \lambda = \frac{3 \times 10^8}{4 \times 10^9} = 0.075 \text{ m}$$

$$\text{NNBW} = 140 \times \left(\frac{\lambda}{D_a}\right) = 8°$$

i.e.
$$D_a = \frac{140\,\lambda}{8} = 1.3125 \text{ m}$$

Capture area,
$$A_c = 0.65 \, A$$

Here,
$$A = \frac{\pi D_a^2}{4} = 1.3529 \text{ m}^2$$

$$\therefore \qquad A_c = 0.879 \text{ m}^2$$

$$\therefore \qquad \boxed{\text{Mouth diameter, } D_a = 1.3125 \text{ m} \\ A_c = 0.879 \text{ m}^2}$$

PROBLEM 8.13   A paraboloid reflector is required to produce a beamwidth between the first nulls equal to 2° at an operating frequency of 4.0 GHz. Find out the mouth diameter and power gain.

Solution   frequency, $f = 4$ GHz

$$\text{NNBW} = 2°$$

$$\therefore \qquad \lambda = \frac{3 \times 10^8}{2.5 \times 10^9} = 0.075 \text{ m}$$

but
$$\text{NNBW} = 140 \times \left(\frac{\lambda}{D_a}\right) = 2°$$

$$\therefore \qquad D_a = 140 \times \left(\frac{\lambda}{2}\right) = \left(\frac{140 \times 0.075}{2}\right) = 5.25 \text{ m}$$

Power gain,
$$g_p = 6.4 \left(\frac{D_a}{\lambda}\right)^2 = 31360 \text{ or } 44.96 \text{ dB}$$

$$\therefore \qquad \boxed{D_a = 5.25 \text{ m}, \; g_p = 44.96 \text{ dB}}$$

PROBLEM 8.14   A paraboloid reflector has a radiation characteristic whose half power beam width is 6°. Find out its null-to-null beam width and power gain.

Solution
$$\text{HPBW} = \phi = 6°$$
$$\therefore \qquad \text{NNBW} = \phi_0 = 2\,\phi = 12°$$

but
$$\phi = 70 \times \left(\frac{\lambda}{D_a}\right)$$

or
$$\left(\frac{\lambda}{D_a}\right) = \frac{\phi}{70}$$

i.e.
$$\left(\frac{D_a}{\lambda}\right) = \frac{70}{6} = 11.667$$

Power gain,
$$g_p = 6.4\left(\frac{D_a}{\lambda}\right)^2 = 871.11$$

or
$$g_p = 29.40 \text{ dB}$$

$\therefore$
$$\boxed{\text{NNBW} = 12°, \ g_p = 29.40 \text{ dB}}$$

PROBLEM 8.15   What is power gain of a paraboloid reflector whose mouth diameter is equal to 6λ

Solution   Power gain,
$$g_p = 6.4\left(\frac{D_a}{\lambda}\right)^2$$

Here,
$$D_a = 6\lambda$$

$\therefore$
$$g_p = 6.4\left(\frac{6\lambda}{\lambda}\right)^2 = 230.4$$

$\therefore$
$$\boxed{g_p = 23.62 \text{ dB}}$$

PROBLEM 8.16   Determine half power and null-to-null beamwidths of paraboloid reflector whose aperture diameter is 7λ. Also find out its directivity.

Solution   Aperture diameter, $= 7\lambda$

$$\text{NNBW} = \phi = 70 \times \left(\frac{\lambda}{D}\right) = 70 \times \left(\frac{\lambda}{7\lambda}\right) = 10°$$

Null-to-null beam width,
$$\phi_0 = 2\phi = 20°$$

The directivity of the paraboloid,

$$D = 6.4 \times \left(\frac{7\lambda}{\lambda}\right)^2$$

$$= 6.4 \times 49 = 313.6$$

$\therefore$
$$\boxed{\phi = 10°, \ \phi_0 = 20°, \ D = 313.6}$$

PROBLEM 8.17   The aperture dimensions of a pyramidal horn are 9×4 cm. It is operating at a frequency of 8 GHz. Find out beamwidth power gain, and directivity.

Solution   Frequency, $f = 8$ GHz

$$\therefore \qquad \lambda = \frac{3 \times 10^8}{8 \times 10^9} = 0.0375 \text{ m} = 3.75 \text{ cm.}$$

$$d = 9 \text{ cm, } W = 4 \text{ cm.}$$

Half power beam width $=$ HPBW

$$\phi_E = 56\frac{\lambda}{d} = 56 \times \frac{3.75}{9} = 23.33°$$

$$\phi_H = 67\frac{\lambda}{W} = 67 \times \frac{3.75}{4} = 62.81°$$

Power gain, $$g_p = \frac{4.5\,Wd}{\lambda^2} = 11.52 = 10.61 \text{ dB}$$

Directivity, $$D = \frac{7.5\,Wd}{\lambda^2} = \frac{7.5 \times 9 \times 4}{3.75^2} = 19.2$$

$\therefore$

$$\boxed{\begin{array}{l} \phi_E = 23.33°, \ \phi_H = 62.81° \\ g_p = 10.61 \text{ dB, } D = 19.2 \end{array}}$$

PROBLEM 8.18   Find out the power gain of a square horn antenna whose aperture size is 10λ.

Solution   The power gain, $$g_p = \frac{4.5\,Wd}{\lambda^2}$$

$$= \frac{4.5 \times 10\lambda \times 10\lambda}{\lambda^2} = 450$$

$\therefore$

$$\boxed{g_p = 26.532 \text{ dB}}$$

PROBLEM 8.19   Find out power gain and directivity of a horn whose dimensions are 10×5 cm operating at a frequency of 8 GHz.

Solution   The dimensions of horn are $d = 10$ cm, $w = 5$ cm, $f = 8$ GHz

$$\therefore \qquad \lambda = \frac{3 \times 10^8}{8 \times 10^9} = 0.0375 \text{ m} = 3.75 \text{ cm.}$$

Power gain, $$g_p = \frac{4.5\,Wd}{\lambda^2} = 16 = 12.04 \text{ dB.}$$

Directivity,
$$D = \frac{7.5\,Wd}{\lambda^2} = 26.667 = 14.26 \text{ dB}$$

$$\therefore \qquad \boxed{g_p = 12.04 \text{ dB},\ D = 14.26 \text{ dB}}$$

PROBLEM 8.20 Find out the complementary slot impedance, when the dipole impedance is

(a) $Z_d = 73 + j\,50\ \Omega$

(b) $Z_d = 70\ \Omega$

(c) $Z_d = 800\ \Omega$

(d) $Z_d = 400\ \Omega$

(e) $Z_d = 50 + j\,10\ \Omega$

(f) $Z_d = 50 - j\,30\ \Omega$

(g) $Z_d = 350\ \Omega$

Solution

(a) We have

$$Z_s = \text{slot impedance}$$

$$= \frac{\eta_0^2}{4(R_d^2 + X_d^2)}\,(R_d - jX_d)$$

$$= \frac{35530.6}{(R_d^2 + X_d^2)}\,(R_d - jX_d)$$

If
$$Z_d = R_d + j\,X_d = 73 + j\,50\ \Omega$$

$$\boxed{Z_s = 331.29 - j\,226.9,\ \Omega}$$

(b) If
$$Z_d = 70 + j\,0\ \Omega$$

$$\boxed{Z_s = 507.58}$$

(c) If
$$Z_d = 800 + j\,0\ \Omega$$

$$\boxed{Z_s = 44.413\ \Omega}$$

(d) If
$$Z_d = 400 + j\,0$$

$$\boxed{Z_s = 88.826\ \Omega}$$

(e) If
$$Z_d = 50 + j\,10\ \Omega$$

$$\boxed{Z_s = 683.25 - j\,136.65\ \Omega}$$

(f) If
$$Z_d = 50 - j\,30$$

$$\boxed{Z_s = 522.5 + j\,313.5\ \Omega}$$

(g) If
$$Z_d = 350 \ \Omega$$

$$\boxed{Z_s = 101.51 \ \Omega}$$

PROBLEM 8.21 What is the radiation resistance of Hertzian dipole of length $\dfrac{\lambda}{20}, \ \dfrac{\lambda}{30}, \ \dfrac{\lambda}{40}$.

Solution The radiation resistance of Hertzian dipole of length $d\ell$ is

$$R_r = 80\pi^2 \left(\frac{d\ell}{\lambda}\right)^2 \ \Omega$$

If $d\ell = \dfrac{\lambda}{20}$
$$R_r = 80\pi^2 \left(\frac{\lambda}{20} \times \frac{\ell}{\lambda}\right)^2$$

$\therefore$
$$R_r = 1.973 \ \Omega$$

If $d\ell = \dfrac{\lambda}{30}$
$$R_r = 80\pi^2 \frac{1}{30^2}$$

or
$$R_r = 0.877 \ \Omega$$

If $d\ell = \dfrac{\lambda}{40}$
$$R_r = 80\pi^2 \frac{1}{40^2}$$

or
$$\boxed{R_r = 0.493 \ \Omega}$$

PROBLEM 8.22 Derive an expression for the directivity of a half-wave dipole

Solution For half-wave dipole,

$$E_{(max)} = \frac{60I}{r}$$

But
$$P_r = 73 \ I^2 = \text{watts}$$
For
$$P_r = 1 \ w,$$

$$I = \frac{1}{\sqrt{73}}$$

$\therefore$
$$E_{(max)} = \frac{60}{r} \times \frac{1}{\sqrt{73}}$$

$$g_{d(max)} = \frac{4\pi(\psi)}{P_r}$$

$$= 4\pi(\psi) \quad [\text{as } P_r = 1 \text{ watt}]$$

$$= 4\pi \times \frac{r^2 E^2}{\eta_0} \quad \left[\text{as } \psi = r^2 \frac{E^2}{\eta_0}\right]$$

$$= \frac{4\pi \times r^2}{\eta_0} \frac{60^2}{r^2} \frac{1}{73}$$

$$= \frac{4\pi \times 60 \times 60}{120\pi} \frac{1}{73}$$

$$= \frac{120}{73} = 1.644$$

$$\therefore \quad \boxed{g_d(\text{max}) = D = 1.644}$$

PROBLEM 8.23 Obtain the radiated power of an antenna whose radiation resistance is 300 Ω operates at a frequency of 12 GHz and with a current of 2 amperes. Find out radiated power.

Solution Radiated power, $\quad P_r = I^2 R_r$

$$= 2^2 \times 300$$

$$= 4 \times 300$$

$$\therefore \quad \boxed{P_r = 1200 \text{ watts}}$$

PROBLEM 8.24 Find the effective area of a half-wave dipole operating at 600 MHz.

Solution The effective area of an antenna is

$$A_e = \frac{\lambda^2}{4\pi} g_d$$

As $\quad f = 600$ MHz.

$$\lambda = \frac{3 \times 10^8}{600 \times 10^6}$$

$$= \frac{3}{6} = 0.5 \text{ m}$$

Directivity of half-wave dipole is

$$(g_d)_{\text{Max}} = D = 1.644$$

$$\therefore \quad A_e = \frac{0.5^2}{4\pi} \times 1.644$$

$$\therefore \quad \boxed{A_e = 0.0327 \text{ m}^2}$$

PROBLEM 8.25 Find the effective area of a Hertzian dipole operating at 200 MHz.

Solution As $\quad f = 200$ MHz,

$$\lambda = \frac{3 \times 10^8}{200 \times 10^6} = 1.5 \text{ m}$$

Directivity of Hertzian dipole, $D = 1.5$

$$\therefore \quad A_e = \text{effective area}$$

$$= \frac{\lambda^2}{4\pi} = \frac{1.5^2 \times 1.5}{4\pi} = 0.268 \text{ m}^2$$

$$\therefore \quad \boxed{A_e = 0.268 \text{ m}^2}$$

8.40 POINTS TO REMEMBER

1. Radiation intensity, $RI = \dfrac{r^2 E^2}{\eta_0}$ watts/unit solid angle

2. Directive gain, $g_d = \dfrac{4\pi \times (RI)}{w_r}$

3. Directivity, $D = (g_d)_{\max}$

4. Power gain, $g_p = \dfrac{4\pi \times (RI)}{w_t}$

5. Antenna efficiency, $\eta = \dfrac{g_p}{g_d}$

6. Effective area, $A_e = \dfrac{\lambda^2}{4\pi} g_d$ or $A_e = \dfrac{\text{received power}}{\text{power flow of incident waves}}$

7. Far-field is represented by $\dfrac{1}{r}$ field term

8. Induction field is represented by $\dfrac{1}{r^2}$ field term.

9. Radiation resistance of Hertzian dipole is $80\pi^2 \left(\dfrac{dl}{\lambda}\right)^2 \ \Omega$

10. Electrostatic field is represented by $\dfrac{1}{r^3}$

11. The far-field and induction field have equal magnitudes at $r = \dfrac{\lambda}{2\pi}$

12. Radiation resistance of half-wave dipole is 73 Ω.

13. Radiation resistance of quarter-wave monopole is 36.5 Ω.

14. Horizontal pattern of vertical dipole is a circle.
15. Radiated power flow of a vertical dipole is in the radial direction.
16. Microwave region extends from 1 GHz to 100 GHz.
17. Parabolic dish antennas are popular in microwave region to produce narrow beams.
18. Reflectors are used to modify the radiation of primary antenna.
19. Cassegrain feed mechanism is very popular for low noise applications.
20. HPBW of paraboloid is $\phi = \dfrac{70\lambda}{D_a}$
21. BWFN of paraboloid is $\phi = \dfrac{140\lambda}{D_a}$
22. Directivity, $D = 9.87\left(\dfrac{D}{\lambda}\right)^2$
23. Power gain of paraboloid is $g_p = 6.4\left(\dfrac{D_a}{\lambda}\right)^2$
24. Paraboloids are popular in TV reception and radars.
25. Shaped beam antennas are useful to produce desired shapes of beam.
26. Narrow beams are useful for point-to-point communication and high angular resolution radars.
27. Sector beams are more useful in search radars.
28. CSC beams are useful for ground mapping and airport surveillance.
29. Horn is a flared out waveguide.
30. The efficiency of corrugated horn is more than that of conventional horn.
31. The directivity of horn is $D = \dfrac{7.5\,A}{\lambda^2}$
32. The power gain of horn is $g_p = \dfrac{4.5\,A}{\lambda^2}$
33. Slot antennas are compact and slot arrays are popular in marine radars.
34. The method of moment (MOM) is useful for finding field distribution in slots and other wire antennas.
35. Lens antennas are popular to convert spherical wave front to plane wave front.
36. Polystyrene and lucite are popular lens materials.
37. The shape of the lens is represented by $r = \dfrac{1(n-1)}{(n\cos\phi - 1)}$
38. Microstrip antennas are popular in cellular phones and to install in places like aircraft's body etc.

39. The desired polarization can be obtained by different shapes of microstrip antenna.

40. The characteristic impedance of patch antenna is $z_p = \dfrac{z_0}{n\sqrt{\epsilon_r}}$

8.41 MULTIPLE CHOICE QUESTIONS

1. The common microwave link antenna is
   (a) Dipole
   (b) Log-periodic
   (c) Rhombic
   (d) Parabolic dish

2. The directivity of half-wave dipole is
   (a) 10
   (b) 1
   (c) 1.5
   (d) 1.64

3. The directivity of current element is
   (a) 1.64
   (b) 1.5
   (c) 2.0
   (d) 5.0

4. Radar antenna is
   (a) parabolic dish
   (b) dipole
   (c) horn
   (d) waveguide

5. The radiation resistance of a current element is

   (a) $80\ \pi^2 \left(\dfrac{dI}{\lambda}\right)^2 \Omega$
   (b) $80\ \pi \left(\dfrac{dI}{\lambda}\right)^2 \Omega$

   (c) $80\ \pi^2 \left(\dfrac{dI}{\lambda}\right) \Omega$
   (d) $80\ \pi \left(\dfrac{dI}{\lambda}\right) \Omega$

6. Circularly polarized antenna is
   (a) dipole
   (b) parabolic dish
   (c) Yagi-Uda
   (d) Helical

7. The directivity of isotropic radiator is
   (a) 1
   (b) zero
   (c) more than 1
   (d) ∞

8. Effective length of half-wave dipole is
   (a) $\lambda/2$
   (b) $>\lambda/2$
   (c) $<\lambda/2$
   (d) 0.55λ

9. A Balun is
   (a) a resistor
   (b) an impedance transformer
   (c) an antenna
   (d) frequency converter

10. The polarization of horizontal dipole is
    (a) horizontal
    (b) vertical
    (c) θ-polarization
    (d) circular

11. Crossed dipoles produce polarization of
    (a) linear
    (b) circular
    (c) horizontal
    (d) vertical
12. Antenna is
    (a) transducer
    (b) filter
    (c) regulator
    (d) amplifier
13. The null-to-null beam width in end-fire array is

    (a) $\dfrac{2\lambda}{Nd}$

    (b) $\sqrt{\dfrac{2\lambda}{Nd}}$

    (c) $2\sqrt{\dfrac{\lambda}{Nd}}$

    (d) $2\sqrt{\dfrac{2\lambda}{Nd}}$

14. The null-to-null beam width in broadside array is

    (a) $\dfrac{2\lambda}{Nd}$

    (b) $2\sqrt{\dfrac{2\lambda}{Nd}}$

    (c) $\dfrac{2\lambda^2}{Nd}$

    (d) $\sqrt{\dfrac{Nd}{2\lambda}}$

15. The maximum directive gain of current element is
    (a) 1.76 dB
    (b) 2.15 dB
    (c) 3 dB
    (d) 0 dB
16. The real part of antenna impedance consists of
    (a) R_r only
    (b) R_r and R_l
    (c) R_l only
    (d) zero ohms of resistance
17. Power and field patterns are related as
    (a) $P \propto E^2$
    (b) $P \propto E$

    (c) $P \propto \sqrt{E}$

    (d) $P \propto \dfrac{1}{E}$

18. Antenna efficiency is

    (a) $\dfrac{g_p}{g_d}$

    (b) $\dfrac{g_d}{g_p}$

    (c) g_p

    (d) g_d

19. If the response of a vertical dipole is 1 for a unity normalized input power, the polarization is
    (a) vertical
    (b) horizontal
    (c) circular
    (d) elliptical
20. If the response of any type of antenna is 0.5 for unity normalized power, the polarization of test antenna is
    (a) linear
    (b) horizontal
    (c) vertical
    (d) unpolarized

21. When the array length is high, the null-to-null beam width is
    (a) small
    (b) high
    (c) constant
    (d) infinity

22. The radiation intensity of an isotropic radiator is
    (a) $\dfrac{P_r}{4\pi r^2}$
    (b) $\dfrac{P_r}{4\pi r}$
    (c) $\dfrac{P_r}{4\pi}$
    (d) P_r

23. An omnidirectional antenna is a
    (a) parabolic dish
    (b) dipole
    (c) horn
    (d) Yagi-Uda antenna

24. A good front-to-back ratio
    (a) increases co-channel interference
    (b) reduces co-channel interference
    (c) has no effect on co-channel interference
    (d) none of these

25. Antenna radiation efficiency is high when its length is
    (a) $\dfrac{\lambda}{2}$
    (b) λ
    (c) $3\dfrac{\lambda}{2}$
    (d) ∞

26. For a 100 Ω antenna with $2A$ of current, the radiated power is
    (a) 400 watts
    (b) 200 watts
    (c) 50 watts
    (d) 25 watts

27. For an ideal antenna, the radiation resistance is
    (a) 73 Ω
    (b) 36.5 Ω
    (c) 292 Ω
    (d) input impedance

28. The power gain in dB of isotropic radiator is
    (a) 0
    (b) 1
    (c) 1.5
    (d) 1.64

29. Half-power beam width of optimum flare horn in E-plane, is
    (a) $\dfrac{56\lambda}{d_E}$
    (b) $\dfrac{28\lambda}{d_E}$
    (c) $\dfrac{122\lambda}{d_E}$
    (d) 112°

30. The normalized radiated power of a dipole is
    (a) 1
    (b) 1.5
    (c) $\sin^2 \theta$
    (d) 1.64

31. The directive gain of electric dipole is
    (a) 1.5
    (b) $1.5 \sin^2 \theta$
    (c) 1.64
    (d) 1.0
32. The effective length of half-wave dipole is
    (a) 0.4 λ
    (b) 0.45 λ
    (c) $\dfrac{\lambda}{\pi}$
    (d) 0.55 λ .
33. Directive gain is equal to power gain if
    (a) $\eta = \infty$
    (b) $\eta = 1$
    (c) $\eta = g_p$
    (d) $\eta = g_d$
34. The radiation resistance of an antenna which radiates 10 kW when a current of 10 ampere flows in it, is
    (a) 100 Ω
    (b) 1000 Ω
    (c) 10 Ω
    (d) 100 KΩ
35. If half-power beamwidth of parabolic antenna is 12°, its null-to-null beamwidth is
    (a) 12°
    (b) 60°
    (c) 24°
    (d) 48°
36. If null-to-null beamwidth of a parabolic antenna is 6.5°, its half-power beamwidth is
    (a) 3.25°
    (b) 6.5°
    (c) 13°
    (d) 26°
37. If the mouth diameter of a parabolic antenna is 2.5 m and if it is operating at a frequency of 10 GHz, the power gain in dB is
    (a) 46.19
    (b) 25
    (c) 250
    (d) 100
38. If the mouth diameter of a parabolic antenna is 2.5 m, and if it is operating at $\lambda = 0.25$ m, half-power beamwidth is
    (a) 7.0°
    (b) 14.0°
    (c) 3.5°
    (d) 21°
39. If the mouth diameter of parabolic antenna is 2.5 m, and if it is operating at $\lambda = 0.25$ m, null-to-null beam width is
    (a) 7°
    (b) 14°
    (c) 21°
    (d) 3.5°
40. If parabolic dish diameter increases
    (a) beamwidth becomes small
    (b) beamwidth becomes high
    (c) beamwidth becomes high and sometimes small
    (d) beamwidth remains constant

41. The radiated electric field is

    (a) $\propto \sqrt{P_r}$

    (b) $\propto P_r$

    (c) $\propto \dfrac{1}{P_r}$

    (d) $\propto \dfrac{1}{\sqrt{P_r}}$

42. The radiation resistance of a current element is

    (a) $\propto dl$ (b) $\propto (dl)^2$

    (c) $\propto \dfrac{1}{dI}$ (d) $\propto \dfrac{1}{(dI)^2}$

43. The common mobile antenna is

    (a) dipole   (b) V-antenna

    (c) whip antenna   (d) dish antenna

44. The length of mobile antenna is

    (a) λ (b) $\lambda/2$

    (c) $\lambda/4$ (d) $> \lambda$

45. At $f = 30$ MHz, the length of the mobile whip antenna is

    (a) 0.4572 m   (b) 4.572 m

    (c) 45.72 m   (d). 0.4572 m

8.42 ANSWERS

1. d	2. d	3. d
4. b	5. a	6. a
7. d	8. a	9. c
10. a	11. d	12. a
13. a	14. b	15. a
16. b	17. a	18. a
19. d	20. a	21. c
22. b	23. b	24. a
25. a	26. d	27. a
28. a	29. c	30. b
31. c	32. c	33. b
34. a	35. c	36. a
37. a	38. a	39. b
40. a	41. a	42. b
43. c	44. c	45. a

8.43 EXERCISE PROBLEMS

1. The field amplitude due to half-wave dipole at 10 km is 0.1 V/m. it operates at 100 MHz. Find out dipole length and its radiated power.

2. What is the length of a half-wave dipole at frequencies of 10 MHz, 50 MHz and 100 MHz.

3. Find out maximum effective area of an antenna at a frequency of 2 GHz when the directivity is 100.

4. Obtain the gain of an antenna whose area is 12 m^2 and operating at a frequency of 6 GHz.

5. Find out the radiated power of an antenna if a current of 10 amp. exists and its radiation resistance is 32.0 Ω.

6. What is the radiation resistance of an antenna if it radiates a power of 120 W and when the current in it is 10 amp.

7. Find out directivity, efficiency and effective area of an antenna if its $R_r = 80 \ \Omega$, $R_l = 10 \ \Omega$. The power gain is 10 dB and antenna operates at a frequency of 100 MHz.

8. If the transmitting power is 10 kW, find out power density at distances of 10 km, 50 km, 100 km assuming the radiator is isotropic.

9. If the current element is z-directed, find out far-field components of H.

10. Derive an expression for distant field θ-field component of E for a dipole of length L.

11. (a) Find the current required to radiate power of 50 W at 60 MHz from 0.1 λ Hertzian dipole.

    (b) Determine its radiation resistance in the element.

12. Find out the radiation efficiency of a Hertzian dipole of length 0.03 λ at a frequency of 100 MHz if the loss resistance is 0.01 Ω.

13. The power gain of transmitting antenna $A_p = 20$. The input power is $P_{in} = 200$ W. Find out effective isotropic radiated power EIRP in dB and dB$_m$

14. Find out power density at a point at 12 km from the transmitting antenna. Its power gain is 10 and input power is 100W.

15. The radiation resistance of a transmitting antenna dipole is 80 Ω, loss resistance is 10 Ω, directive gain is 15 and input power is 1 kW. Find antenna efficiency and radiated power.

16. The capture area of a receiving antenna is 10 cm^2 and available power density is 10 μW/cm^2. Find out capture power.

17. What is bandwidth in percentage of an antenna operating at a frequency of 100 MHz if the dB frequencies are 300 MHz and 350 MHz.

18. The diameter of parabolic reflector is 2.0 m. It radiates a power of 100 W at an operating frequency of 3 GHz. Its efficiency is 60% and its aperture efficiency is 60%. Find out antenna power gain and bandwidth.

19. What is the free space path loss when the transmitting and receiving antennas are separated by 100 km, while operating at a frequency of 10 GHz.

20. A uniformly illuminated parabolic reflector whose aperture size is 2 m is operated at 6 GHz. Find out null-to-null beamwidth and power gain with reference to dipole of half-wave length. Assume that the antenna is lossless.

21. A 2 m parabolic reflector operating at $f = 6$ GHz radiates a power of 100 W. It has efficiency of 60% and its aperture efficiency is 60%. Find out antenna power gain in dB.

22. For the 3 m parabolic reflector operating at 8 GHz radiates a power of 10 W. It has an efficiency of 55% and its aperture efficiency is 60%. Find out the receiver power gain. Also find out EIRP.

23. The power radiated by an antenna is 100 W and dissipated power is 10 W. The antenna has a directional gain of 250. Find out antenna efficiency and power gain.

Chapter 9

Microwave Measurements

Measurement of parameters is a means of analysis.

9.1 INTRODUCTION

The main objective of this chapter is to provide the procedures of the measurement of different parameters of microwave devices. The methods to obtain the characteristics of the microwave active devices like reflex klystron and Gunn oscillator are presented.

The main and common microwave test and measurement equipment are:

- Microwave sources
- Oscilloscopes
- Spectrum analyzer
- Frequency meters/wavemeters
- Slotted lines
- Power meters
- Detectors
- Noise meters
- Coaxial slotted lines
- Waveguide slotted lines
- Microammeters
- Network analyzer
- VSWR meters

- Matched terminations
- Shorts
- Directional couplers
- Attenuators
- Isolators
- Flanges
- Waveguide terminations
- Coaxial to waveguide adopters
- Horns
- Klystron power supplies
- Gunn power supplies
- Reflex klystrons
- Gunn diodes
- Smith charts etc.

Most of the devices and components are described in the preceding chapters. The details of the other equipment are briefly presented here.

9.2 MICROWAVE SOURCES

These are known as microwave signal generators. These can be sweep generators or reflex klystrons. In the laboratories, Gunn oscillators are also often used. It is possible to generate continuous waves, *AM* and *FM* modes, pulse and sweep frequencies.

9.3 OSCILLOSCOPES AND SAMPLING OSCILLOSCOPES

These provide the display of time domain information.

9.4 FREQUENCY METERS / WAVEMETERS

A wavemeter is a frequency meter and consists of an adjustable resonant circuit. It is calibrated to provide the resonant frequency in terms of the setting of the tuning adjustment.

The resonant circuit consists of the following.

> ➢ Lumped constants
> ➢ Coaxial two-wire lines
> ➢ Cavities

The wavemeters are of three types

> ➢ Absorption type
> ➢ Reaction type
> ➢ Transmission type

9.5 ABSORPTION TYPE WAVEMETER

This type of meter indicates the current induced in it. The wavemeter is loosely coupled to the oscillations whose frequency is to be determined and is adjusted for optimum response.

9.6 REACTION TYPE WAVEMETER

In this type of wavemeter, the adjustment of the wavemeter corresponding to the frequency to be measured is obtained from the reaction produced by the wavemeter on the system. For example, when a wavemeter is loosely coupled to the tank circuit of the oscillator, the current drops abruptly when the coupled wavemeter is tuned to resonance with the frequency of the oscillator.

9.7 TRANSMISSION TYPE WAVEMETER

In this type of meter, wavemeter is used as a coupling device in a system that transmits power from the source to a load or indicator. In this, appreciable transmission of energy to the load occurs only when the wavemeter is tuned to the frequency of the source. These types of meters are widely used at microwave frequencies.

9.8 COAXIAL WAVEMETERS

These are of two types.

> ➤ Transmission wavemeter and
> ➤ Absorption wavemeter

The transmission wavemeter has two input lines into cavity. It allows only resonant frequency to propagate through it and the other frequencies are reflected back.

An absorption wavemeter absorbs energy from the line around the resonant frequency. The resonant frequency of the wavemeter cavity depends on the cavity dimensions. These dimensions are adjusted with the help of the shorted plunger in the end of the line.

9.9 SPECTRUM ANALYZER

It is used to display a wave in frequency domain.

The spectrum analyzer is of two types.

> ➤ The real-time spectrum analyzer
> ➤ Swept-tuned frequency spectrum analyzer

The real time spectrum analyzer displays transient response of a system.

Both the above types display periodic and random signals. The main disadvantage of the spectrum analyzer is that it does not display phase relation on the cathode ray tube.

The real time spectrum analyzer is characterized by limited bandwidth, number of filters and maximum switching time.

Moreover, it is expensive. Its main advantage is, it provides display of periodic random waves with their transient response.

On the other hand, swept tuned spectrum analyzer is often used at microwave frequencies. It is basically, a superheterodyne receiver. This type of spectrum analyzer is simple and economical. It displays the frequency domain of periodic waves. But it does not display the transient response.

9.10 SALIENT FEATURES OF SPECTRUM ANALYZER

> ➤ They provide two amplitude scales, namely, linear and longitudinal.
> ➤ They are useful for the measurements of relative and absolute measurements.
> ➤ They provide different modes of operation. The modes are wideband, narrow band, amplitude versus time display etc.
> ➤ Their usefulness depends on frequency of operation, resolution, stability, number of input levels, Frequency and time domain response and calibration of different parameters.

9.11 APPLICATIONS OF SPECTRUM ANALYZER

These are widely used at microwave frequencies to measure

> ➤ frequency
> ➤ distortion
> ➤ signal to noise ratio
> ➤ power and voltage levels of each frequency
> ➤ power
> ➤ frequency response of a system and to display variations of amplitude with time and also to identify types of waves and types of modulation etc.

9.12 POWER METERS

Power meters are extremely useful for many measurements. They are available with a family of sensors covering different range of frequencies. In general, the meters have an internal power reference source (1 mW) to calibrate the meter for absolute power. Some times micro-ammeters are used in place of power meters due to their low cost.

9.13 NETWORK ANALYZER

It is an instrument that measures scattering matrix, impedance and admittance parameters of linear networks through sine wave testing. The measurements are done by configuring its various components around the device under test. The network analyzer detects separated signals and form the desired signal ratio and display the results.

9.14 PRECAUTION IN MICROWAVE MEASUREMENTS

The precautions to be taken in microwave measurements are given below.

> ➤ Microwave measuring equipment is costly and they should be handled with care.
> ➤ Proper storage should be provided to avoid damage and loss.
> ➤ Proper filtering and adapters should be used to interconnect the equipment and devices.
> ➤ Proper impedance matching of the equipment is required to avoid mismatching. Any mismatching results in reflection.
> ➤ The sources must be isolated from load to avoid damage due to reflected power.
> ➤ Connections between waveguides should be tight to avoid leakage.
> ➤ Coaxial connections should be carefully connected and tightened to protect the threads.
> ➤ The low loss coaxial lines suitable for the frequency of operation should be used.

EXPERIMENT 1

9.15   REFLEX KLYSTRON CHARACTERISTICS

I. AIM

1. To obtain the reflex klystron output and frequency characteristic.
2. To find mode number.
3. To find transit time.
4. To find electronic tuning range.
5. To find (ETS) electronic tuning sensitivity.

II. THEORETICAL CONCEPTS

> Klystron is a vacuum tube microwave oscillator. Its operation depends on the principle of velocity of modulation and transit time.

The transit time and mode number, power, frequency characteristics, electronic tuning sensitivity, output efficiency and klystron mode characteristics are some of the important parameters in reflex klystron.

Transit Time

The transit time and the frequency are related by

$$t = \frac{n + 3/4}{f_0}$$

Here n is zero or a natural number, each value of n corresponds to a different transit time. These transit times are called modes. The modes are represented by n.

Mode Number

Mode number is given by

$$N_n = n + 3/4$$

If $n = 0$, it gives ¾ mode and if $n = 1$, it gives 1 ¾ mode.

Power and Frequency Characteristics

The mode of operation depends on the repeller voltage. The variations of power with repeller voltage are called power characteristics. The variation of frequency with repeller voltage are called frequency characteristics.

Frequency of Reflex Klystron

The frequency of klystron depends on the cavity dimensions. These dimensions can be varied by flexing a portion of the cavity wall and also by changing the space of the cavity grids. This method of change of frequency is known as mechanical tuning.

The klystron frequency can also be varied by reflector voltages. This type of variation of frequency is known as electronic tuning. Electronic tuning sensitivity is defined as

$$ETS \equiv \frac{f_2 - f_1}{V_2 - V_1} \text{ MHz/Volt}$$

Here, f_1 and f_2 are frequencies expressed in MHz corresponding to half of its value at the dip, V_1 and V_2 are repeller voltages.

Efficiency of Klystron

The efficiency of klystron is defined as

$$\eta \equiv \frac{P_o}{V_b \, I_b} \times 100\%$$

Here P_o is output power
V_b = beam voltage
I_b = beam current

III. EQUIPMENT REQUIRED

- Klystron power supply
- Reflex klystron, klystron mount
- Variable attenuator
- Frequency meter
- Crystal detector and detector mount
- SWR meter or micro-ammeter
- Waveguides and cooling fan
- Flanges,
- Coaxial to waveguide adaptors
- Waveguide to coaxial adaptors
- Isolator

IV. EXPERIMENTAL SETUP

The setup is shown in fig. 9.1.

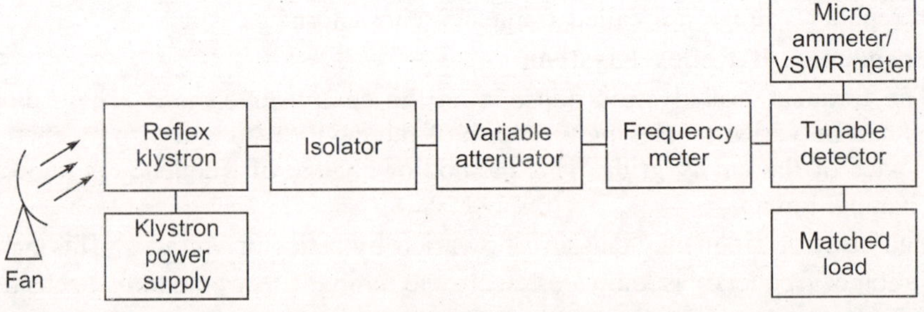

Fig. 9.1 Microwave bench measurement setup for klystron characteristics

V. PROCEDURE

1. Assemble the equipment as in fig. 9.1.
2. Adjust the variable attenuator to a maximum position.
3. Adjust the beam voltage for optimum value for klystron to have *CW* operation.
4. Adjust the repeller voltage to maximum negative value in steps of 1 V.
5. Record microammeter readings for each repeller voltage.
6. Measure the frequency by tuning frequency meter corresponding to a dip in the output.
7. The frequency meter is detuned each time while measuring output power and tabulate the results.
8. Plot the variation of power with repeller voltage.
9. Plot the variation of frequency with repeller voltage.
10. Find out the mode number, transit time.

VI. RESULTS

Table 9.1 Reflex klystron characteristics

S. No.	Repeller Voltage	Microammeter Reading (μA)	Frequency Meter Reading in GHz

The mode numbers are calculated from

$$\frac{N_2}{N_1} = \frac{V_1}{V_2} = \frac{(n+1) + 3/4}{(n + 3/4)}$$

Here, N_1 and N_2 are mode numbers, V_1 and V_2 are repeller voltages and the transit time is determined from

$$t_t = \frac{n + 3/4}{f_0}$$

The transit time of each mode is found from

$$t_{t1} = \frac{N_1}{f_{01}}, \ t_{t2} = \frac{N_2}{f_{02}}$$

Typical Output and Frequency Characteristics

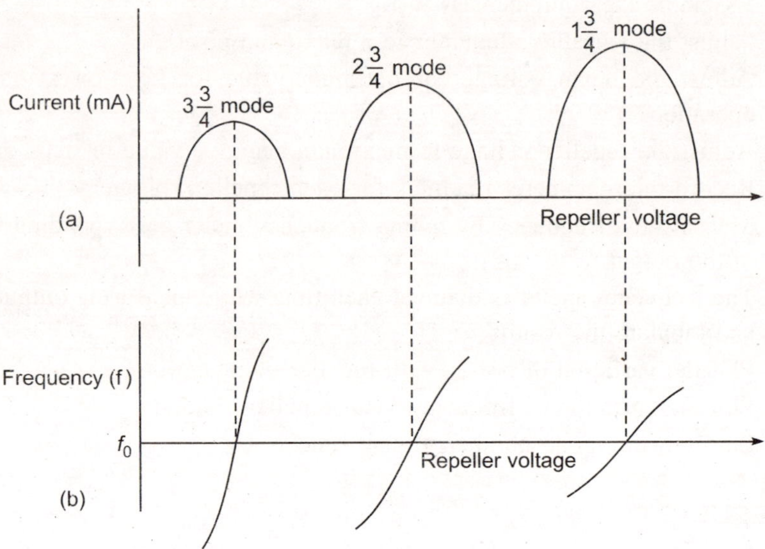

Fig. 9.2 (a) Variation of current with repeller voltage *(b)* variation of frequency with repeller voltage

VII. EXPERIMENT WITH MODULATED SOURCE

1. Assemble components as in fig. 9.1.
2. Apply klystron power supply with 1000 cycle square wave modulation.
3. Tune klystron with optimum beam voltage.
4. Obtain peak reading in VSWR by adjusting modulation and repeller voltage.
5. Tune the klystron for a maximum reading in VSWR meter.
6. Adjust repeller voltage in steps of 1 V and note power output and frequency.
7. Plot variation of power output and frequency with repeller voltage.
8. Calculate the mode number and transit time.

Table 9.2   Reflex klystron characteristics with modulation source

S. No.	Repeller Voltage	Microammeter Reading (μA)/Power	Frequency Meter Reading in GHz

VIII. DETERMINATION OF MODES USING OSCILLOSCOPE

1. Assemble the components and equipment as shown in fig. 9.3

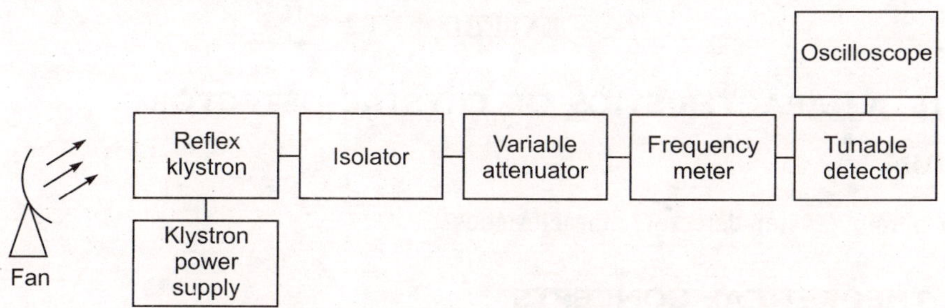

Fig. 9.3 Experimental setup for mode characteristics of reflex klystron

2. Keep variable attenuator at maximum attenuation position.
3. Switch on the klystron power supply and oscilloscope.
4. Keep mode selector switch in FM-MOD.
5. Set beam voltage in the klystron power supply at 300 V and keep amplitude knob of *FM* modulator at maximum position.
6. Change the reflector voltage and observe the variation of output power with reflector voltage.
7. Vary the reflector voltage and amplitude of *FM* modulation position and observe the klystron modes on the CRO.

IX. PRECAUTIONS

1. An isolator or attenuator should be used between the klystron and the other equipment in the setup to avoid loading of the klystron.
2. While measuring frequency, frequency meter should be detuned each time.
3. The negative repeller voltage should be applied first before anode voltage is applied.
4. Before switching on power supply, the control knobs of klystron power supply should be kept as below:

Meter switch : OFF

Mode switch : *AM*

Beam voltage knob : Fully anti-clockwise

Reflector voltage : Fully clockwise

AM – Amplitude : Fully clockwise

AM – Frequency knob : Mid position

5. The control knob of VSWR meter should kept as below

Meter switch : Normal

Input Switch : Low impedance position

Range dB switch : 40/50 dB

Gain control knob : Fully clockwise

6. Cooling fan should be used to avoid heating of klystron tube.

EXPERIMENT 2

9.16 CHARACTERISTICS OF CRYSTAL DETECTOR

I. AIM

To obtain crystal detector characteristics.

II. THEORETICAL CONCEPTS

The detector is a microwave device used to sense the presence of microwave power. It is basically a crystal. It consists of a movable probe placed on a mount. The crystal detector is a sensitive, non-linear and non-reciprocal device. It rectifies the detected signal and produces a current proportional to input power. The current is proportional to the square of the voltage and hence the crystal detector is called square law detector. This property exists for low powers (< 10 mW). At high power levels, it behaves like a linear detector.

The equivalent circuit of crystal detector is shown in fig. 9.4.

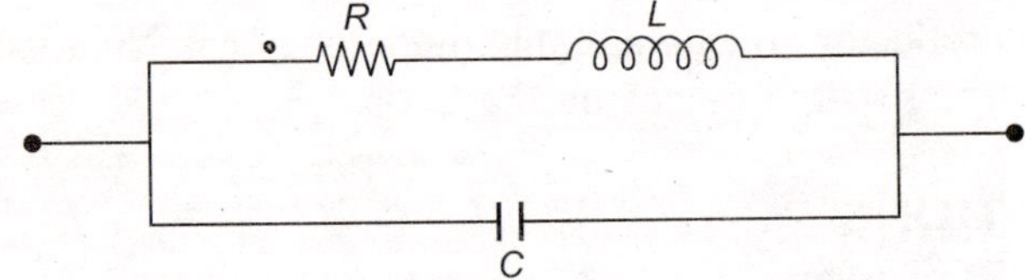

R = load resistance, L = load inductance, C = junction capacitance

Fig. 9.4   The equivalent circuit of crystal detector

III. EQUIPMENT REQUIRED

- Reflex klystron
- Klystron power supply
- Frequency meter
- Tunable crystal detector
- Variable attenuator
- Microammeter / VSWR meter
- Waveguide stands
- Cooling fan
- Flanges
- Coaxial to waveguide adaptors
- Waveguide to coaxial adaptors
- Matched terminators

IV. EXPERIMENTAL SETUP

Method 1 – Crystal detector characteristics using slotted wvaeguide

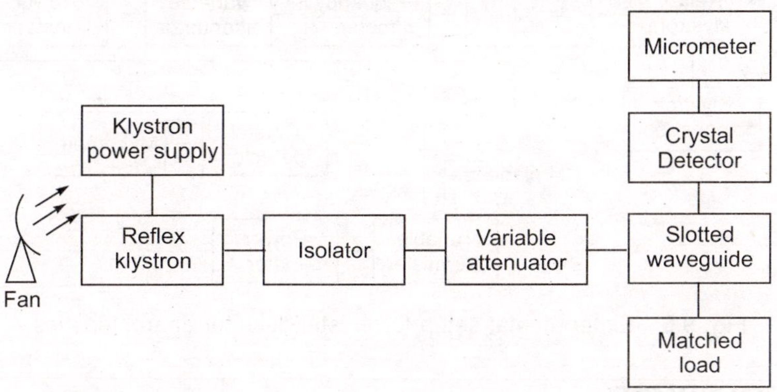

Fig. 9.5   Experimental setup for crystal detector characteristics

V. PROCEDURE (Method 1)

1. Assemble the devices and components as in the fig. 9.5.
2. The reflex klystron is adjusted to operate at the top of the mode curve.
3. Measure the guide wavelength.
4. Slide the probe to obtain a convenient minimum in the output meter. Let this be the reference position.
5. Slide the probe carriage towards next minimum and note the reading.
6. Plot $(2\pi L/\lambda_g)$ Vs current. Here, L is the distance from the reference point.
7. The results are presented in tabular form (table 9.3).

Table 9.3   Crystal detector characteristics

S. No.	Current, $I\ (\mu A)$	Slotted Line Location,	L	Normalized Current, I/I_{max}	$\sin(2\pi L/\lambda_0)$

8. The graph between $\sin(2\pi L/\lambda_g)$ and current gives a straight line on a log sheet. The slope of the line is equal to x. x is the exponent in the current and voltage relation of crystal detector.

Method 2 – Crystal detector characteristics using H-plane Tee junction

VI. EXPERIMENTAL SETUP

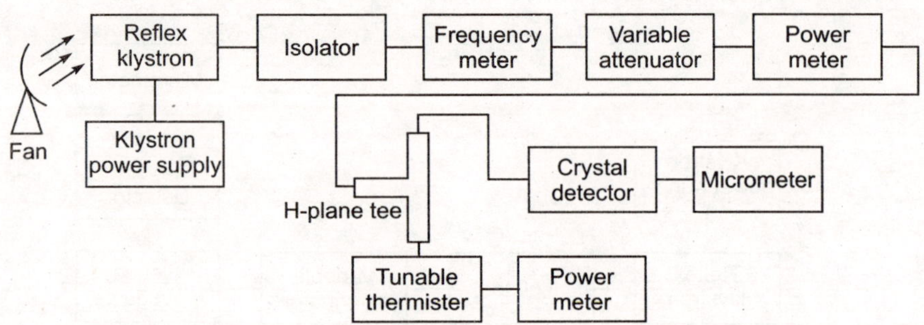

Fig. 9.6   Experimental setup for crystal detector characteristics

VII. PROCEDURE

1. Assemble the setup as in fig. 9.6.
2. Adjust the klystron for a convenient frequency.
3. Keeping the minimum attenuation in the attenuator, adjust the source for maximum power output.
4. Measure power at the input Tee junction.
5. Adjust the variable attenuator to make the micro-ammeter reading zero.
6. Increase input power by varying attenuator in steps of 0.1 mW as indicated by power meter.
7. Note the corresponding current.
8. Set another frequency and repeat steps (3-7).
9. Tabulate the results (table 9.4).

Table 9.4   Crystal detector characteristics

S. No.	f_1		f_2	
	Input Power	*Output Power*	*Input power*	*Output Power*

10. Plot the variation of output current with power or plot the variation of output current with the change of attenuation in dB.

Method 3

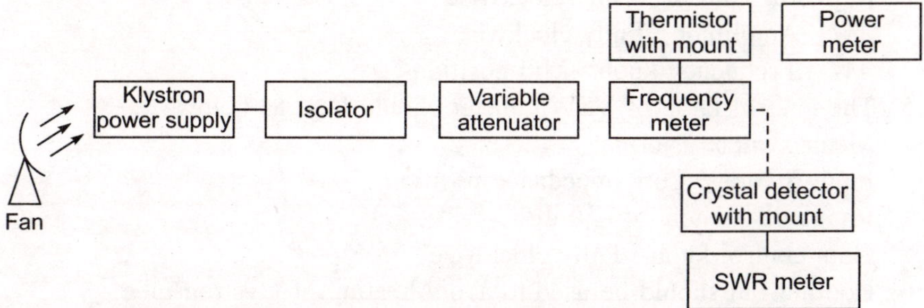

Fig. 9.7   Experimental setup for crystal detector characteristics

1. Assemble the equipment as shown in fig. 9.7.
2. Set the source frequency in a convenient position f_1.
3. Adjust the variable attenuator to read zero dBm.
4. Connect crystal detector and SWR meter in place of thermistor mount and power meter.
5. Tune the detector to note a significant signal level in SWR meter.
6. Select the scale on SWR meter to read zero dBm by using gain control knob.
7. Increase the attenuation in steps of 1 dB and note SWR meter reading.
8. Change the frequency to f_2 and repeat the above steps.
9. Tabulate the results (table 9.5).
10. Plot the variation of SWR meter reading with attenuator reading.

Table 9.5   Crystal detector characteristics

S. No.	f_1		f_2	
	Input Power	*Output Power*	*Input power*	*Output Power*

VIII. PRECAUTIONS

1. An isolator or attenuator should be used between the klystron and the other equipment in the setup to avoid loading of the klystron.
2. While measuring frequency, frequency meter should be detuned each time.
3. The negative repeller voltage should be applied first before anode voltage is applied.
4. Before switching on power supply, the control knobs of klystron power supply should be kept as below :
    Meter switch : OFF
    Mode switch : *AM*

Beam voltage knob : Fully anti-clockwise
Reflector voltage : Fully clockwise
AM – Amplitude : Fully clockwise
AM – Frequency knob : Mid position

5. The control knob of VSWR meter should kept as below
Meter switch : Normal
Input Switch : Low impedance position
Range dB switch : 40/50 dB
Gain control knob : Fully clockwise

6. Cooling fan should be used to avoid heating of klystron tube.

EXPERIMENT 3

9.17 MEASUREMENT OF GUIDE WAVELENGTH AND SOURCE FREQUENCY

I. AIM

To measure the guide wavelength and frequency.

II. THEORETICAL CONCEPTS

The waveguide is a metallic hollow pipe which guides the electromagnetic wave.

Salient Features of Waveguides

1. It supports *TE* and *TM* waves.
2. It does not support *TEM* waves.
3. The electric and magnetic fields in a rectangular waveguide exist as in fig. 9.8.

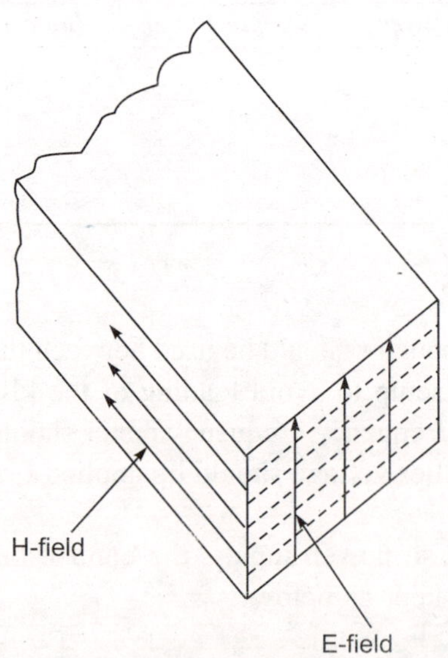

Fig. 9.8 *E* and *H* fields in a waveguide

4. The waves are characterized by a propagation constant given by

$$\gamma_g = \sqrt{\omega^2 \mu \in - \left(\frac{m\pi}{a}\right)^2 - \left(\frac{n\pi}{b}\right)^2}$$

Here, a is wide wall dimension

$\quad b$ is narrow wall dimension

$\quad m, n$ represent mode numbers

1. When the wave propagates in the waveguide, it can have an infinite types of patterns. These patterns are called modes.
2. The *TE* and *TM* waves are represented by TE_{mn} and TM_{mn}. The first subscript indicates the number of half-wave variations of electric or magnetic field across the wide dimension of the waveguide and second subscript indicates the number of half-wave variations of electric or magnetic field across the narrow wall.
3. If there is no variation across the narrow dimension, the mode is represented by TE_{10} mode. This is called dominant mode.
4. The waveguide dimensions depend on the operating frequency range. For example, X-band waveguide has inner dimension of 2.286 cm $\times$ 1.016 cm.
5. They are used as transmission lines at microwave frequencies.
6. They can be used as a high pass filters, radiators and antenna feed elements.
7. The cut-off wavelength in rectangular waveguide is

$$\lambda_c = \frac{2}{\sqrt{\left(\frac{m}{a}\right)^2 + \left(\frac{n}{b}\right)^2}}$$

8. The cut-off wavelength for the dominant mode, TE_{10} is $\lambda_c = 2a$.
9. The modes other than the dominant mode are called high order modes.
10. The cut-off frequencies in a rectangular waveguide is $f_c = \dfrac{\upsilon_0}{2a}$.
11. All the frequencies above cut-off frequencies are propagated.
12. The cut-off frequency depends on the propagation mode.
13. If two or more modes propagate they may interact in an unwanted manner.
14. It is preferred to allow only one mode preferably dominant mode at a time.
15. When a discontinuity generates a high order mode, the mode dies out quickly as it is below cut-off.
16. In some components and circuits higher order modes are used.
17. The propagation characteristics in a rectangular waveguide are consolidated as

(1) Propagation constant,

$$\gamma_g = \sqrt{\left(\frac{m\pi}{a}\right)^2 + \left(\frac{n\pi}{b}\right)^2 - \omega^2 \mu \in} \; , \; \left(\frac{1}{m}\right)$$

$$= \alpha_g \; \text{if} \; \left(\frac{m\pi}{a}\right)^2 + \left(\frac{n\pi}{b}\right)^2 > \omega^2 \mu \in$$

(2) Phase constant,

$$\beta_g = \sqrt{\omega^2 \mu \in - \left(\frac{m\pi}{a}\right)^2 - \left(\frac{n\pi}{b}\right)^2} \; , \; \text{(rad/m)}$$

(3) Cut-off frequency,

$$f_c = \frac{1}{2\pi\sqrt{\mu \in}} \sqrt{\left(\frac{m\pi}{a}\right)^2 + \left(\frac{n\pi}{b}\right)^2} \; , \; \text{(Hz)}$$

(4) Cut-off wavelength,

$$\lambda_c = \frac{2}{\sqrt{\left(\frac{m}{a}\right)^2 + \left(\frac{n}{b}\right)^2}} \; , \; \text{(m)}$$

(5) Phase velocity,

$$\upsilon_p = \frac{\omega}{\sqrt{\omega^2 \mu \in - \left(\frac{m\pi}{a}\right)^2 - \left(\frac{n\pi}{b}\right)^2}} \; , \; \text{(m/s)}$$

(6) Guide wavelength,

$$\gamma_g = \frac{2\pi}{\sqrt{\omega^2 \mu \in - \left(\frac{m\pi}{a}\right)^2 - \left(\frac{n\pi}{b}\right)^2}} \; , \; \text{(m)}$$

or

$$\lambda_g = \frac{\lambda}{\left[1 - \left(\frac{\lambda}{\lambda_c}\right)^2\right]^{1/2}} \; , \; \text{(m)}$$

III. EQUIPMENT REQUIRED

Klystron power supply, Reflex klystron, Cooling fan, Frequency meter, Isolator, Variable attenuator, Slotted waveguide section, detector mount, VSWR meter and short circuit plunger.

IV. EXPERIMENTAL SETUP

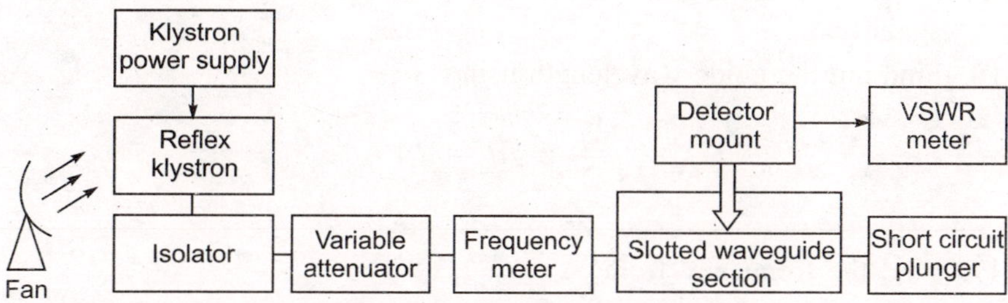

Fig. 9.9 Experimental setup for measuring guide wavelength and source frequency

V. PROCEDURE

1. Setup the components and equipment as shown in fig. 9.9.
2. Set variable attenuator at maximum position.
3. Keep the control knobs of VSWR meter as below
   Range dB – 50 dB position
   Input switch – Crystal low impedance
   Meter switch – Normal position
   Gain (course & fine) – Mid position
4. Keep the control knobs of klystron power supply as given below
   Meter switch – OFF ; MOD switch – *AM*
   beam voltage knob : Fully anti-clockwise,
   Reflector voltage – fully clockwise
   AM frequency knob – Around mid position
5. Switch on klystron power supply and VSWR meter.
6. Turn the switch of power supply to beam voltage position and set beam voltage at 300 V with the help of beam voltage knob.
7. Adjust the reflector voltage to get some deflection in VSWR.
8. Maximize the deflection with AM amplitude knob and AM frequency control knobs of the power supply.
9. Tune the plunger of klystron mount for maximum deflection.
10. Tune the reflector voltage knob for maximum deflection in VSWR meter.
11. Tune the probe connected to the detector for maximum deflection in VSWR meter.

12. Tune the frequency meter knob to obtain a dip on the VSWR scale and note down the frequency meter.

13. Detune the frequency meter now.

14. Move the probe along the slotted line till a minimum deflection position in VSWR meter is obtained. If required VSWR meter dB switch may be kept at higher position. Record the probe position corresponding to minimum deflection (d_1).

15. Move the probe to get next minimum position and record the probe position again (d_2).

16. Find out the guide wavelength using

$$\frac{\lambda_g}{2} = (d_2 - d_1)$$

$$\lambda_g = 2(d_2 - d_1)$$

17. Find the frequency from

$$f_{\lambda g} = v_0 \sqrt{\frac{1}{\lambda_g^2} + \frac{1}{\lambda_c^2}}$$

Here, $v_0 = 3 \times 10^8$ m/sec

λ_c = cut-off wavelength

= $2a$, a is inner broad dimension of the waveguide

18. Verifying the frequency obtained with the frequency obtained in the frequency meter.

19. Repeat the procedure at different frequencies.

VI. OBSERVATIONS AND RESULTS

Cut-off wavelength $\lambda_c = 2a$

Table 9.6   Results for guide wavelength and source frequency

S. No.	Frequency Setting	d_1	d_2	λ_g	f_M	$f_{\lambda g}$
1.	f_1					
2.	f_2					
3.	f_3					
-	-					
-	-					
-	-					

VII. PRECAUTIONS

1. An isolator or attenuator should be used between the klystron and the other equipment in the setup to avoid loading of the klystron.

2. While measuring frequency, frequency meter should be detuned each time.

3. The negative repeller voltage should be applied first before anode voltage is applied.

4. Before switching on power supply, the control knobs of klystron power supply should be kept as below :

   Meter switch : OFF

   Mode switch : *AM*

   Beam voltage knob : Fully anti-clockwise

   Reflector voltage : Fully clockwise

   AM – Amplitude : Fully clockwise

   AM – Frequency knob : Mid position

5. The control knob of VSWR meter should kept as below

   Meter switch : Normal

   Input Switch : Low impedance position

   Range dB switch : 40/50 dB

   Gain control knob : Fully clockwise

6. Cooling fan should be used to avoid heating of klystron tube.

EXPERIMENT 4

9.18 *V-I* CHARACTERISTICS OF GUNN DIODE

I. AIM

To study the *V-I* characteristics of Gunn diode.

II. THEORETICAL CONCEPTS

Gunn diode is a negative resistance device. It is used as amplifier and oscillator.

Salient Features

- Gunn diode discovered by Gunn in 1963.
- It is used as amplifier and oscillator.
- They can be used up to 100 GHz frequency.
- Its power output is about 10 W at lower microwave frequency range.
- It does not have a *p-n* junction.
- It is an example of transfer electron device (TED).
- It is a wide band device.
- It has low noise characteristics.
- It is a bulk effect device.
- Its *V-I* characteristics is given by

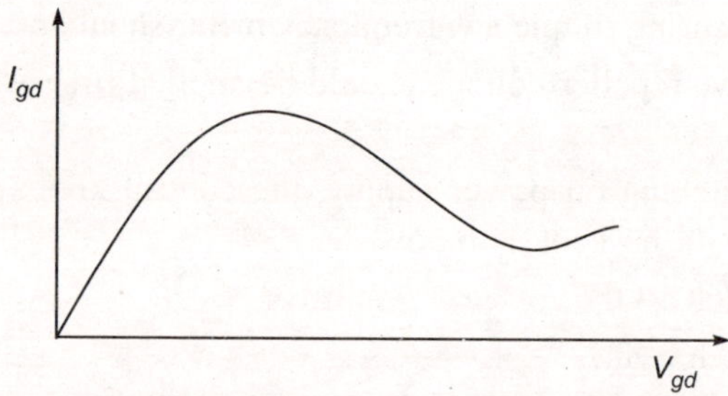

Fig. 9.10   *V-I* Characteristics of a Gunn diode

III. EQUIPMENT REQUIRED

- Gunn oscillator
- Gunn power supply
- PIN modulator
- Isolator
- Frequency meter
- Variable attenuator
- Detector mount
- Waveguide stands
- SWR meter
- Cables and accessories

IV. EXPERIMENTAL SETUP

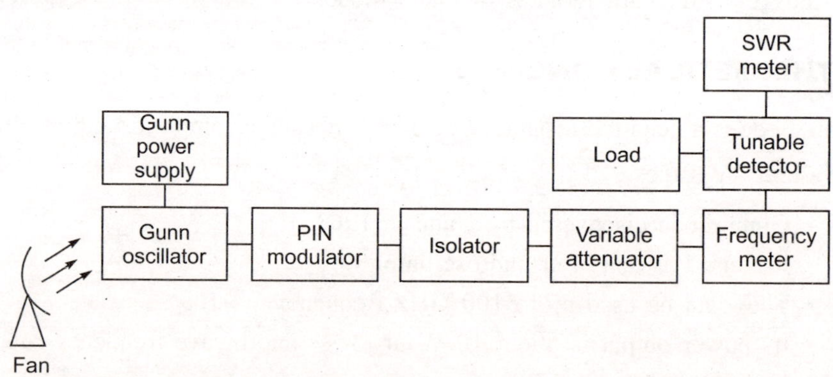

Fig. 9.11   Measurement setup for *V-I* characteristics of Gunn diode

V. PROCEDURE

1. Assemble equipment as shown in fig. 9.11.
2. Set the variable attenuator for maximum attenuation.

3. Adjust the micrometer of Gunn oscillator for required frequency of operation, say f_1.
4. Switch ON the Gunn power supply VSWR meter and cooling fan.
5. Vary the voltage using the Gunn bias knob and note the corresponding the Gunn diode current. These measurements are made using the panel meter and meter switch.
6. Tabulate the results.
7. Plot the voltage and current reading on the graph as shown in fig. 9.12.
8. Note the threshold voltage. This voltage corresponds to the maximum current.

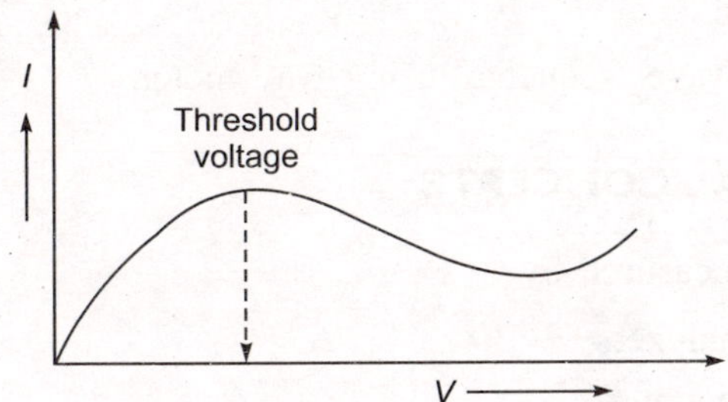

Fig. 9.12   Characteristics of Gunn oscillator

9. Repeat steps (4-8) for two more frequencies f_1, f_2.

Table 9.7   Gunn diode characteristics

S. No.	f_1		f_2		f_3	
	V	I	V	I	V	I

VI. PRECAUTIONS

1. Do not keep Gunn bias knob position at threshold position for more than 10-15 seconds.
2. Reading should be obtained as fast as possible. Otherwise due to excessive heating. Gunn diode may burn.
3. Before application of the Gunn power supply, keep Gunn bias and PIN bias knobs fully anti-clockwise.
4. In VSWR meter, the meter switch should be kept in normal position. Input switch at low impedance position. Keep range switch in 40 dB and gain knob in fully clockwise.

5. Threshold voltage is kept for not more than about 10 seconds to avoid burning of Gunn diode.

6. An isolator or attenuator should be used between the Gunn oscillator and the other equipment in the setup to avoid loading of the Gunn oscillator.

7. While measuring frequency, frequency meter should be detuned each time.

8. Cooling fan should be used to avoid heating of Gunn diode.

EXPERIMENT 5

9.19 FREQUENCY MEASUREMENTS

I. AIM

To measure microwave frequency using wavemeter.

II. THEORETICAL CONCEPTS

The frequency is measured by

- Spectrum analyzers
- Frequency counters
- Null beam method
- Wavemeters and
- Slotted lines etc.

A wavemeter has a resonant cavity. It is analogous to the tuned resonant circuit. There are two types of cavities.

1. Transmission type
2. Absorption type

The transmission type cavity wavemeter passes only the signal to which it is tuned. The absorption type cavity wavemeter absorbs only the frequency to which it is tuned. The absorption type wavemeter is commonly used in laboratory measurements.

III. EQUIPMENT REQUIRED

- Reflex klystron
- Klystron power supply
- Attenuator
- Wavemeter
- Slotted line
- Detector
- VSWR meter

IV. EXPERIMENTAL SETUP

A typical measurement setup is shown in fig. 9.13.

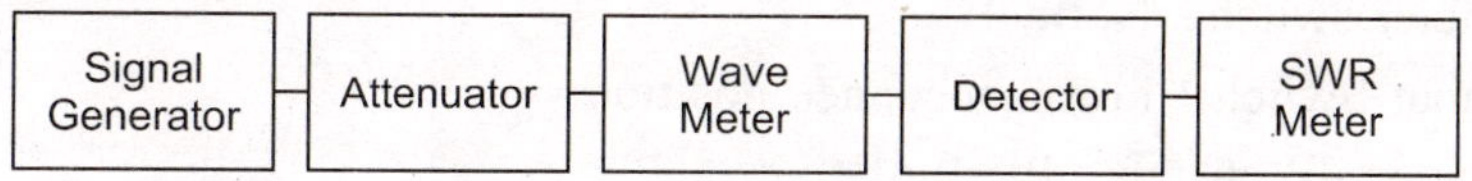

Fig. 9.13   Measurement of frequency with wavemeter

V. PROCEDURE

1. Assemble components as in fig. 9.13.
2. Adjust the power level to give full scale reading on the output meter.
3. Then tune the wavemeter slowly till a dip in the power level is absorbed.
4. Read the frequency from the wavemeter dial.
5. The above steps are repeated for different input frequencies.
6. Tabulate the results as in the table 9.8.

Table 9.8   Measurement of frequency

S. No.	Frequency from Signal Generator	Measured Frequency

VI. PRECAUTIONS

1. An isolator or attenuator should be used between the klystron and the other equipment in the setup to avoid loading of the klystron.
2. While measuring frequency, frequency meter should be detuned each time.
3. The negative repeller voltage should be applied first before anode voltage is applied.
4. Before switching on power supply, the control knobs of klystron power supply should be kept as below :

   Meter switch : OFF
   Mode switch : *AM*
   Beam voltage knob : Fully anti-clockwise
   Reflector voltage : Fully clockwise

AM – Amplitude : Fully clockwise

AM – Frequency knob : Mid position

5. The control knob of VSWR meter should kept as below

Meter switch : Normal

Input Switch : Low impedance position

Range dB switch : 40/50 dB

Gain control knob : Fully clockwise

6. Cooling fan should be used to avoid heating of klystron tube.

EXPERIMENT 6

9.20 VSWR MEASUREMENT

I. AIM

To measure VSWR of a given unit under test.

II. THEORETICAL CONCEPT

VSWR means Voltage Standing Wave Ratio. Standing waves indicate the quality of transmission. The well matched loads have no reflections and VSWR is one.

$$\text{VSWR} \equiv \frac{V_{max}}{V_{min}}$$

III. EQUIPMENT REQUIRED

- Klystron power supply
- Reflex klystron
- Isolator
- Frequency meter
- Variable attenuator
- Slottedline
- Klystron mount
- Detector with probe
- VSWR meter
- BWC cable
- SS tuners
- Matched terminals
- Movable short or unit under test
- Cooling fan

IV. EXPERIMENTAL SETUP

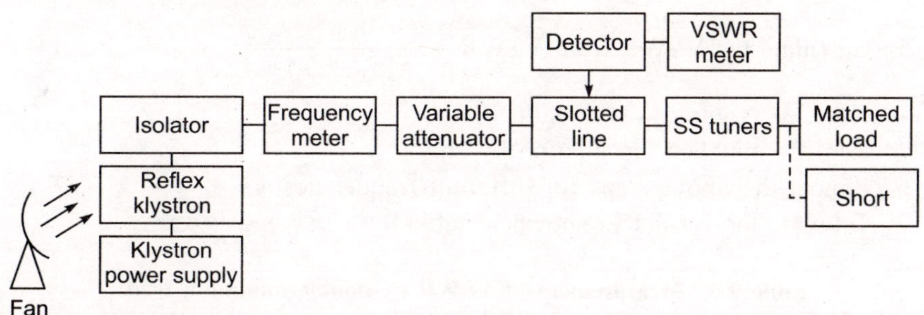

Fig. 9.14   Experimental setup for measurement of VSWR

V. PROCEDURE

For Low VSWR Values

1. Connect the components as in fig. 9.14.
2. Keep variable attenuator at maximum
3. Switching ON klystron power supply, VSWR meter and cooling fan.
4. Set beam voltage to 300 V.
5. Adjust the reflector voltage to obtain deflection in VSWR meter.
6. Tune the output with the help of reflector voltage, amplitude and frequency of *AM*.
7. Tune klystron to get maximum deflection in the meter.
8. Adjust attenuation, gain cosntrol knob and dB switch to get deflection in the scale of VSWR meter.
9. Move the probe to get maximum deflection in VSWR meter.
10. Set the VSWR meter gain control knob or attenuator till the meter indicates 1.0 on normal scale of 0 to ∞.
11. Keeping the control knobs fixed, move the probe to the next minimum position. Note the VSWR from the scale.
12. Now change SS tuner probe and repeat the steps 9 to 11. Note VSWR each position.
13. Change dB range in VSWR meter corresponding to the value of VSWR.

Measurement of High Values of VSWR (Double Minimum Method)

1. Set a convenient frequency when the matched load is in place.
2. The probe is so inserted to a depth where the minimum is read without difficulty.
3. The probe is moved to a point where the power is twice the minimum. Let the position be d_1.

4. The probe is moved to the twice power point the other side of the minimum. Let this point be d_2.

5. Calculate the VSWR from $\text{VSWR} = \dfrac{\lambda_g}{\pi(d_1 - d_2)}$

   Here, λ_g is guide wavelength.
   Replace matched termination by short.

6. Repeat the above steps for different frequencies.
7. Tabulate the results as shown in table 9.9.

Table 9.9   Measurement of VSWR by double minimum feed

Frequency	Matched Load			Short		
	d_1	d_2	VSWR	d_1	d_2	VSWR
f_1						
f_2						
f_3						

VI. MEASUREMENT OF VSWR BY CRO

1. Connect detector output to CRO.
2. Move the probe along the slotted line for a typical frequency.
3. Measure V_{max} and V_{min}.
4. Change the frequency and repeat the above steps.
5. Tabulate the results as shown in table 9.10.

6. Find VSWR from $\dfrac{V_{max}}{V_{min}}$.

Table 9.10   Measurement of VSWR using CRO

Frequency	Matched Load			Short		
	V_{max}	V_{min}	VSWR	V_{max}	V_{max}	VSWR
f_1						
f_2						
f_3						

VII. PRECAUTIONS

1. An isolator or attenuator should be used between the klystron and the other equipment in the setup to avoid loading of the klystron.
2. While measuring frequency, frequency meter should be detuned each time.
3. The negative repeller voltage should be applied first before anode voltage is applied.

4. Before switching on power supply, the control knobs of klystron power supply should be kept as below :

    Meter switch : OFF

    Mode switch : *AM*

    Beam voltage knob : Fully anti-clockwise

    Reflector voltage : Fully clockwise

    AM – Amplitude : Fully clockwise

    AM -- Frequency knob : Mid position

5. The control knob of VSWR meter should kept as below

    Meter switch : Normal

    Input Switch : Low impedance position

    Range dB switch : 40/50 dB

    Gain control knob : Fully clockwise

6. Cooling fan should be used to avoid heating of klystron tube.

EXPERIMENT 7

9.21   MEASUREMENT OF ATTENUATION

I. AIM

To measure attenuation

II. THEORETICAL CONCEPTS

At microwave frequencies, attenuation takes place due to

1. energy dissipation within the device as heat.
2. the insertion loss of the device. This occurs due to mismatch.

Attenuation is defined as the reduction in power through a device. It is the power loss between source and a load when a device is inserted between the two.

The attenuation, α is defined as

$$\alpha(\text{dB}) \equiv 10 \log_{10} \frac{p_i}{p_0}$$

Here, p_i = input power

p_0 = output power

The attenuation measurements are made when the input and output ports of the device are matched without reflections.

Insertion loss measurements are made when the device is not matched with the presence of reflections.

III. EQUIPMENT REQUIRED

Microwave power source

- Klystron power supply
- Reflex klystron
- Variable standard attenuator
- Component under test
- Tuned detector
- Indicating meter
- Waveguide section
- Waveguide stands
- Cables, waveguide stands, cooling fan etc.

IV. EXPERIMENTAL SETUP

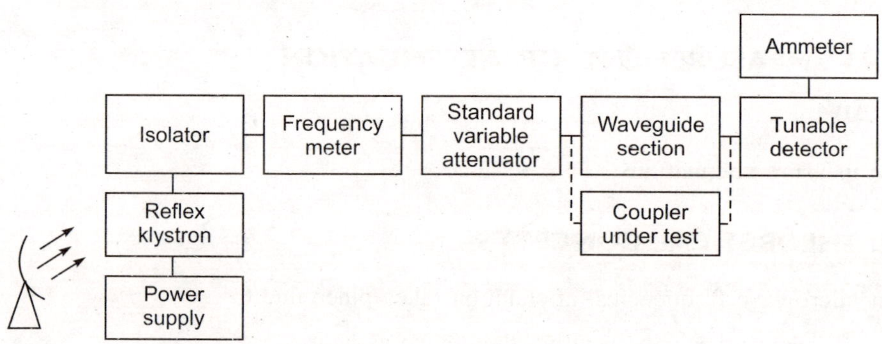

Fig. 9.15   Attenuation measurement setup

V. PROCEDURE

1. Assemble as in fig. 9.15.
2. Set a typical frequency
3. Keep the variable standard attenuation at maximum value.
4. Set the input to have maximum in the ammeter.
5. Replace waveguide by component under test.
6. Adjust the standard variable attenuator to obtain the same reading in the ammeter.
7. Record the difference in readings of standard attenuator.
8. Set another frequency and repeat steps (3 − 6) and tabulate the results as shown in table. 9.11.

Table 9.11   Results on attenuation

S. No.	Frequency	Initial Reading of Standard Attenuator (1)	Final Reading in the Standard Attenuator (2)	Attenuation of Coupler Under Test (1 − 2)
1.	f_1			
2.	f_2			
3.	f_3			
-	-			
-	-			
-	-			

VI. PRECAUTIONS

1. An isolator or attenuator should be used between the klystron and the other equipment in the setup to avoid loading of the klystron.
2. While measuring frequency, frequency meter should be detuned each time.
3. The negative repeller voltage should be applied first before anode voltage is applied.
4. Before switching on power supply, the control knobs of klystron power supply should be kept as below :

   Meter switch : OFF

   Mode switch : *AM*

   Beam voltage knob : Fully anti-clockwise

   Reflector voltage : Fully clockwise

   AM – Amplitude : Fully clockwise

   AM – Frequency knob : Mid position
5. The control knob of VSWR meter should be kept as below

   Meter switch : Normal

   Input Switch : Low impedance position

   Range dB switch : 40/50 dB

   Gain control knob : Fully clockwise
6. Cooling fan should be used to avoid heating of klystron tube.

EXPERIMENT 8

9.22 MEASUREMENT OF PARAMETERS OF DIRECTIONAL COUPLER

I. AIM

To find out coupling factor and directivity of a given directional coupler.

II. THEORETICAL CONCEPTS

Definition

> A waveguide directional coupler is a four-port waveguide junction device that samples part of the EM wave power through the main waveguide.

Salient Features of Directional Couplers

1. It is a passive four-port device.
2. It consists of a primary guide with ports 1 and 2 and a secondary guide with port 3 and 4.
3. A typical directional coupler is shown in fig. 9.16.

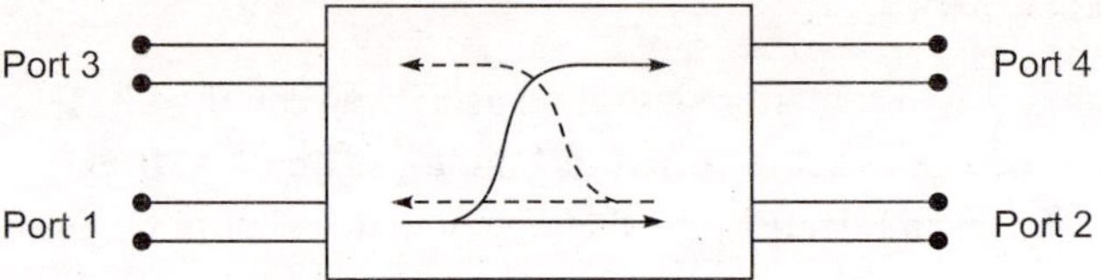

Fig. 9.16   Directional coupler

4. It is made of two connected waveguides. One of the waveguides curves away (fig. 9.16).
5. The waveguides are coupled through holes between them.
6. The directional coupler is said to be consisting of main arm and an auxiliary arm (fig. 9.17).

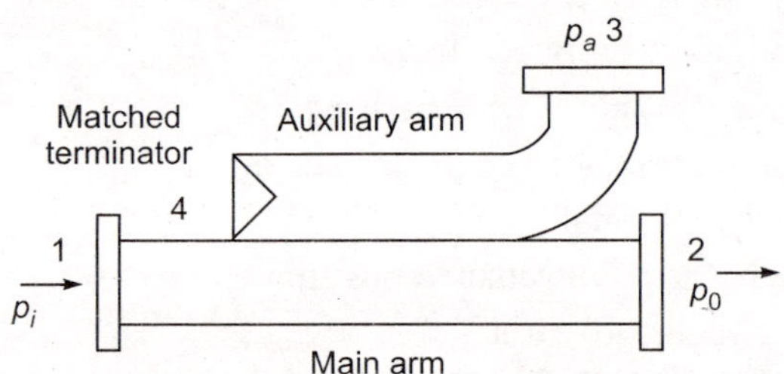

Fig. 9.17   Directional coupler

7. The amount of power coupled to the auxiliary arm depends on the number of holes and their size.
8. The matched termination absorbs the power reaching it without reflections.
9. The coupler is used to find the power reflection coefficient in a waveguide and to find out incident and reflected power values.
10. The powers at ports 4 and 2 have a phase difference of 90°. Similarly, the powers at port 3 and 1 have a phase difference of 90° when the propagation is in reverse direction.
11. The guides 1-2 and 3-4 are identical. Any one of them can be used as primary and the other acts the auxiliary guide.

12. The directional couplers are described by coupling factor directivity and VSWR.

13. It is a reciprocal device.

The Parameters of Directional Couplers

1. Coupling factor (C)

It is defined as the ratio of input power and the output power at auxiliary arm. That is,

$$C \equiv 10 \log_{10} \equiv \frac{p_i}{p_a}, \text{ dB}$$

Here, p_i = input power to the primary guide

$\quad\quad p_a$ = power output at auxiliary guide.

2. Directivity (D)

Definition :   It is defined as the ratio of power in the auxiliary arm due to power in forward direction to the power at the same port due to power in the reverse direction. That is,

$$D(\text{dB}) \equiv 10 \log_{10} \frac{p_{af}}{p_{ar}}$$

Here, p_{af} = power in the auxiliary arm due to power in forward direction

$\quad\quad p_{ar}$ = power in the auxiliary arm to power in the reverse direction.

If D is more, the isolation of the auxiliary arm from leakage output due to reverse power leakage is more.

Applications of Directional Couplers

1. They are used in the measurement of parameters of antennas like slot coupled Tee junctions, horns, etc.
2. They are used to sample the incoming power.
3. They are widely used in impedance bridges for microwave measurements.
4. They are used for power monitoring.
5. They are used in microwave mixers.
6. They are used as input and output couplers in balanced microwave amplifier circuits.

III. EQUIPMENT REQUIRED

- Reflex klystron
- Klystron mount
- Isolator
- Directional coupler

- Matched load
- Crystal detector
- Micro-ammeter

IV. EXPERIMENT SETUP

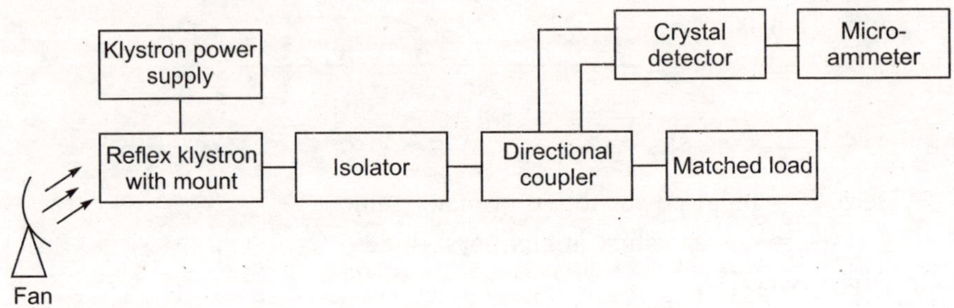

Fig. 9.18 Experimental setup for measuring C and D of directional coupler

V. FORMULAE

$$\text{Coupling, } C = 20 \log_{10} \frac{I_{3f}}{I_1} \text{ in dB}$$

$$\text{Directivity, } D = 20 \log_{10} \frac{I_{3f}}{I_{3r}} \text{ in dB}$$

Here, $I_1 = I_{max}$ with matched load at 2

I_{3f} = current at port 3 with forward power

I_{3r} = current at port 3 with reverse power.

VI. PROCEDURE

1. Connect devices as shown in the experimental setup. Port 4 is permanently matched terminated.
2. Fix beam voltage of 200 volts and a reflector voltage of 150 V/s in klystron power supply.
3. Connect matched load at port 2 and measure I_{3f} at port 3.
4. Measure I_1 by connecting crystal detector and micro-ammeter to the port 2 (i.e. at the port 1 of directional coupler).
5. Connect matched load at port 3 and measure I_2 at port 2.
6. Connect match load at port 1 and connect input at port 2 and measure output at port 3 i.e. I_{3r}.

VII. RESULTS

Calculate the coupling and directivity as a fraction and also in dB from the following.

$$\text{Coupling, } C = \frac{p_1}{p_{3f}} = \frac{I}{I_{3f}}$$

$$\text{Coupling in dB} = 20 \log_{10} \left(\frac{I}{I_{3f}} \right)$$

$$\text{Directivity, } D = \frac{p_{3f}}{p_{3r}} = \left(\frac{I_{3f}}{I_{3r}} \right)$$

$$\text{Directivity in dB} = 20 \log_{10} \left(\frac{I_{3f}}{I_{3r}} \right)$$

VIII. OBSERVATIONS

Beam voltage = 200 V
Reflector voltage = 150 V (say)

Table 9.12   Results on coupling and directivity

I_1	I_{3f}	I_{3r}	I_1/I_{3f}	I_{3f}/I_{3r}	C in dB	D in dB

IX. PRECAUTIONS

1. An isolator or attenuator should be used between the klystron and the other equipment in the setup to avoid loading of the klystron.
2. While measuring frequency, frequency meter should be detuned each time.
3. The negative repeller voltage should be applied first before the anode voltage is applied.
4. Before switching on power supply, the control knobs of klystron power supply should be kept as below :
   Meter switch : OFF
   Mode switch : *AM*
   Beam voltage knob : Fully anti-clockwise
   Reflector voltage : Fully clockwise
   AM – Amplitude : Fully clockwise
   AM – Frequency knob : Mid position

5. The control knob of VSWR meter should kept as below
   Meter switch : Normal
   Input Switch : Low impedance position
   Range dB switch : 40/50 dB
   Gain control knob : Fully clockwise
6. Cooling fan should be used to avoid heating of klystron tube.

EXPERIMENT 9

9.23 STUDY OF MAGIC TEE

I. AIM

To study magic tee

II. THEORETICAL CONCEPTS

It is a combination of E-plane Tee and H-plane Tee.

Salient features

1. It is a four-port device and it is also called magic Tee because of its unusual characteristics.
2. Its four arms are two side arms, shunt arm and series arm. The shunt arm is called H-arm and series arm is called E-arm. The side arms are called collinear arms.
3. The electric fields in the shunt and series arms are perpendicular to each other and hence they are said to be cross-polarized.
4. There is no coupling between shunt and series arm.
5. In this device, E-arm can see only side arm and similarly H-arm can see only side arm.
6. The magic Tee will be matched if the shunt and series arm are matched.
7. If a signal is given at the shunt arm port, power divides equally into the side arms and they are in phase. There is no coupling to the series arm.
8. If a signal is given to the series arm port, power divides equally into the side arms and they are 180° out of phase. There is no coupling to H-arm.
9. If a signal is given to a side arm, it divides equally into series and shunt arms. There is no coupling to the 2nd side arm.
10. If signals are given to side arms, they will add in phase in H-arm and they add in E-arm with 180° out of phase.
11. H-arm is called sum arm and E-arm is called difference arm.
12. Magic Tee is used to produce sum and difference of signals simultaneously.
13. A typical magic Tee is shown in fig. 9.19.

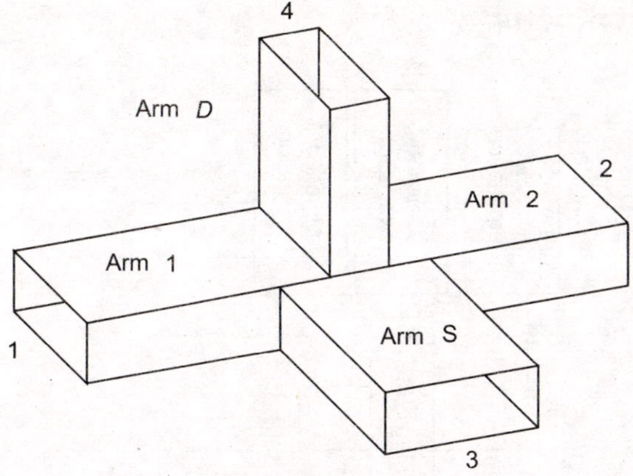

Fig. 9.19 Magic Tee

14. The scattering matrix of magic Tee is

$$S = \begin{bmatrix} 0 & 0 & s_{13} & s_{14} \\ 0 & 0 & s_{23} & s_{24} \\ s_{31} & s_{32} & 0 & 0 \\ s_{41} & s_{42} & 0 & 0 \end{bmatrix}$$

Applications

The magic Tee is used for

- ➢ mixing
- ➢ duplexing
- ➢ producing sum and difference signals.
- ➢ impedance measurements
- ➢ to couple two transmitters to the antenna without loading each other.

III. EQUIPMENT REQUIRED

- Microwave source
- Isolator
- Variable attenuator
- Frequency meter
- Slotted line
- Tunable probe
- Detector mount
- Magic tee
- Matched terminations
- Waveguide stand
- VSWR meter
- Accessories

IV. Experimental Setup

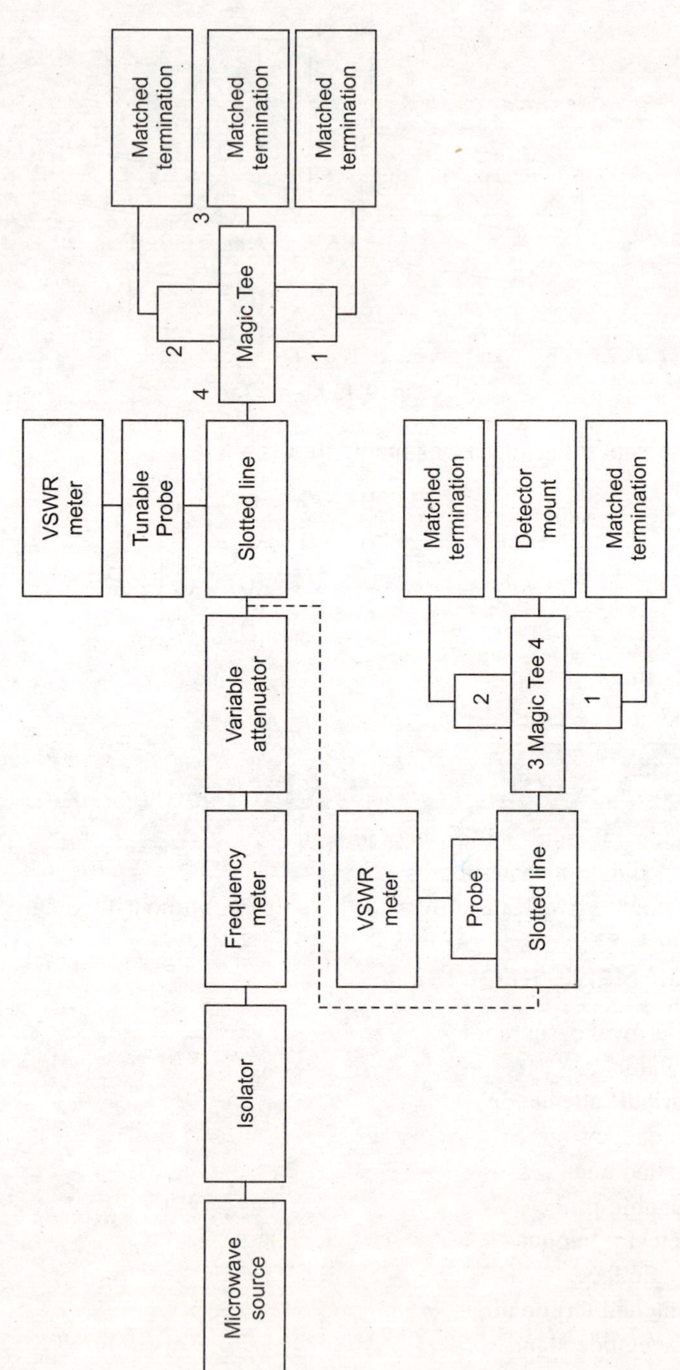

Fig. 9.20 Setup for the study of Magic Tee

V. PROCEDURE

Measurement of Isolation and Coupling Coefficient

1. Connect the detector mount to slotted line as in fig. 9.20.
2. Tune the microwave source for particular frequency of operation and tune the detector mount for maximum output.
3. With the help of variable attenuator and gain control knob of VSWR meter, set a reference power level in the VSWR meter and note down. Let it be P_3.
4. Without disturbing the position of variable attenuator and gain control knob. Note down the reading of VSWR meter. Let it be P_4.
5. Determine the isolation between port 3 and 4 as $P_3 - P_4$ in dB.
6. Determine the coupling coefficient from equation given below.
7. The same experiment can be repeated for other ports also.
8. Repeat the above experiment for other frequencies.

$$C_{ij} = 10^{-\alpha/20}$$

Here, $\alpha = 10 \log 10 \left(\dfrac{p_i}{p_j} \right)$

RESULTS

Table 9.13   Results for coupling

S. No.	Frequency	p_3	p_4	p_1	p_2

Table 9.14   Results for isolation

	$\alpha = 10 \log_{10} \dfrac{p_4}{p_3}$	$C_{43} = 10^{-\alpha/20}$	$\alpha = 10 \log 10 \dfrac{p_1}{p_3}$	$\alpha = 10 \log_{10} \dfrac{p_2}{p_3}$	C_{31}	C_{32}
f_1						
f_2						
f_3						

EXPERIMENT 10

9.24 STUDY THE ISOLATOR AND CIRCULATOR

I. AIM

To measure insertion loss and isolation of a isolator and a circulator.

II. THEORETICAL CONCEPTS

Isolators

An isolator is a passive device which transmits the wave in the forward direction and does not allow transmission in the reverse direction. It means, it does not exhibit any attenuation in the transmission from port 1 to port 2 and it attenuates totally while transmitting in the reverse direction. It is also called uniline. It behaves like a *p-n* diode. It is used to load network.

Salient Features

1. Isolator is usually made of ferrite.
2. It is a non-reciprocal two-port device.
3. It passes microwave signals with low loss in the forward direction and absorbs energy in the reverse direction.
4. The measurements are made in both directions.
5. It exhibits low insertion loss in the forward direction and high attenuation in the reverse direction.
6. It is described by back-to-front ratio. The back-to-front radio is the ratio of reverse attenuation in decibels to the forward insertion loss in decibels.
7. A resonant absorption isolator consists of a section of a rectangular waveguide with ferrite material placed halfway to the center for the waveguide along the axis of the guide.
8. The isolator is popularly known as uniline.
9. Isolator is used to improve the frequency stability in microwave generators like klystrons and magnetrons etc.
10. The isolator is kept between the generator and load to prevent reflected power.
11. It protects the generators from reflected power.
12. An ideal isolator has insertion loss of zero dB. A practical isolator has an insertion loss of about 0.5 dB in the forward direction.
13. An ideal isolator has infinite insertion loss. But practically, it provides an isolation of about 25 dB.
14. It behaves like a diode.
15. It is also similar to a gyrator in construction except that it uses a 45° twist section and 45° Faraday's rotation.
16. Its band width is 10 percent.
17. Typical transmission loss is less than 1 dB in the forward direction and it is about 30 dB in the reverse direction.

Circulators

A circulator is a microwave passive multiport device in which the incident wave at port 1 is coupled to port 2 only, incident wave at port 2 is coupled to port 3 only and so on. The ideal circulator is a matched device. It means, when all the ports are terminated in a matched load except one port, the input impedance of the other ports is equal to characteristic impedance of the input line.

A four-port circulator is shown in fig. 9.21.

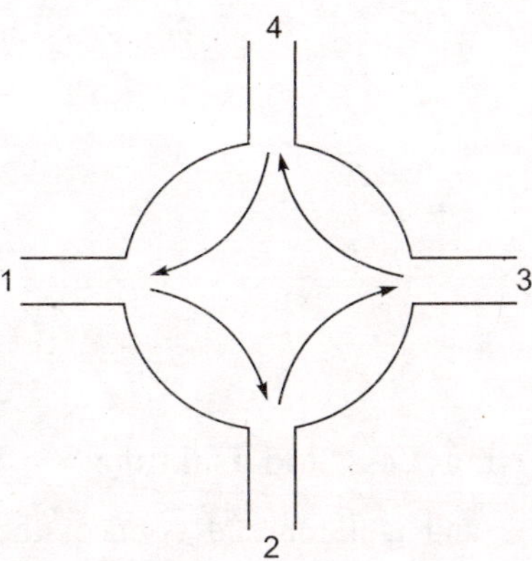

Fig. 9.21   A four-port circulator

A four-port circulator is also constructed using a 3 dB side hole directional coupler and rectangular waveguide and non-reciprocal phase shifters. This is a more compact circulator.

Applications

- It is used to separate the input and output in negative resistive applications.
- It is used to couple a transmitter and receiver to a common antenna.

Salient Features

1. It is usually made of ferrite.
2. It is a mutliport device.
3. In this device, if energy is fed into first port, it goes only to the second port. If energy is fed into second port, it goes only to third port. If energy is fed into the last port, it goes to the first.
4. There exists isolation or attenuation from any port to all ports other than the next to it.
5. The circulators containing three or four ports are the most common.
6. A circulator consists of two nonreciprocal phase shifter waveguide sections mounted side by side with two slots in the walls.
7. Two magic tees are connected into 180° phase shift to create a four-port circulator.
8. Circulator can handle 100 or more kilowatts of average power.
9. Their bandwidth is more than one octane.

III. EQUIPMENT REQUIRED

- Microwave source
- Isolator

- Circulators
- Frequency meter
- Variable attenuator
- Slotted line
- Tunable probe
- Detector mount
- VSWR meter
- Test isolation
- Circulation

IV. PROCEDURE

Measurements of Insertion Loss and Isolation

1. Connect the probe and isolator and connect the detector mount to the slotted section. Connect the output of the detector mount to VSWR meter.

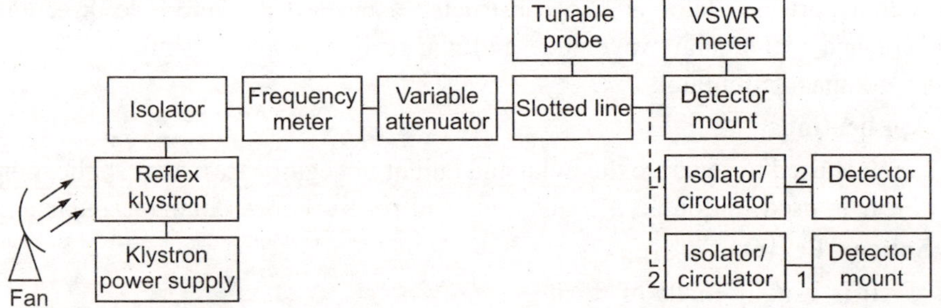

Fig. 9.22 Setup for measurement loss & isolation of isolator & circulator

2. Tune the source for maximum output particular frequency of operation. Tune the detector mount for maximum output in the VSWR meter.
3. Set any reference level of power in VSWR meter with the help of variable attenuator and gain control knob of VSWR meter. Let it be P_1.
4. Remove the detector mount from slotted line without disturbing the position of setup. Connect the isolator/circulator between slotted line and detector mount as in fig. 9.22. Keeping input port to slotted line and detector at its output port. A matched termination should be placed at the third port in case of circulator.
5. Now the reading in the VSWR meter. If necessary change range – dB switch from high to low position. Let it be P_2.
6. Find insertion loss from $P_1 - P_2$.
7. For measurement of isolation, the isolator or circulator has to be connected in reverse i.e. output port to slotted line and detector to input port with another port terminated by matched termination (in case circulator) after

setting a reference level without isolator or circulator in the setup as described in insertion loss measurement. Let same P_1 level is set.

8. Note the reading of VSWR meter inserting the isolator or circulator as given in step 7.
9. Find isolation as $P_1 - P_3$.
10. The same experiment can be done for other ports of circulator.
11. Repeat the above experiment for other frequencies if required.

Table 9.15   Results for insertion loss and isolation

S. No.	Frequency	P_1	P_2	P_3	Insertion Loss	Isolation
	f_1					
	f_2					
	f_3					

V. PRECAUTIONS

1. An isolator or attenuator should be used between the klystron and the other equipment in the setup to avoid loading of the klystron.
2. While measuring frequency, frequency meter should be detuned each time.
3. The negative repeller voltage should be applied first before anode voltage is applied.
4. Before switching on power supply, the control knobs of klystron power supply should be kept as below :
   Meter switch : OFF
   Mode switch : *AM*
   Beam voltage knob : Fully anti-clockwise
   Reflector voltage : Fully clockwise
   AM – Amplitude : Fully clockwise
   AM – Frequency knob : Mid position
5. The control knob of VSWR meter should kept as below
   Meter switch : Normal
   Input Switch : Low impedance position
   Range dB switch : 40/50 dB
   Gain control knob : Fully clockwise
6. Cooling fan should be used to avoid heating of klystron tube.

EXPERIMENT 11

9.25   CALIBRATION OF ATTENUATORS

I. AIM

To study the attenuators (fixed and variable type)

II. THEORETICAL CONCEPTS

Definition

An attenuator is a device which reduces the power of the signal when the signal passes through it.

Attenuation is the real part of propagation constant and is the reduction of the power and is expressed in positive decibels for convenience. The attenuation of the attenuator is

$$\alpha = 10 \log_{10} \frac{p_i}{p_0}$$

Here, p_i = input power

p_0 = output power

Types of Attenuators

There are

- Fixed attenuators
- Step attenuators
- Continuously variable attenuators

Fixed Attenuators

In these, the attenuation is reduced by a fixed amount. The 3 dB, 10 dB, 20 dB attenuators are available. These attenuators are made of film resistors around a center conductor and disks. The film resistor attenuators are useful at wide range of frequencies as they exhibit constant values. In this, the material depth is less than the skin depth. The attenuators are specified by frequency and attenuation in dB.

The fixed attenuators are also called pads. For example, 6 dB pad means the output power of attenuator is one-fourth of the input.

Continuously Variable Attenuators

The attenuation is given by

$$\alpha = 2\pi \sqrt{\left(\frac{1}{\lambda_c}\right)^2 - \left(\frac{1}{\lambda}\right)^2}$$

Here, α is attenuation

λ_c is cut-off wavelength

λ = free space wavelength

Salient Features of Attenuator

1. It reduces input power given to it.
2. It is a two-port reciprocal network.
3. Its scattering matrix is

$$S = \begin{bmatrix} 0 & s_{12} \\ s_{21} & 0 \end{bmatrix}$$

4. $s_{11} = s_{22} = 0$.
5. It has transmission coefficients.
6. Fixed attenuation uses a tapered edge of resistive vane which is made of lossy material.
7. A variable attenuator consists of a tapered resistive card. The depth of penetration into the waveguide is adjustable.
8. Accuracy of the attenuator increases if the resistive card is replaced by a piece of a metal film with glass coating.
9. The variable attenuator is frequency sensitive and some phase shift is introduced while changing the attenuation.
10. Variable attenuators are constructed by placing lossy resistive card in the waveguide—its surface parallel to the electric field.
11. The variation of attenuation depends on the mount of card insertion.
12. The card is tapered to minimize the reflection.

III. EQUIPMENT REQUIRED

- Microwave source
- Isolator
- Frequency meter
- Variable attenuator
- Slotted line
- Tunable probe
- Detector mount
- Matched termination
- VSWR meter
- Test fixed and variable attenuator and accessories

IV. PROCEDURE

A. Input VSWR Measurement

1. Connect the equipment as shown in fig. 9.23.

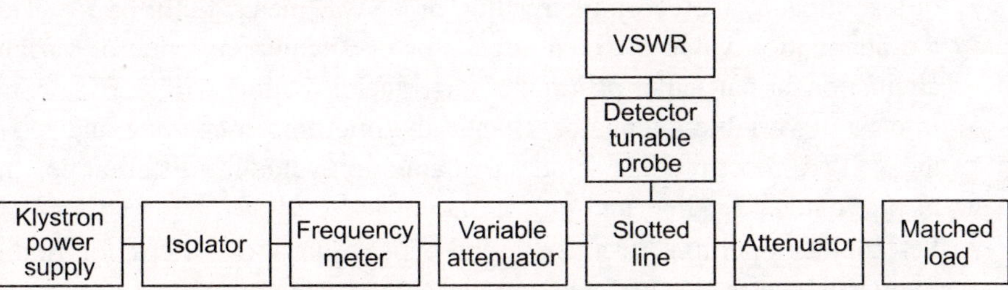

Fig. 9.23   Setup for VSWR measurement of attenuator

2. The klystron for maximum power at a frequency of operation.
3. Measure the VSWR with the help of tunable probe, slotted line and VSWR meter as described in the experiment of low and medium VSWR.
4. Repeat at the above step for other frequencies.

B. Insertion Loss/Attenuation Measurement

1. Connect the setup as in fig. 9.24.

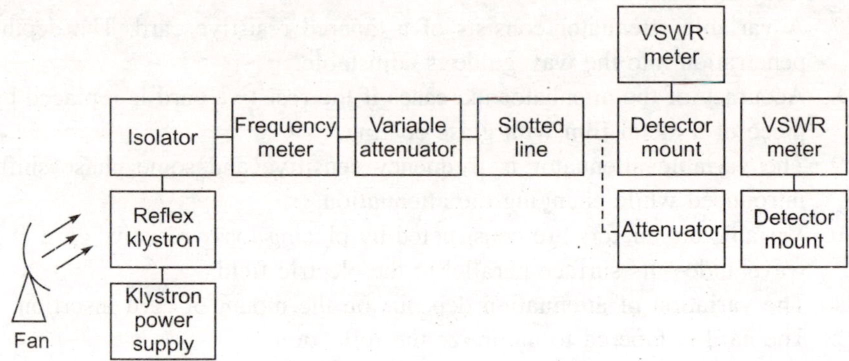

Fig. 9.24 Setup for insertion loss & attenuation measurement of attenuator

2. Connect the detector mount to the slotted line, and tune the detector mount also for maximum deflection on VSWR meter (detector mount's output should be connected to VSWR meter).
3. Keep any reference level on the VSWR meter with the help of variable attenuator and gain control knob of VSWR meter. Let it be P_1.
4. Disconnect the detector mount from the slotted line, without disturbing any position on the setup. Place the test variable attenuator to the slotted line and detector mount to other port of test variable attenuator. Keep the micrometer reading of test variable attenuator to zero and record the reading of VSWR meter. Let it be P_2. Then the insertion loss of test attenuator will be $P_1 - P_2$ dB.
5. For measurement of attenuation of fixed and variable attenuator after step 4 of the above measurement, carefully disconnect the detector mount from the slotted line without disturbing any position obtained up to step 3. Place the test attenuator to the slotted line and detector mount to the other port of test attenuator. Record the reading of VSWR meter. Let it be P_3. Then the attenuation value of fixed attenuator or attenuation value of variable attenuator for particular position of micrometer reading will be $P_1 - P_3$ dB.
6. In case of variable attenuator, change the micrometer reading and record the VSWR meter reading. Find out attenuation value for different position of micrometer reading and plot a graph.
7. Set another operating frequency and repeat the above steps for finding frequency sensitivity of fixed and variable attenuator.

V. PRECAUTIONS

1. An isolator or attenuator should be used between the klystron and the other equipment in the setup to avoid loading of the klystron.
2. While measuring frequency, frequency meter should be detuned each time.
3. The negative repeller voltage should be applied first before anode voltage is applied.
4. Before switching on power supply, the control knobs of klystron power supply should be kept as below :
   Meter switch : OFF
   Mode switch : *AM*
   Beam voltage knob : Fully anti-clockwise
   Reflector voltage : Fully clockwise
   AM – Amplitude : Fully clockwise
   AM – Frequency knob : Mid position
5. The control knob of VSWR meter should kept as below
   Meter switch : Normal
   Input Switch : Low impedance position
   Range dB switch : 40/50 dB
   Gain control knob : Fully clockwise
6. Cooling fan should be used to avoid heating of klystron tube.

EXPERIMENT 12

9.26 MEASUREMENT OF UNKNOWN IMPEDANCE

I. AIM

To find out unknown impedance using slotted line.

II. THEORETICAL CONCEPTS

The impedance of an antenna or a transmission line is a parameter. It is always expressed in terms of load and characteristic impedances.

If $Z_L \neq Z_0$, reflection takes place. The magnitude of the reflection coefficient, VSWR, phase, relative position of standing wave patterns with respect to short-circuit, characteristic impedance are important parameters of the load.

The input impedance of a transmission line is defined as

$$Z_{\text{in}} = \frac{V_s}{I_s}$$

and it is given by

$$Z_{\text{in}} = \frac{V_L \cos h\gamma \ell + Z_o I_L \sin h\gamma \ell}{I_L \cos h\gamma \ell + \dfrac{V_L}{Z_o} \sin h\gamma \ell}$$

Here, $\gamma = \alpha + j\beta$, If $\alpha = 0$, $\gamma = j\beta$

For a line of length,

$$Z_L = Z_0 \left(\frac{1 - j\rho \tan \beta \Delta\ell}{\rho - j \tan \beta \Delta\ell} \right)$$

Here, Z_L = the impedance at the receiving end

Z_0 = characteristic impedance

Z_{in} = input impedance

$\beta \Delta\ell$ = electrical distance in wavelength

III. EQUIPMENT REQUIRED

- Gunn power supply
- Gunn oscillator
- Isolator
- Frequency meter
- Variable attenuator
- Slotted line
- Tunable probe
- Detector
- VSWR meter
- Short
- Unknown load
- Impedance.

IV. EXPERIMENTAL SETUP

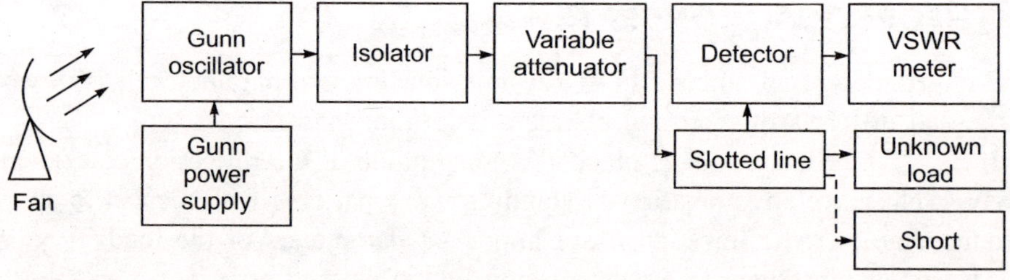

Fig. 9.25 Experimental setup for the measurement of unknown impedance

V. PROCEDURE

1. Assemble the components as in fig. 9.25.
2. Set variable attenuator for maximum attenuation.
3. Connect the unknown load.

4. Switch on the Gunn power supply and set the micrometer of Gunn oscillator for required operating frequency.

5. Keep the mode switch power supply to square wave with modulation frequency of 1 KHz.

6. Move the probe along the slotted line for a minimum. Let it be a reference value at X_1.

7. Obtain next successive minimum at X_2.

8. Replace load by a short.

9. Move the probe to a new minimum at X_s.

10. Find out $X_s - X_2$ or $X_s - X_1$. It is positive if the minimum is shifted towards load. It is negative if minimum is shifted towards generator.

11. Find out $(X_s - X_1)/\lambda_g$.

12. Find out the load impedance from $Z_L = Z_0 \left(\dfrac{1 - \rho \tan \beta \Delta \ell}{\rho - j \tan \beta \Delta \ell} \right)$.

    Here, ρ = reflection coefficient, $\beta \Delta \ell = 2\pi (X_2 - X_1)/\lambda_g$.

VI. CALCULATION OF IMPEDANCE USING SMITH CHART

1. Note VSWR and shift in minimum X in term of wavelength.

   $$\text{VSWR} = V_{max}/V_{min} \text{ and } |\rho| = \frac{\text{VSWR} - 1}{\text{VSWR} + 1}.$$

2. Draw a VSWR circle with a radius of 1/VSWR.

3. Locate a point at a distance of X from $(0, 0)$ moving anticlockwise direction on the circumference. Join this point to the centre of Smith chart.

4. The intersection of VSWR and the line gives the normalized load impedance.

5. Calculate the load impedance from the product of normalized impedance and characteristic impedance of slotted line.

6. The characteristic impedance is given by $Z_0 = \dfrac{\eta_0}{\left[1 - \left(\dfrac{\lambda_c}{\lambda_0} \right)^2 \right]^{\frac{1}{2}}}$.

VIII. PRECAUTIONS

1. An isolator or attenuator should be used between the klystron and the other equipment in the setup to avoid loading of the klystron.

2. While measuring frequency, frequency meter should be detuned each time.

3. The negative repeller voltage should be applied first before anode voltage is applied.

4. Before switching on power supply, the control knobs of klystron power supply should be kept as below :

   Meter switch : OFF

   Mode switch : *AM*

   Beam voltage knob : Fully anti-clockwise

   Reflector voltage : Fully clockwise

   AM – Amplitude : Fully clockwise

   AM – Frequency knob : Mid position

5. The control knob of VSWR meter should kept as below

   Meter switch : Normal

   Input Switch : Low impedance position

   Range dB switch : 40/50 dB

   Gain control knob : Fully clockwise

6. Cooling fan should be used to avoid heating of klystron tube.

EXPERIMENT 13

9.27 MICROWAVE POWER MEASUREMENT

I. AIM

To measure microwave power

II. THEORETICAL CONCEPTS

Microwave power measurements are divided into three power ranges.

1. Low power (less than 1 mW)
2. Medium power (between 1 mW and 10 W)
3. High power (greater than 10 W)

Power at microwave frequencies is detected by

- Bolometer
- Thermo coupler
- Microwave crystal

Bolometers: These are used to detect power. Their operation depends on thermal principles. The bolometer resistance varies with temperature. The temperature is in turn influenced by the power absorbed. The power absorbed is proportional to the resistance.

Thermo-Couplers: Thermo-couplers are also used as power detectors. The voltage across a thermo-couplers changes with the difference in temperature at the junctions. The voltage is proportional to the power.

Microwave Crystals: The microwave crystals are nothing but rectifiers which are basically nonlinear. The rectified output is proportional to the power.

Low Power Measurement (1 to 10 mW) by Bolometer or Thermister

Bolometer is a temperature sensitive device. When microwave power is applied to bolometer, its temperature changes which in turn changes its resistance. The change of resistance is measured using a bridge circuit like Wheatstone Bridge (fig. 9.26).

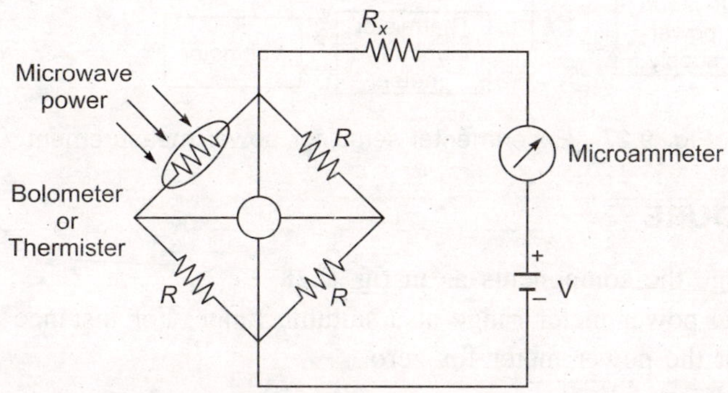

Fig. 9.26 Low power measurement by bridge circuit using bolometer

The bolometer is connected in one of the arm of the bridge. Initially the bridge is balanced in the absence of incident of power. When microwave power is applied, imbalances of the bridge takes place due to change in resistance. This change in resistance is measured and hence the power is measured by rebalancing with R_x.

The power is measured from

$$P = \frac{1}{4}\,(I_1^2 - I_2^2)R_b$$

Here I_1 = Initial current

I_2 = changed current through bolometer

R_b = bridge resistance

III. EQUIPMENT REQUIRED

- Reflex klystron
- Klystron power supply
- Frequency meter
- Variable attenuator
- Power meter
- Thermistor and its mount
- Voltmeter
- Waveguide stands
- Waveguide to coaxial adaptors

IV. EXPERIMENTAL SETUP

The measurement setup is shown in fig. 9.27.

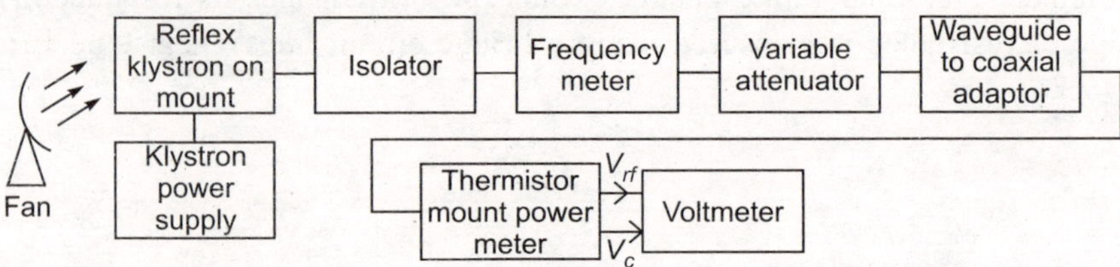

Fig. 9.27   Experimental setup for power measurement

V. PROCEDURE

1. Arrange the components as in fig. 9.28.
2. Set the power meter range at a suitable range. For instance at 1 mW.
3. Adjust the power meter for zero.
4. Note V_{rf} and V_c.
5. Keep maximum attenuation.
6. Tune reflex klystron and adjust for a specified frequency.
7. Decrease the attenuator slowly.
8. Adjust the attenuator to read 3 mW in power meter.
9. Note V_{rf} (rf bridge voltage) and V_c (corresponding bridge voltage).

10. Calculate *rf* power from $p_0 = $ microwave power $+ \dfrac{V_{rf}^2}{4R}$.

11. Repeat the steps 10 for power setting of 2 mW, 1 mW and 0.5 mW. Use calibration factor for thermister mount to determine exact power.
12. Tune reflex klystron and set two more frequencies and repeat steps 8 to 12.
13. Tabulate the results as in the table 9.16.

Table 9.16   Power measurements

S. No.	Frequency	3 mw V_{rf}	V_c	Power Meter Reading	Total Power
1.	f				

VI. PRECAUTIONS

1. An isolator or attenuator should be used between the klystron and the other equipment in the setup to avoid loading of the klystron.
2. While measuring frequency, frequency meter should be detuned each time.

3. The negative repeller voltage should be applied first before anode voltage is applied.

4. Before switching on power supply, the control knobs of klystron power supply should be kept as below :

   Meter switch : OFF

   Mode switch : *AM*

   Beam voltage knob : Fully anti-clockwise

   Reflector voltage : Fully clockwise

   AM – Amplitude : Fully clockwise

   AM – Frequency knob : Mid position

5. The control knob of VSWR meter should kept as below

   Meter switch : Normal

   Input Switch : Low impedance position

   Range dB switch : 40/50 dB

   Gain control knob : Fully clockwise

   Cooling fan should be used to avoid heating of klystron tube.

EXPERIMENT 14

9.28 MEASUREMENT OF DIELECTRIC CONSTANT

I. AIM

To measure the dielectric constant of a given material.

II. THEORETICAL CONCEPTS

The materials are characterized by their dielectric constant, permeability and conductivity. The dielectric constant may be complex and is given by

$$\epsilon = \epsilon' - j\epsilon''$$

Here, ϵ is the dielectric constant and ϵ'' is the loss factor.

For non-metallic materials, $\mu = \mu_0$.

The energy lost in the dielectric is indicated by the loss tangent ($\tan \delta$). It is given by $\tan \delta = \dfrac{\epsilon''}{\epsilon'}$.

At microwave frequencies, the dielectric constant is determined from the propagation characteristics of electromagnetic wave through the dielectric material.

III. EQUIPMENT REQUIRED

- Microwave source
- Isolator
- Variable attenuator
- Frequency meter

- Slotted line
- Tunable probe
- Phase shifter
- Movable short
- VSWR meter
- Cables and accessories

IV. MEASUREMENT SETUP

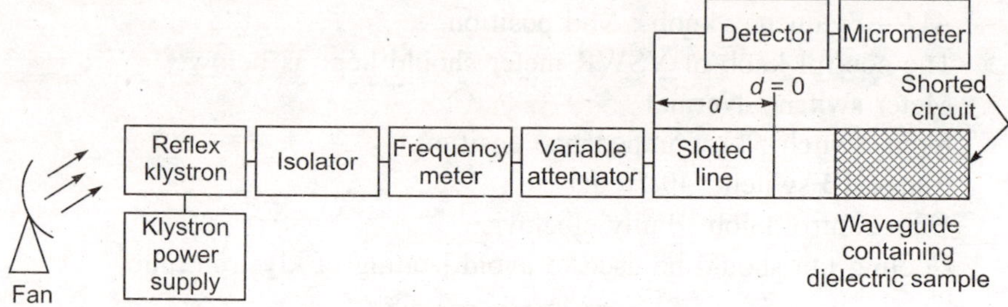

Fig. 9.28   Setup for the measurement of dielectric constant

V. PROCEDURE

1. Assemble the equipment as shown in fig. 9.28.
2. Connect the short at the load end of the waveguide.
3. Measure guide wavelength λ_g.
4. Obtain the standing waves in the waveguide for short in place.
5. Measure ΔX_s using the twice maximum method for high standing wave ratio. The distance to be measured are indicated in the diagram. Measurements are made with the slotted section connected to the waveguide which contains the sample.
6. Note the position of the minimum X_s.
7. Insert the dielectric sample in the waveguide so that it is against the short circuit.
8. Measure ΔX_s by the twice a minimum method and note the position of minimum $X_\in$.

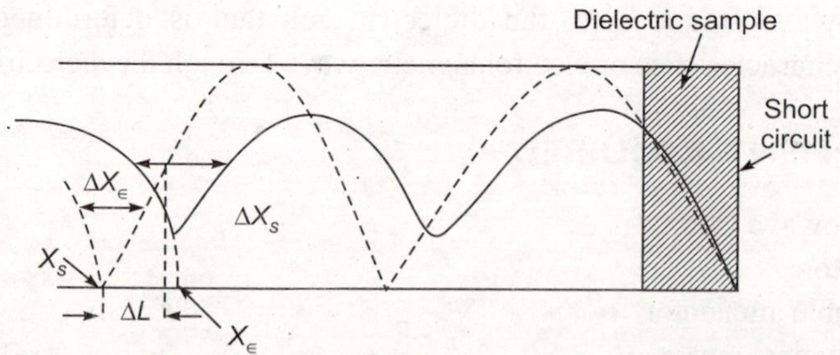

Fig. 9.29   Standing waves in the waveguide with and without the sample

9. Measure the difference between the minima ΔL.

10. Calculate

$$\frac{\tan x}{x} = \frac{\tan \beta_{d_1} w_1}{\beta_{d_1} w_1} = \frac{\lambda_g}{2\pi d} \tan \beta(\Delta L + w) \tag{1}$$

Here, $x = \beta d_1\, w_1$, w_1 is the thickness of the dielectric sample.

βd_1 = Phase constant in the guide with the dielectric

$$= \frac{2\pi}{\lambda}\left[\left(\epsilon_r \mu_r - \left(\frac{\lambda}{\lambda_c}\right)\right)^2\right]^{1/2}$$

The equation (1) has infinite number of solutions. The measurement is repeated with two more samples of thickness w_2 and w_3. βd_2 and βd_3 are determined. The correct solution is obtained from the intersection of the three variations.

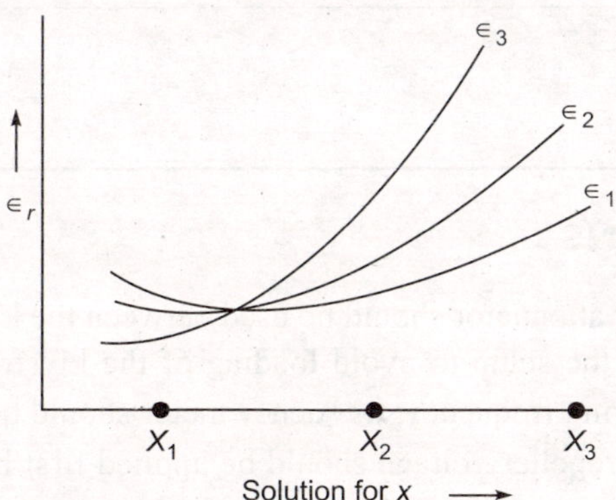

Fig. 9.30 Calculating of dielectric constant using three-point method

11. Find the dielectric constant from

$$\epsilon_r = \left(\frac{x\lambda}{2\pi d}\right)^2 + \left(\frac{\lambda}{\lambda_c}\right)^2$$

Here, λ = free space wavelength

   λ_c = cut-off wavelength of the waveguide of dominant mode

   = 2 × broad dimension of the waveguide

12. Find $\epsilon'' = \epsilon_r \tan \delta$

$$= \frac{(\Delta x_\epsilon - \Delta x_s)}{d}\left(\frac{\lambda}{\lambda_g}\right)^2$$

13. Place the dielectric sample in the waveguide so that it is against the short circuit.

14. Measure the position of minima in the slotted line with respect to the reference plane ($d = 0$). Let it be d_2.
15. Find VSWR.
16. Repeat steps (1) to (6) with sample having different length.
17. Find guide wavelength λ_g.
18. Find cut-off wavelength from $\lambda_\in = 2a$.
19. Tabulate the results.

VI. RESULTS

Table 9.17   Measurement for dielectric constant

S. No.	Sample	Sample Thickness $s(\ell)$	minima position of standing wave d_1	d_2	β	$\beta\ell_t = \beta(\ell_t + d_1 - d_2)$	Solution of x $x = \beta(\ell_\in + d_1 - d_2)$

VII. PRECAUTIONS

1. An isolator or attenuator should be used between the klystron and the other equipment in the setup to avoid loading of the klystron.
2. While measuring frequency, frequency meter should be detuned each time.
3. The negative repeller voltage should be applied first before anode voltage is applied.
4. Before switching on power supply, the control knobs of klystron power supply should be kept as below :

    Meter switch : OFF

    Mode switch : *AM*

    Beam voltage knob : Fully anti-clockwise

    Reflector voltage : Fully clockwise

    AM – Amplitude : Fully clockwise

    AM – Frequency knob : Mid position
5. The control knob of VSWR meter should kept as below

    Meter switch : Normal

    Input Switch : Low impedance position

    Range dB switch : 40/50 dB

    Gain control knob : Fully clockwise
6. Cooling fan should be used to avoid heating of klystron tube.

9.29 POINTS TO REMEMBER

- Reflex klystron is a microwave source.
- A typical klystron is 2K25.
- Isolator is used in the measurements to protect source from reflections.
- Modes in reflex klystron are determined from the variation of output power with repeller voltage.
- The frequency of reflex klystron depends on beam voltage, repeller voltage and cavity dimensions.
- Attenuator is also called pad.
- Frequency meters operate on the basis of absorption or reaction or transmission of waves.
- Oscilloscope provides time-domain information.
- Spectrum analyzer provides frequency domain information.
- Network analyzer is a device which is useful to measure both amplitude, phase of a microwave signal, passive and active microwave network and component parameters, impedance, gain, transmission and reflection parameters.
- Micro-ammeters are used for power measurements in place of power meter due to low cost.
- The transit time and frequency in reflex klystron are related by

$$T_t = \frac{n + 3/4}{f_0}.$$

- Mode number is given by $N_n = n + 3/4$.

- Electronic turning sensitivity is given by $\text{ETS} = \dfrac{(f_2 - f_1)}{(V_2 - V_1)}$ MHz/volt.

- Klystron efficiency is $\eta = \dfrac{p_0}{V_a I_a} \times 100\%$.

- Variable attenuator is used to control the incoming power level.
- The repeller voltage is negative.

- The mode numbers are calculated from $\dfrac{N_2}{N_1} = \dfrac{V_1}{V_2} = \dfrac{(n+1) + 3/4}{(n + 3/4)}$.

- The minimum value of VSWR is 1.0.
- The maximum value of VSWR is ∞.
- The crystal detector is used to sense the presence of microwave power.
- The equivalent circuit of a crystal detector is

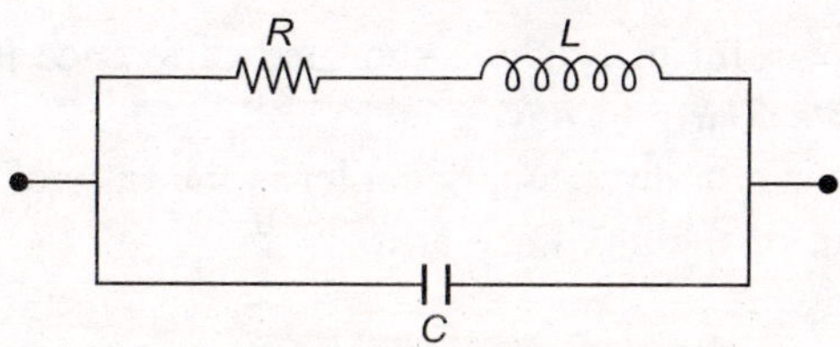

- Matched load absorbs incident power.
- Shorts reflect incident power.
- Bolometers are used to measure power.
- Wavemeters are used to measure frequency.
- Waveguide slotted line consists of longitudinal slot in the centre of the broad wall of a rectangular waveguide.
- VSWR meter can be used to measure VSWR, reflection coefficient and power.
- VSWR can be measured by CRO, double minimum method or by VSWR meter.
- VSWR is given by $\dfrac{V_{max}}{V_{min}}$.
- Input impedance of a transmission line is $Z_{in} = \dfrac{V_L \cosh\gamma\ell + Z_o I_L \sinh\gamma\rho}{I_L \cosh\gamma\ell + \dfrac{V_L}{Z_o}\sinh\gamma\ell}$.
- Microwave frequency can be measured from $f = v_0 \sqrt{\dfrac{1}{\lambda_g^2} + \dfrac{1}{\lambda_c^2}}$.
- Reflection is measured from $|\rho| = \dfrac{VSWR - 1}{VSWR + 1}$.
- Gunn diode is used as a microwave oscillator.
- Gunn diode is a negative resistance device.
- Reflex klystron uses the concepts of velocity modulation.
- Null-to-null beam width is approximately twice that of half-wave beam width.
- Attenuation is given by $\alpha(dB) = 10 \log_{10} \dfrac{p_i}{p_0}$.
- Free space wavelength is measured from $\dfrac{1}{\lambda^2} = \dfrac{1}{\lambda_g^2} - \dfrac{1}{\lambda_c^2}$.
- Guide wavelength for the dominant mode is $\lambda_g = \dfrac{\lambda}{\left[1 - \left(\dfrac{\lambda}{2a}\right)^2\right]^{\frac{1}{2}}}$.
- Magic Tee is useful to produce sum and difference patterns.
- Magic Tee is a four-port device.
- Coupling factor of a directional coupler is the ratio of power in main arm power and out of the auxiliary arm.

> Directivity of the directional coupler is the ratio of power in auxiliary arm in the forward direction to the power in auxiliary arm in the reverse direction.

> Reflex klystrons are used as local oscillators in commercial, military, air borne, Doppler radar and in missiles.

> The equivalent circuit of reflex klystron is

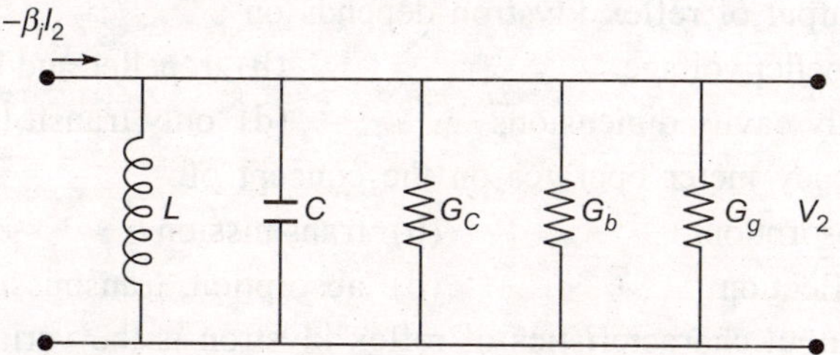

> The modes of operation of Gunn diode are stable amplification mode, LSA oscillation mode and bias-circuit oscillation mode.

> Gunn diode is a TED device.

9.30 MULTIPLE CHOICE QUESTIONS

1. Reflex klystron consists of
   (a) single cavity
   (b) two cavity
   (c) multi-cavity
   (d) no cavities

2. Reflex klystron is
   (a) an amplifier
   (b) oscillator
   (c) detector
   (d) rectifier

3. Reflex klystron contains
   (a) no feedback
   (b) oscillations
   (c) detector
   (d) rectifier

4. Reflex klystron is a
   (a) low power generator
   (b) high power oscillator
   (c) low gain amplifier
   (d) not an oscillator

5. Typical range of efficiency of reflex klystron is
   (a) 80 to 90 %
   (b) 70 to 80 %
   (c) 10 to 20 %
   (d) 20 to 30 %

6. The reflex klystrons are used in
   (a) microwave broadcast transmitters
   (b) short wave broadcast transmitters
   (c) as local oscillators
   (d) as amplifier in TV broadcast transmitters

7. The efficiency of reflex klystron is
   (a) p_0/p_{ac}
   (b) $p_0/V_a I_a$
   (c) p_{dc}/p_{ac}
   (d) p_{ac}/p_{dc}

8. Variable attenuator is used to
   (a) amplify power
   (b) control power
   (c) rectify power
   (d) convert dc to ac

9. The output of reflex klystron depends on
   (a) repeller voltage
   (b) repeller and beam voltage
   (c) only cavity dimensions
   (d) only transit time

10. Frequency meter operates on the concept of
    (a) absorption
    (b) transmission
    (c) reflection
    (d) absorption, transmission and reaction

11. The output characteristics of reflex klystron is the variation of
    (a) power with repeller voltage
    (b) power with beam voltage
    (c) power with frequency
    (d) power with transit time

12. The frequency characteristics is variation of
    (a) frequency with power
    (b) frequency with repeller voltage
    (c) frequency with beam voltage
    (d) frequency with transit time

13. Bolometer is used to measure
    (a) low power
    (b) high power
    (c) frequency
    (d) low voltage

14. Waveguide stands are used to
    (a) connect the equipment properly
    (b) align the components
    (c) measure VSWR
    (d) measure power

15. Matched load
    (a) reflects power
    (b) absorbs power
    (c) transmits power
    (d) scatter power

16. Shorted load results in reflection coefficient to be
    (a) -1
    (b) ∞
    (c) 0
    (d) $-\infty$

17. Pure reactive loads makes magnitude of the reflection coefficient to be
    (a) 0
    (b) ∞
    (c) 1
    (d) -1

18. Open circuit loads makes reflection coefficient to be
    (a) -1
    (b) 1
    (c) 0
    (d) ∞

19. The reflection coefficient in terms of Z_L and Z_0 is
    (a) $\dfrac{Z_L - Z_0}{Z_L + Z_0}$
    (b) $\dfrac{Z_0 - Z_L}{Z_0 + Z_L}$

(c) $\dfrac{Z_L + Z_0}{Z_L - Z_0}$ (d) Z_L/Z_0

20. If $Z_L = 200\ \Omega$, $Z_0 = 100\ \Omega$, reflection coefficient is
    (a) 2   (b) 0.5
    (c) 0.333   (d) ∞

21. If the reflected voltage is 25 and incident voltage is 50, the reflection coefficient is
    (a) 0.5   (b) 2
    (c) ∞ (d) 0

22. The minimum VSWR is
    (a) 10   (b) 1.0
    (c) 0   (d) -1.0

23. If the reflection coefficient is 0.25, and input voltage is 50 V, the reflected voltage is
    (a) 200 V   (b) 2 V
    (c) 12.5 V   (d) 100 V

24. If V_{max} is 3 V and V_{min} is 1.5 V in the standing wave pattern, VSWR is
    (a) 2   (b) 0.5
    (c) 2.5   (d) 1

25. If the reflection coefficient is 0.377, VSWR is
    (a) 3.77   (b) 37.7
    (c) 2.21   (d) 377

26. If VSWR is 1.67, VSWR in dB is
    (a) 4.44 dB   (b) 44.4 dB
    (c) 444 dB   (d) 2.22 dB

27. Zero reflection takes place when Z_L is
    (a) Z_0 (b) $0\ \Omega$
    (c) $\infty\ \Omega$ (d) $50\ \Omega$

28. VSWR is 1.0 when
    (a) $Z_L \neq Z_0$ (b) $\rho = 0$
    (c) $\rho = 1$ (d) $\rho = -1.0$

29. VSWR meter is useful to measure directly
    (a) current   (b) reflection coefficient
    (c) VSWR   (d) impedance

30. Isolator is used to
    (a) protect source   (b) provide free transmission
    (c) provide free reflection   (d) match load impedance

31. Isolator is a
    (a) three-port device   (b) four-port device
    (c) two-port device   (d) one-port device

32. Guide wavelength can be measured by
    (a) microwave bench setup
    (b) VSWR meter
    (c) micro-ammeter
    (d) klystron

33. Cut-off wavelength for dominant mode in rectangular waveguide is
    (a) more than free space λ
    (b) less than free space λ
    (c) ∞
    (d) zero

34. The reflection coefficient is

    (a) $|\rho| = \dfrac{S-1}{S+1}$
    (b) $|\rho| = \dfrac{1-S}{1+S}$

    (c) $\rho = \dfrac{1+S}{1-S}$
    (d) $\rho = \dfrac{1+S}{S-1}$

35. In double minimum method, SWR is
    (a) $\lambda_g/(d_1 - d_2)$
    (b) $\lambda_g/\pi(d_1 - d_2)$

    (c) $\pi(d_1 - d_2)$
    (d) $\dfrac{\pi(d_1 - d_2)}{\lambda_g}$

36. If power is fed at port 3 (H-arm) of Magic Tee, power at port 4 (E-arm)
    (a) exist
    (b) does not exist
    (c) half the input power exists
    (d) exists with opposite phase to that of input power

37. If arm 1 and 2 of Magic Tee fed with two signals p_1 and p_2, the signal at port 3 is
    (a) $p_1 + p_2$
    (b) $p_1 - p_2$
    (c) 0
    (d) $-p_1 - p_2$

38. If arm 1 and 2 fed with two signals (p_1 and p_2), the signal at port 4 is
    (a) difference of (p_1 and p_2)
    (b) ($p_1 + p_2$)
    (c) 0
    (d) $-p_1 - p_2$

39. When the arm 1 and 2 are terminated the isolation in the matched load between E and H arm of magic Tee is

    (a) $10 \log_{10} \dfrac{p_4}{p_3}$
    (b) $10 \log_{10} \dfrac{p_3}{p_4}$

    (c) $10 \log_{10} \dfrac{p_1}{p_2}$
    (d) $10 \log_{10} \dfrac{p_2}{p_1}$

40. H-plane Tee junction is called
    (a) shunt junction
    (b) series junction
    (c) shunt-series junction
    (d) series-shunt junction

41. E-plane Tee junction is called
    (a) series junction
    (b) series-shunt junction
    (c) shunt-series junction
    (d) shunt junction

42. In hollow rectangular waveguide *TEM*
    (a) exists
    (b) does not exist
    (c) exists with high cut-off frequency
    (d) exists with low cut-off frequency
43. The cut-off frequency of *TEM* wave is
    (a) zero
    (b) 1.0
    (c) ∞
    (d) moderate
44. In microwave measurements, the slot in the broad wall of the rectangular waveguide can be
    (a) displaced from centre line
    (b) at the centre
    (c) inclined
    (d) inclined and displaced
45. The characteristic impedance of a transmission line to match a 75 Ω line with 50 Ω line is
    (a) 61.2 Ω
    (b) 50 Ω
    (c) 75 Ω
    (d) 0.666 Ω
46. If $Z_L = R_L$ and $R_L > Z_0$, VSWR is

    (a) $\sqrt{R_L Z_0}$
    (b) $\dfrac{R_L}{Z_0}$

    (c) $\dfrac{Z_0}{R_L}$
    (d) $\sqrt{\dfrac{R_L}{Z_0}}$

47. If $R_L = 100\ \Omega$, $Z_0 = 50\ \Omega$, VSWR is
    (a) 2
    (b) 0.5
    (c) $\sqrt{2}$
    (d) 1.41
48. The minimum impedance on a transmission line is

    (a) $\dfrac{Z_0}{\text{VSWR}}$
    (b) $\dfrac{\text{VSWR}}{Z_0}$

    (c) $Z_0 \times \text{VSWR}$
    (d) $\dfrac{Z_L}{\text{VSWR}}$

49. SWR is

    (a) $\dfrac{I_{max}}{I_{min}}$
    (b) $\sqrt{\dfrac{I_{max}}{I_{min}}}$

    (c) $\dfrac{I_{min}}{I_{max}}$
    (d) $\dfrac{V_{min}}{V_{max}}$

50. The relation between different velocities is
    (a) $v_0^2 = v_p v_g$
    (b) $v_g^2 = v_0 v_p$
    (c) $v_p = v_g v_0$
    (d) $v_p = \dfrac{v_0}{v_g}$

9.31   ANSWERS

1. a	18. b	35. b
2. b	19. a	36. b
3. b	20. c	37. a
4. a	21. a	38. a
5. d	22. b	39. a
6. c	23. c	40. b
7. b	24. a	41. d
8. b	25. c	42. b
9. b	26. a	43. a
10. d	27. a	44. b
11. a	28. b	45. a
12. b	29. c	46. b
13. a	30. a	47. a
14. b	31. c	48. a
15. b	32. a	49. a
16. a	33. a	50. a
17. c	34. a	

9.32   EXERCISE PROBLEMS

1. If the coupling factor is 0 dB, find power coupled to the auxiliary arm in a directional coupler.

2. Input power to a directional coupler is 10 mW and the power coupled to the auxiliary arm is 1 mW. Find the coupling factor.

3. The coupling factor of a directional coupler is 20 dB and the directivity 30 dB. It has an insertion loss of 0 dB. Find the coupled power, if input power is 10 mW.

4. If the incident and reflected powers are 3 mW and 0.1 mW from a given load. Find VSWR.

5. An input power of 10 mW is given to the collinear port 1 of a H-plane Tee junction. Find the power delivered through each port. Assume the junction is lossless and other ports are matched terminated.

6. What is the power delivered to the loads of 50 Ω and 75 Ω connected to ports 1 and 2 if 20 mW is given to the matched port 3 of H-plane Tee junction.

7. In a 3 dB directional coupler, find the power coupled through the auxiliary arm if the input power is 50 mW.

8. If the forward current in the auxiliary arm of a directional coupler is 50 μA and the reverse current is 5 μA, find the directivity of the coupler.

Objective Questions and Answers

CHAPTER I

INTRODUCTION TO MICROWAVES AND APPLICATIONS

1. Microwave frequency range is from 10 KHz – 100 GHz   – Yes / No.
2. Microwave frequency range is from 0.3 GHz – 100 GHz   – Yes / No.
3. Wavelength of microwaves is very high   – Yes / No.
4. Microwaves offer large bandwidth   – Yes / No.
5. The frequency of microwave in free space is v_0/λ – Yes / No.
6. Poynting vector is $H \times E$ – Yes / No.
7. Poynting vector is $E \times H$ – Yes / No.
8. Microwaves are received by _______
9. Microwaves are produced by _______
10. $\gamma = \beta + j\alpha$ – Yes / No.
11. $\gamma = \alpha + j\beta$ – Yes / No.
12. The unit of γ is __________
13. The unit of attenuation constant is __________
14. The unit of phase constant is __________
15. Static electric field is __________
16. Static magnetic field is __________
17. The time-varying electric field is __________
18. The time varying magnetic field is __________
19. The velocity of the propagation of microwave is __________
20. The unit of Poynting vector __________
21. The unit of electric field is __________
22. The unit of magnetic field is __________
23. X-rays are part of electromagnetic spectrum   – Yes / No.
24. Visible light is part of electromagnetic spectrum   – Yes / No.
25. Pico means __________
26. Peta means __________
27. Attenuation of microwaves in free space is __________

28. Phase constant of microwaves is a function of _______________
29. Wavelength is directly proportional to frequency   – Yes / No.
30. Propagation constant is expressed in dB/m   – Yes / No.

Answers for Chapter 1

1. No	2. Yes	3. No
4. Yes	5. Yes	6. No
7. No	8. Antennas	9. Antennas
10. No	11. Yes	12. m^{-1}
13. dB/m	14. rad/m	15. $-\nabla V$
16. B/μ	17. $-\nabla V - \dfrac{\partial A}{\partial t}$	18. $\in (V \times E)$
19. $\dfrac{\omega}{\beta}$	20. Watt/m^2	21. Volt/m
22. Amp/m	23. Yes	24. Yes
25. 10^{-12}	26. 10^{18}	27. Zero
28. $f, \mu, \in, \sigma$	29. No	30. Yes

CHAPTER 2

MICROWAVE TUBES FOR MICROWAVE SIGNAL GENERATION

1. Conventional vacuum tubes are used at microwave frequencies
    – Yes / No.
2. The conventional tubes are useful for frequencies below 100 MHz
    – Yes / No.
3. The high frequency operation in conventional vacuum tube triode is effected by transit time effect   – Yes / No.
4. Undesirable lead inductance and interelectrode capacitance exist in vacuum tube triode   – Yes / No.
5. The lead inductances in vacuum tubes are _________
6. The interelectrode capacitances in a vacuum triode are _______
7. The transit time is ________
8. The gain-bandwidth product in a triode is _________
9. The limitations of conventional tubes at microwave frequencies are overcome by _________
10. Klystron is a vacuum tube   – Yes / No.
11. Klystron is used as an amplifier   – Yes / No.
12. The klystron should have only one cavity   – Yes / No.
13. Two-cavity klystron is used as an amplifier   – Yes / No.
14. The velocity of the electron in klystron is ___________

15. The velocity modulator in klystron means ____________
16. Klystron can be used up to 100 GHz   – Yes / No.
17. Bandwidth of klystron is limited due to the use of ________
18. Noise figures of klystron is approximately ________
19. The gain is large in multicavity klystron   – Yes / No.
20. The reflex klystron is a single cavity tube   – Yes / No.
21. The reflex klystron is an __________
22. Two-cavity klystron is not usually constructed   – Yes / No.
23. The klystron oscillates if the output power is feedback to the input cavity and if the loop gain has a magnitude of unity with a phase shift of multiple of 2π – Yes / No.
24. The reflex klystron is a low power oscillator   – Yes / No.
25. The reflex klystron can be used as a local oscillator   – Yes / No.
26. The electrons in reflex klystron are modulated by ________
27. The interaction of electrons and RF field in TWT is ____________
28. The wave in TWT is propagating   – Yes / No.
29. The wave in klystron is propagating   – Yes / No.
30. The bandwidth of TWT is about ________
31. The efficiency of TWT is about ____________
32. At higher frequencies, the inductance and capacitance of resonant circuit must be decreased   – Yes / No.
33. In slow wave structures the wave velocity is small   – Yes / No.
34. In microwave ovens, magnetrons are used   – Yes / No.
35. Magnetron is a higher power device   – Yes / No.
36. Magnetron is a cross-field tube   – Yes / No.
37. Magnetron oscillators usually operate in the ______ mode.
38. The cyclotron angular frequency in the magnetron is ________
39. AMPLITRON is nothing but ________
40. CARCINOTRON is nothing but ____________
41. BWA has two terminals   – Yes / No.
42. BWO has one terminal   – Yes / No.
43. BWO can be tuned over a wide range of frequency by varying ______
44. BWO is a continuous oscillator   – Yes / No.
45. In BWO, the beam is hollow and located in close proximity to the inside of the helix   – Yes / No.
46. In BWA, helix wave depends on RF input   – Yes / No.
47. TWYSTRON is a device that provides the advantages of ________
48. In TWYSTRON, input is located in the ____________
49. In TWYSTRON, input is located in TWT   – Yes/ No.
50. Traveling wave magnetron is used in the wavelength regions of ______

Answers for Chapter 2

1. No
2. Yes
3. Yes
4. Yes
5. L_p, L_g and L_k
6. C_{gp}, C_{gk}, C_{pk}
7. d/v_0
8. g_n/C
9. using re-entrant cavities and slow wave tubes
10. Yes
11. Yes
12. No
13. Yes
14. $\sqrt{2eV/m}$
15. the velocity of electron varies in accordance with RF input voltage
16. Yes
17. cavity resonator
18. 20 dB
19. Yes
20. Yes
21. oscillator
22. Yes
23. Yes
24. Yes
25. Yes
26. cavity gap voltage
27. continuous
28. Yes
29. No
30. 0.8 GHz
31. 20% to 40%
32. Yes
33. Yes
34. Yes
35. Yes
36. Yes
37. π
38. eB/m
39. backward wave amplifier
40. backward wave oscillator
41. Yes
42. Yes
43. the helix-cathode voltage
44. Yes
45. Yes
46. Yes
47. TWT and klystron
48. klystron
49. Yes
50. centimetre-to-millimeter

CHAPTER 3

MICROWAVE SEMICONDUCTOR DEVICES

1. The operation of semiconductor devices depends on the energy differences between valence and conduction bands – Yes / No.
2. The common semiconductors of microwave solid-state devices are _____
3. The microwave bipolar transistors operate in class _____
4. The operation of microwave bipolar transistor depends on _____
5. In microwave transistor, emitter strip width and base thickness are _____
6. The operation of FET depends on _____ of semiconductor layer of GaAs.

7. FET is used as an ______________
8. Tunnel diode is ______________ resistance device.
9. Esaki diode is nothing but ____________
10. The impurity concentration in tunnel diode is______________
11. The equivalent circuit of tunnel diode is____________
12. Tunnel diode is useful as ____________
13. The transient response of tunnel diode is limited by ______________
14. TED means ____________
15. TED has two junctions   – Yes / No.
16. TED is an ______________ device.
17. TED is used an ______________
18. The output power of TED is ______________ than that of BJT.
19. An example of TED is ______________
20. LSA diode is a TED   – Yes / No.
21. Gunn diode is a negative resistance device   – Yes / No.
22. Gunn diode is used as oscillator   – Yes / No.
23. Gunn diode does not have any junction   – Yes / No.
24. Gunn diode is a low noise device   – Yes / No.
25. LSA diode means ______________
26. LSA diode is a positive resistance device   – Yes / No.
27. LSA diode provides high power output   – Yes / No.
28. IMPATT diode means ____________
29. The operation of IMPATT diode depends on ______________
30. IMPATT diode is used up to 100 GHz   – Yes / No.
31. IMPATT diode is used as amplifier   – Yes / No.
32. IMPATT diode is popular in digital communication systems   – Yes / No.
33. The frequency of the IMPATT diode is ____________
34. TRAPATT diode means ____________
35. TRAPATT diode is a high efficiency diode oscillator   – Yes / No.
36. TRAPATT diode is an efficiency in the range of ______________
37. The avalanche zone velocity in TRAPATT diode is ____________
38. TRAPATT diode is used in low power Doppler radars   – Yes / No.
39. TRAPATT diode is in alternatives   – Yes / No.
40. TRAPATT diode is useful over a frequency range of __________
41. BARITT diode means ____________
42. BARITT diode is a low noise oscillator   – Yes / No.
43. BARITT diode is more useful as an amplifier   – Yes / No.
44. BARITT diode has p-n structure   – Yes / No.
45. BARITT diode is high noise device   – Yes / No.
46. BARITT diode is useful at low frequencies   – Yes / No.
47. BARITT diode is a microwave device   – Yes / No.

48. PIN diode means ______________
49. PIN diode is used as phase shifter – Yes / No.
50. PIN diode is used an amplitude modulator – Yes / No.

Answers for Chapter 3

1. yes
2. GaAs, Inp and Si
3. *C*
4. thickness of the *p-n* junction depletion region and transit time
5. very small
6. conductivity
7. oscillator
8. negative
9. tunnel diode
10. high
11. 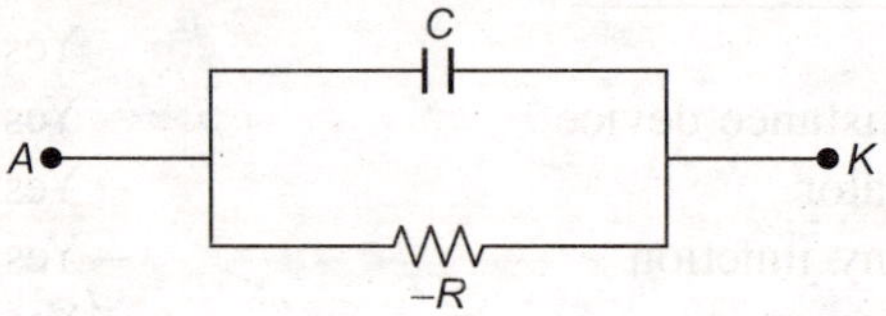

12. oscillator
13. shunt capacitance
14. transferred electric device
15. no
16. active
17. oscillator
18. higher
19. gunn
20. yes
21. yes
22. yes
23. yes
24. yes
25. limited space charge accumulation device
26. no
27. yes
28. impact
29. reverse biased voltage characteristics of *p-n* junction and phase delay of the applied *RF* signal
30. yes
31. yes
32. yes
33. $1/2\tau$
34. trapped plasma avalanche triggered transit diode
35. yes
36. 20 – 40%
37. J/eNA
38. yes
39. yes
40. a few hundred MHz to several GHz
41. barrier injected transit time diode
42. yes
43. yes
44. no
45. no
46. no
47. yes
48. *p*-intrinsic *n* diode
49. yes
50. yes

CHAPTER 4

SCATTERING MATRIX PARAMETERS

1. Scattering matrix is a square matrix only — Yes / No.
2. S matrix gives information about reflection coefficient and transmission coefficient — Yes / No.

3. s_{11} is defined as $\left.\dfrac{x_1}{y_1}\right|_{x_2} = 0$ — Yes / No.

4. s_{21} is defined as $\left.\dfrac{y_1}{x_1}\right|_{x_2} = 0$ — Yes / No.

5. Scattering coefficients can be expressed in terms of eigenvalues of the characteristic equation — Yes / No.
6. For a matched junction, $s_{11} = 0$ — Yes / No.
7. For a matched junction, $|s_{21}| = 0$. — Yes / No.
8. For a matched junction, $|s_{21}| = 1$. — Yes / No.
9. For a matched junction, $s_{ii} = 0$ — Yes / No.
10. For a matched junction, $|s_{ii}| = 1$ — Yes / No.
11. For a reciprocal network, $s_{ij} = s_{ji}$ — Yes / No.
12. For a reciprocal network, $(s)_t \neq s$ — Yes / No.
13. S-matrix exhibits phase shift property — Yes / No.

14. Return loss is $\dfrac{p_i}{p_0}$ — Yes / No.

15. Insertion loss is $\dfrac{p_i}{p_r}$ — Yes / No.

16. Return loss $= 10 \log \dfrac{1}{|s_{11}|}$ — Yes / No.

17. Transmission loss is $\left(\dfrac{p_i - p_s}{p_0}\right)$ — Yes / No.

18. For a symmetric junction, $z_{11} = z_{22}$ — Yes / No.
19. For a reciprocal junction, $z_{12} \neq z_{21}$. — Yes / No.
20. Unitary property for S-matrix does not exist — Yes / No.

Answers for Chapter 4

1. yes 2. yes
3. no 4. yes

5. yes	6. yes
7. no	8. yes
9. yes	10. yes
11. yes	12. no
13. yes	14. yes
15. yes	16. no
17. yes	18. yes
19. no	20. yes

CHAPTER 5

MICROWAVE PASSIVE COMPONENTS

1. Attenuator _________________ power
2. Attenuator can be of ________________
3. Attenuation of attenuator is ______________
4. If the input power is 350 mW and output power of an attenuator is 35 mW, attenuation in dB is ____________
5. Step attenuator reduces power by ____________
6. Fixed attenuators are often called ____________
7. The continuously variable attenuator provides the output power in ___
8. An example of variable attenuator is ______________
9. Attenuation means ______________ through the device.
10. The attenuation can be often by ____________
11. A line is said to be matched if it is terminated in __________
12. Mismatch can be eliminated by making Z_L = _________
13. Mismatch results in ____________
14. The high reflection coefficient increases ____________
15. Matched load means ________________
16. The shorted load means ____________
17. The open load means ___________
18. The microwave directional coupler is a device that ____________
19. Coupling factor of directional coupler is ______________
20. The directivity of directional coupler is ______________
21. Hybrid Tee is a _________ port junction
22. An E-plane Tee is the electrical equivalent of connecting the arm in __
23. An H-plane Tee is the electrical equivalent of connecting the arm in __
24. In hybrid Tee junction, ________________
25. If E arm in hybrid Tee is fed with a signal, it is split in opposite phase between arm 1 and 2 of the main guide   – Yes / No.
26. If H arm in hybrid Tee is fed with a signal, it is split with the same phase between arm 1 and 2 of the main guide   – Yes / No.

27. In hybrid Tee, the signal entering E arm cannot exit through H-arm
                                                                – Yes / No.
28. In hybrid Tee, the signal entering H-arm cannot pass through E-arm
                                                                – Yes / No.
29. Magic Tee is nothing but _____________
30. If two signals of the phase and magnitude are applied to arm 1 and arm 2 of hybrid Tee, they exit from H-arm   – Yes / No.
31. If two signals of opposite phase are applied to arm 1 and 2 of hybrid Tee, they exit from E-arm   – Yes / No.
32. A ferrite circulator is a device with four or more ports   – Yes / No.
33. A four-port circulator is made by two magic Tees connected with $180°$ phase shifter   – Yes / No.
34. A cavity is a ________________ circuit
35. Q of a cavity is the ratio of ________________
36. Multicavity coupled filters are formed by combining cavities with _____
37. Wavemeters are of ____________
38. The resonant frequency of a cavity is ________________
39. Q of cavity is ____________
40. In re-entrant cavity resonators, the walls are connected   – Yes / No.
41. Wavemeters are used to ____________
42. Choke joint permit ________________ electrical continuity across connection of waveguide.
43. Mixers are non-linear devices   – Yes / No.
44. Detectors convert microwave energy to ____________
45. Microwave circulator is a __________ waveguide junction.
46. In a microwave circulator, wave can travel only from n^{th} port to ___
47. A circulator consists of two 3 dB directional couplers and a rectangular waveguide with two ________________
48. The S-matrix of a perfectly matched, lossless and non-reciprocal circular is _________
49. Microwave isolator is ________________
50. An ideal isolator completely absorbs the power for propagation in one direction and provides lossless transmission in the opposite direction
                                                                – Yes / No.
51. An example of uniline is ________________
52. Isolator improves ________________ of generator
53. The S-matrix of directional coupler is ____________
54. A directional coupler is a __________ port waveguide junction.
55. The waveguide corner is a useful to ____________
56. Bend is useful to ____________
57. Twist is useful to ________________
58. The S-matrix of ideal hybrid ring is ____________

59. The S-matrix of magic Tee is __________
60. H-plane Tee is called _______
61. E-plane Tee is called __________
62. The S-matrix of E-plane Tee is ______
63. The S-matrix of H-plane Tee is ________
64. The S-matrix of H-plane is ________
65. Energy transfer takes place in transmission lines by __________
66. Z_0 of a lossless line is __________
67. Pair of transmission lines are not useful at high frequencies due to ___
68. Skin depth is __________________ to frequency.
69. Transmission lines are useful as ______________
70. The wave in coaxial lines is __________
71. The cut-off frequency of TEM is ____________
72. Z_0 of a coaxial line is __________
73. The standard value of Z_0 of coaxial line is ________
74. Coaxial lines can be used up to a frequency of 100 GHz.   – Yes / No.
75. At high frequencies, the attenuation in coaxial lines is high due to _____
76. The reflection coefficient is ________
77. The time taken for a wave to propagate through a lossless line is ______
78. The minimum value of VSWR is ________
79. If VSWR $= \infty$, $|\rho| = $ __________
80. VSWR $= \dfrac{Z_L}{R_0}$ if ______________
81. Phase constant in a transmission line is ____________
82. The input impedance of half-wave transmission line is __________
83. 1 W in dB is __________
84. 1 mW in dB is ________
85. 1 μW in dB is __________
86. 10 mW in dB is ________
87. Hollow rectangular waveguide ______________ support TEM
88. The cut-off wavelength in a rectangular waveguide is ________
89. In TE wave, if the direction of propagation is Z, $E_z = $ __________
90. In TM wave, if the direction of propagation is x, $H_x = $ ____________
91. In circular waveguide, the dominant mode is ____________
92. In ridge waveguide the phase velocity is ____________
93. The bandwidth in ridge waveguide is __________
94. The scattering matrix of a continuously variable attenuation is _______
95. In a continuously variable attenuator, $s_{11} = s_{12} = $ ________
96. A matched load ___________ power
97. Bends in circular waveguides ____________
98. Multistep twists are used in __________________ applications
99. Isolator is made of ______________

100. Directional coupler has ___________ ports
101. The directional coupler is _________ device
102. The equivalent circuit of H-plane Tee is ____________

Answers for Chapter 5

1. reduces
2. fixed or variable type
3. $10 \log_{10} p_{in}/p_{out}$
4. 10 dB
5. fixed amounts
6. pads
7. required quantities
8. piston attenuator
9. reduction of power
10. resistive material
11. Z_0
12. Z_0
13. reflection
14. VSWR
15. $Z_L = Z_0$
16. $Z_L = 0 \ \Omega$
17. $Z_L = \infty \ \Omega$
18. samples part of the power
19. $10 \log_{10} p_i/p_{out}$ dB
20. $10 \log_{10} \dfrac{p_{aux} \text{ (forward)}}{p_{aux} \text{ (reverse)}}$ dB
21. 3
22. series
23. parallel
24. E and H plane Tee junctions are combined
25. Yes
26. Yes
27. Yes
28. No
29. hybrid Tee
30. Yes
31. Yes
32. Yes
33. Yes
34. resonant
35. energy stored per cycle to energy loss per cycle
36. irises and screws
37. transmission, reaction or absorption type
38. $(f_{high} + f_{low})/2$
39. $f_r/(f_{high} - f_{low})$
40. Yes
41. select or reject specific frequencies
42. zero
43. Yes
44. dc signals
45. multiport
46. $(n+1)^{th}$ port in one direction
47. non-reciprocal phase shifters
48. $S = \begin{bmatrix} 0 & s_{12} & s_{13} & s_{14} \\ s_{21} & 0 & s_{23} & s_{24} \\ s_{31} & s_{32} & 0 & s_{34} \\ s_{41} & s_{42} & s_{43} & 0 \end{bmatrix}$
49. a non-reciprocal transmission device
50. Yes
51. isolator
52. frequency stability

53. $S = \begin{bmatrix} 0 & p & 0 & jq \\ p & 0 & jq & 0 \\ 0 & jq & 0 & p \\ jq & 0 & p & 0 \end{bmatrix}$

54. 4

55. change the direction of the guide

56. change the direction of the guide

57. change the direction of the guide

58. $S = \begin{bmatrix} 0 & s_{12} & 0 & s_{14} \\ s_{21} & 0 & s_{23} & 0 \\ 0 & s_{32} & 0 & s_{34} \\ s_{41} & 0 & s_{43} & 0 \end{bmatrix}$

59. $S = \begin{bmatrix} 0 & 0 & s_{13} & s_{14} \\ 0 & 0 & s_{23} & s_{24} \\ s_{31} & s_{32} & 0 & 0 \\ s_{41} & s_{42} & 0 & 0 \end{bmatrix}$

60. shunt Tee

61. series Tee

62. $S = \begin{bmatrix} s_{11} & s_{12} & s_{13} \\ s_{12} & s_{11} & -s_{13} \\ s_{13} & -s_{13} & s_{33} \end{bmatrix}$

63. $S = \begin{bmatrix} \dfrac{1}{2} & \dfrac{1}{2} & \dfrac{1}{\sqrt{2}} \\ \dfrac{1}{2} & \dfrac{1}{2} & -\dfrac{1}{\sqrt{2}} \\ \dfrac{1}{\sqrt{2}} & -\dfrac{1}{\sqrt{2}} & 0 \end{bmatrix}$

64. $S = \begin{bmatrix} \dfrac{1}{2} & -\dfrac{1}{2} & \dfrac{1}{\sqrt{2}} \\ -\dfrac{1}{2} & \dfrac{1}{2} & \dfrac{1}{\sqrt{2}} \\ \dfrac{1}{\sqrt{2}} & \dfrac{1}{\sqrt{2}} & 0 \end{bmatrix}$

65. E and H fields

66. $\sqrt{L/C}$

67. radiation losses

68. inversely

69. stubs

70. *TEM*

71. zero

72. $\dfrac{138}{\sqrt{\epsilon_r}} \log \left(\dfrac{D}{d} \right)$

73. 50 Ω

74. No

75. low skin depth

76. $\rho = V_r/V_i$

77. $\sqrt{LC}$

78. one

79. 1	80. $Z_L > R_0$
81. $2\pi/\lambda$	82. $Z_i = Z_L$
83. 0	84. 0
85. 0	86. -20
87. does not	88. $2a$
89. 0	90. 0
91. TE_{11}	92. low
93. high	94. $\begin{bmatrix} 0 & s_{12} \\ s_{21} & 0 \end{bmatrix}$
95. 0	96. absorbs
97. does not exist	98. broadband
99. ferrite	100. 4
101. reciprocal	102. 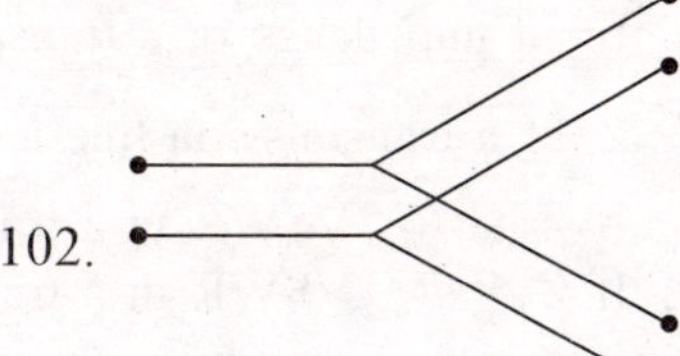

CHAPTER 6

MICROWAVE TRANSMISSION LINES

1. The transmission line carries ___________
2. The equivalent circuit of a transmission line is ___________
3. The primary constants of a transmission line are ___________
4. The secondary constants of a transmission line are ___________
5. The minimum value of VSWR is ___________
6. The maximum value of VSWR is ___________
7. The transmission line is used as impedance transformer   – Yes / No.
8. The condition for a lossless line is ___________
9. The condition for a distortionless line is ___________
10. The characteristic impedance, Z_0 is $\sqrt{Z/Y}$.
11. The reflection coefficient of a line in terms of Z_0, Z_L is ___________
12. VSWR in terms reflection coefficient is ___________
13. The reflection coefficient in terms of VSWR is ___________
14. The equivalent circuit of open line of length, $\ell < \dfrac{\lambda}{4}$ is ___________
15. The equivalent circuit of shorted line of length, $\ell < \dfrac{\lambda}{4}$ is ___________

16. The open wire transmission line is not susceptible to external noise
    – Yes / No.

17. The units of G and C in the equivalent circuit of a transmission line are
    _______________________ respectively.

18. R in the equivalent circuit of a transmission line _______________

19. The definition of Z_0 of a transmission line is _______________

20. For RF lines, $\omega L \gg G$ – Yes / No.

21. For RF lines, $\omega C \gg G$ – Yes / No.

22. For RF lines, $\beta_\ell =$ _______________

23. For a lossless line $\alpha = 0$ – Yes / No.

24. For a lossless line, $Z_0 = \sqrt{L/C}$ – Yes / No.

25. For a distortionless line, $\alpha_\ell =$ _______________

26. Input impedance of a transmission line is defined as _______________

27. Z_i of a transmission line is $\sqrt{(Z_i)_{oc}\,(Z_i)_{sc}}$ – Yes / No.

28. If $Z_0 > R_L$, VSWR in a transmission line is R_L/Z_0 – Yes / No.

29. If $Z_0 < R_L$, VSWR in a transmission line is R_L/Z_0 – Yes / No.

30. Losses in a transmission line are _______________

31. Matched transmission line means _______________

32. Mismatched transmission line means _______________

33. A Smith chart is a circular graph – Yes / No.

34. A Smith chart is used to find VSWR – Yes / No.

35. A Smith chart is used to plot normalized impedance – Yes / No.

36. The stub is a transmission line – Yes / No.

37. The stub cancels the resistive part of the transmission line – Yes / No.

38. The impedance of a stub is purely resistive – Yes / No.

39. The impedance of a stub is purely reactive – Yes / No.

40. If $Z_L = 100 + j200\ \Omega$, Z_0 is $50\ \Omega$, the normalized impedance is ______

41. In a hollow rectangular waveguide TEM exists – Yes / No.

42. In a hollow rectangular waveguide TE, TM wave exist – Yes / No.

43. Circular waveguides produce circularly polarized fields – Yes / No.

44. The cut-off frequency of TEM wave is _______________

45. The dominant mode in a rectangular waveguide is represented by ____

46. The guide wavelength is less than that of free space wavelength
    – Yes / No.

47. The group velocity, $v_g = v_0^2/v_p$ – Yes / No.

48. The phase velocity in a waveguide is _______________

49. $v_0^2 = v_g v_p$ – Yes / No.

50. If $Z_L = 150\ \Omega$, $Z_0 = 75\ \Omega$, VSWR = _______________

51. If $Z_0 = 50\ \Omega$, $Z_L = 40\ \Omega$, VSWR = _______________

52. The strip line is a three conductor transmission line – Yes / No.

53. The microstrip line is a two conductor transmission line   – Yes / No.
54. The strip lines are used over a frequency range of 10 KHz to 100 KHz
    – Yes / No.
55. The strip lines are used for frequencies of 100 MHz to 100 GHz
    – Yes / No.
56. The velocity of propagation in a strip line is v_0/ϵ_r – Yes / No.
57. If $\epsilon = 2$, the velocity of propagation in a strip line is more than that of free
    space   – Yes / No.
58. Microstrip line is symmetrical transmission line   – Yes / No.
59. Strip line is symmetrical transmission line   – Yes / No.
60. Attenuation in microstrip line depends on frequency   – Yes / No.
61. Pair of lines are used in   – Yes / No.
62. Twisted pairs and coaxial cables are used in   __________________

Answers for Chapter 6

1. power
2. a distributed network
3. R, L, G and C
4. Z_0, r
5. 1.0
6. ∞
7. yes
8. $R = 0$ and $G = 0$
9. $\dfrac{R}{L} = \dfrac{G}{C}$
10. $\sqrt{Z/Y}$
11. $(Z_L - Z_0)/(Z_L + Z_0)$
12. $(1 + |\rho|)/(1 - |\rho|)$
13. $(S - 1)/(S + 1)$
14. capacitor
15. inductor
16. no
17. mho/m and F/km
18. loop resistance/unit length
19. V_f/I_f
20. yes
21. no
22. $\omega\sqrt{L/C}$
23. yes
24. $\sqrt{L/C}$
25. $\sqrt{RG}$
26. $Z_i = V_s/I_s$
27. yes
28. no
29. yes
30. copper, dielectric and radiation losses
31. $Z_L = Z_0$
32. $Z_L \neq Z_0$
33. yes
34. yes
35. yes
36. yes
37. no
38. no
39. yes
40. $2 + j\,4$
41. no
42. yes
43. yes
44. zero

45. TE_{10}	46. no
47. yes	48. $\dfrac{\lambda_g}{\lambda_0} v_0$
49. yes	50. 2
51. 1.25	52. yes
53. yes	54. no
55. yes	56. no
57. no	58. no
59. yes	60. yes
61. telephony	62. computer networks

CHAPTER 7

MICROWAVE INTEGRATED CIRCUITS

1. Passive and active devices can be fabricated by MICs   – Yes / No.
2. Substrate materials are ___________
3. VLSI circuits contains more than ________________ components.
4. MMIC means _______________
5. HIC means _______________
6. HIC consists of one _______________
7. One application of MMIC is in _________
8. Advantages of MMIC are ___________
9. MMIC materials are substrates, conductors, dielectrics and __________
10. Resistive materials have ___________
11. Specifications in the design of planar resistors are ___________
12. The resistance of planar resistor is ________________
13. The capacitance of metal-oxide-metal capacitor is __________
14. One application of HIC is _______________
15. MMIC fabrication consists of oxidation, film deposition, epitaxial growth and _______________
16. Dielectrics withstand __________ voltages.
17. Two shapes of planar inductor films are _______________
18. Typical thickness of thick film is __________
19. Transmission lines can be fabricated by MICs   – Yes / No.
20. MOSFETs are fabricated in MMICs   – Yes / No.

Answers for Chapter 7

1. yes	2. ceramic, glass etc.
3. 1 million	4. monolithic microwave integrated circuit

5. hybrid integrated circuit

6. MMIC, one IC, DCs and a film IC

7. aircraft

8. low cost, small size

9. resistive materials

10. low

11. sheet resistivity, thermal resistance and bandwidth

12. $\ell/\omega t \sigma_s$

13. $C = \in \ell w/h$

14. phased arrays

15. lithography

16. high

17. circular and square spiral

18. a few thousand angstroms

19. yes

20. yes

CHAPTER 8

MICROWAVE ANTENNAS

1. An antenna is a transducer   – Yes / No

2. An antenna is a sensor of *EM* waves   – Yes / No

3. An antenna acts as an impedance matching

4. Effective length of a wire antenna is always greater than actual length   – Yes / No

5. Directive gain = Power gain for an antenna   – Yes / No

6. The units of radiation intensity are _______________

7. Directivity is _______________________

8. Efficiency of an antenna is _______________

9. Efficiency of an antenna in terms of directive and power gains is _______________

10. Effective area is _______________

11. The radiation fields are nothing but far-fields   – Yes / No

12. The far-field is indicated by the presence of _______

13. The induction field is indicated by the presence of _______

14. The electrostatic field is indicated by the presence of _______

15. The radiation resistance of an isolated half-wave dipole is _______

16. The radiation resistance of quarter-wave monopole is _______

17. The current distribution in a half-wave dipole is _______

18. The current distribution in alternating current element is _______

19. The current distribution in very short dipoles is _______

20. The radiation pattern of vertical and horizontal dipoles are identical   – Yes / No

21. The directivity of current element is _______

22. The directivity of half-wave dipole is _______

23. The patterns of half-wave dipole and quarter-wave monopole are identical   – Yes / No

24. If a current element is x-directed, vector magnetic potential is

    ____

25. Radiation resistance of short monopole is _________________

26. Radiation resistance of short dipole is _________________

27. The radiated fields of z-directed half-wave dipole consists of $E_\theta, E_r, H_r, H_\theta$, terms

    – Yes / No

28. The radiated fields of z-directed dipole consists of only E_θ, E_r and H_ϕ

    – Yes / No

29. At *LF* and *VLF*, polarization often used is _________________

30. dB_i means _________________

31. dB_m means power gain in dB _________________

32. If the signal level is 1 mW, power gain is

    (a) 0 dBm   (b) 1 dBm

    (c) 10^{-3} dBm   (d) 10 dBm

33. Marconi antenna has a physical length of

    (a) $\lambda/4$ (b) $\lambda/2$

    (c) $3\lambda/2$ (d) λ

34. For a 300 Ω antenna operating with $5A$ of current, the radiated power is

    (a) 7500 W   (b) 750 W

    (c) 75 W   (d) 1500 W

35. Effective area of antenna is a function of frequency   – Yes / No

36. Antenna used in mobile communications is _________________

37. If a current element is z-directed, vector magnetic potential is

    ____

38. If vector magnetic potential has only A_z, E_ϕ is _________________

39. Radiation resistance of current element is _________________

40. Radiation resistance of quarter wave monopole is _________________

41. Directional pattern of a short dipole in the horizontal plane is a

    ____

42. Directional pattern of a horizontal half-wave centre fed dipole is

    ____

43. Effective length of a dipole is always _________________ than the actual length

44. The directivity in dB of half-wave dipole is _________________

45. The directivity in dB of current element is _________________

46. Effective area of a hertzian dipole operating at 100 MHz is

    ____

47. Ideally reflector size is infinitely large   – Yes / No.

48. The polarization and the position of the primary antennas control the radiating properties of complete system   – Yes / No.

49. Reflector is called as primary antenna   – Yes / No.
50. Microwave frequency range is _______________
51. Corner reflector is better than plane reflectors in collimating *EM* energy
       – Yes / No.
52. Bandwidth of corner reflector is more when elements are cylindrical dipoles rather than thin wires   – Yes / No.
53. The grid-wired corner reflector reduces the weight of the antenna system
       – Yes / No.
54. Efficiency of corner reflector is reduced when spacing of feed element becomes small   – Yes / No.
55. Multiple lobes are produced when the spacing of feed element from the vertex is large   – Yes / No.
56. In corner reflectors, the spacing of the feed point should be greater than the length of the sides   – Yes / No.
57. If the main beam is narrow, the directivity is said to be small
       – Yes / No.
58. Collimation of *EM* energy means generation of parallel rays
       – Yes / No.
59. Parabolic reflector is different from paraboloid   – Yes / No.
60. Dish antenna and paraboloid are one and the same   – Yes / No.
61. The gain of an antenna with a paraboloid reflector depends on (D_a/λ) and the illumination   – Yes / No.
62. In Cassegrain feed, the size of the hyperboloid reflector depends on its distance from horn feed, mouth diameter of horn and frequency
       – Yes / No.
63. The size of hyperboloid reflector is small if its distance from the feed antenna is small   – Yes / No.
64. Cassegrain feed is best suited for _______________
65. The disadvantage of Cassegrain feed is the obstruction of *EM* energy by hyperbolic reflector   – Yes / No.
66. If HPBW is 10° in the radiation of pattern of paraboloid, NNBW is

    ____
67. The power gain of paraboloid is given by _______________
68. Capture area of paraboloid is _______________ where $K = 0.65$ for dipole feed and *A* is actual area.
69. If the actual area of paraboloid reflector is 10 m^2, its capture area is

    ____
70. Sector beams are used in _______________ antennas.
71. Cosec beams are used for _______________
72. Narrow beams are used for point-to-point communication purposes
       – Yes / No.

73. For height finding, the antenna beam is ___________

74. In pyramidal Horn, flaring is done in only one plane — Yes / No.

75. Power gain of horns is greater than that of paraboloid reflectors
— Yes / No.

76. Directivity of horns is greater than that of waveguide — Yes / No.

77. Power gain of a horn is more than its directivity — Yes / No.

78. Feed system with corrugated horn reduces spillover efficiency
— Yes / No.

79. Feed system with corrugated horn reduces cross-polarization
— Yes / No.

80. Horizontal slot produces vertical polarized radiation fields — Yes / No.

81. Horizontal dipole produces horizontal polarized radiation fields
— Yes / No.

82. If impedance of dipole is inductive, slot impedance is capacitive
— Yes / No.

83. If the impedance of the slot is capacitive, the impedance of complementary
dipole is inductive — Yes / No.

84. From slot antenna, in a conducting plane, its complementary dipole is
formed by interchanging air and metallic regions in the slot
— Yes / No.

85. Impedance of the slot antenna can be changed by changing feed point
— Yes / No.

86. Back radiation from a slot in a conductive plane can be avoided by

____

87. Slot gain is increased by array of slots — Yes / No.

88. The radiation pattern of annular slot antenna is _______________

89. Array of slots are used in _______________

90. An array of slots when excited with appropriate amplitude and phase is
suitable in _______________

91. Dipole of small length to diameter ratio increases the bandwidth
— Yes / No.

92. Slot of small length to width ratio increases the band width
— Yes / No.

93. Notch antennas are used in aircraft — Yes / No.

94. Notch antennas are used in edges of wing surface of aircraft
— Yes / No.

95. Notch antenna is broad band — Yes / No.

96. The purpose of dielectric filling of notch is _______________

97. Microstrip antennas are used because of _______________

98. Microstrip antennas are used for frequencies above ___________

99. The bandwidth of microstrip antenna is _______________

100. In microstrip antennas, B.W. can be increased by _____________ the thickness of the strip

101. If $\in_r$ of substrate is high in microstrip antenna, B.W. increases
– Yes / No.

102. If reactive component is added in microstrip antenna, B.W. is increased
– Yes / No.

103. If reactive component is added in microstrip antennas VSWR is increased
– Yes / No.

104. The radiation beam of microstrip antenna is _______________

105. The characteristic impedance Z_0 of microstrip antenna is _______________

106. Trihedral forms of corner reflectors are used as _________

107. Rod reflectors are nothing but parasitic elements   – Yes / No.

108. The length of the rod reflector is greater than $\lambda/2$ – Yes / No.

109. Rod reflector is an active radiating element   – Yes / No.

110. In cassegrain feed the dimension of the hyperboloid depends on its distance from primary feed antenna   – Yes / No.

111. In cassegrain feed the dimension of the hyperboloid depends on mouth diameter of the horn   – Yes / No.

112. In cassegrain feed the dimension of the hyperboloid depends on frequency of operation   – Yes / No.

113. Flare angle of the horn is related to axial length   – Yes / No.

114. The directivity of the paraboloid is greater than that of horn
– Yes / No.

115. The size of the horn becomes large if the flare angle is small
– Yes / No.

116. Horn antenna is called as secondary antenna when used with paraboloid
– Yes / No.

117. The disadvantage of lens antenna at low frequencies is ______

118. The material of lens antenna is _________________

119. Lens are preferred over parabolic reflectors at _____________

120. Lens is used to correct the curved wavefront   – Yes / No.

121. The refractive index of lens material is different from unity – Yes / No.

122. In fanned beams, the directivity is poor in one of the principal planes
– Yes / No.

123. If the beamwidth is small, target resolution is high
– Yes / No.

124. Fanned beams are used for _______________

125. For feed systems using corrugated horns, the aperture efficiency is

____

126. Babinet's principle is applicable in electromagnetic problems
– Yes / No.

127. Babinet's principle is valid in optics – Yes / No.

128. For a slot in conducting sheet, there exists a complementary dipole
– Yes / No.

129. The gain of the horn antenna is ______________

130. Vertical slot in the narrow wall of rectangular waveguide does not radiate
– Yes / No.

131. Longitudinal centred slot in the broad wall of a rectangular waveguide does not radiate – Yes / No.

132. The equivalent circuit of an inclined slot in the narrow wall of a rectangular waveguide is a ______________

133. Resonant length of the slot is ______________

134. Method of moments is useful to solve ______________

135. Patch antennas are ______________

136. Patch is made of dielectric material – Yes / No.

137. Pyramidal horn is nothing but rectangular horn – Yes / No.

138. Conical horn is excited conveniently by a circular waveguide
– Yes / No.

139. For lossless antenna directivity is the same as gain – Yes / No.

140. Aperture efficiency is given by ______________

141. A slot can be excited by a waveguide – Yes / No.

142. A slot can be excited by an energized cavity – Yes / No.

143. A slot can be excited by a transmission line – Yes / No.

144. The efficiency of patch antenna is ______________

145. The equivalent circuit of symmetrical vertical slot in the broad wall of a rectangular waveguide is ______________

146. For producing circular polarized waves the shape of the patch antenna is

____

Answers for Chapter 8

1. yes

2. yes

3. yes

4. no

5. no

6. Watts/unit solid angle

7. Maximum directive gain

8. $\dfrac{w_r}{(w_r + w_1)}$

9. g_p/g_d

10. $\dfrac{\lambda^2}{4\pi} g_d$

11. yes

12. $\dfrac{1}{r}$ term

13. $\dfrac{1}{r^2}$ term

14. $\dfrac{1}{r^3}$ term

15. 73 Ω

16. 36.5 Ω

17. Sinusoidal

18. Constant

19. Triangular

20. no

21. 1.5

22. 1.64

23. no

24. X-directed

25. $100\left(\dfrac{1}{\lambda}\right)^2 \Omega$

26. $200\left(\dfrac{1}{\lambda}\right)^2 \Omega$

27. no

28. yes

29. Vertical

30. power gain of the antenna in dB relative to isotropic antenna

31. compared to 1 mW

32. a

33. a

34. a

35. yes

36. whip antenna

37. z-directed

38. Zero

39. $80\pi^2\left(\dfrac{dI}{\lambda}\right)^2 \Omega$

40. 36.5 Ω

41. Circle

42. figure of eight

43. Less

44. 2.15

45. 1.64

46. 1.07 m^2

47. yes

48. yes

49. no

50. 1 GHz – 100 GHz

51. yes

52. yes

53. yes

54. yes

55. yes

56. no

57. no

58. yes

59. no

60. yes

61. yes

62. yes

63. yes

64. low noise receiver applications

65. yes

66. 20°

67. $6.4\left(\dfrac{D}{\lambda}\right)^2$

68. KA

69. 6.5 m^2

70. surface search from shipborne

71. airport surveillance

72. yes

73. sharp in elevation

74. no

75. no
76. yes
77. no
78. yes
79. yes
80. yes
81. yes
82. yes
83. yes
84. yes
85. yes
86. boxing the slot suitably
87. yes
88. narrow beam
89. aircraft
90. scanning radars without antenna movement
91. yes
92. yes
93. yes
94. yes
95. yes
96. to eliminate aerodynamic drag in aircraft
97. small size, less weight low cost etc.
98. 100 MHz
99. small
100. increasing
101. yes
102. yes
103. no
104. broad

105. $Z_0 = \eta \sqrt{\dfrac{\mu_r}{\epsilon_r}}\ \Omega$

106. radar targets
107. yes
108. yes
109. no
110. yes
111. yes
112. yes
113. yes
114. yes
115. no
116. no
117. bulkiness
118. Lucite
119. millimeter and submillimeter frequencies
120. yes
121. yes
122. yes
123. yes
124. air search from ground
125. $75 - 80\%$
126. no
127. yes
128. yes
129. moderate
130. yes
131. yes
132. shunt admittance
133. $\lambda/2$
134. integral equations
135. very compact
136. no
137. yes
138. yes
139. yes
140. ratio of effective aperture and physical aperture
141. yes
142. yes
143. yes
144. low
145. series impedance
146. circular

CHAPTER 9

MICROWAVE MEASUREMENTS

1. The principle involved in the reflex klystron is _______________
2. Reflex klystron is _______________
3. Feedback in reflex klystron is _______________
4. Gunn diode is a _______________
5. Gunn diode is used as oscillator — Yes / No.
6. The minimum value of VSWR is _______________
7. The maximum value of VSWR is _______________
8. VSWR in-term of reflex coefficient is _______________
9. Return power loss in a transmission line is _______________
10. The output crystal detector is _______________
11. The crystal detector is used to _______________

12. The relation between λ_g, λ_c and λ is $\lambda_g = \dfrac{\lambda}{\left[1-\left(\dfrac{\lambda}{\lambda_c}\right)^2\right]^{\frac{1}{2}}}$.

13. For TE_{10} mode in a rectangular guide, $\lambda_c = $ _______________
14. Isolator is useful for _______________
15. Attenuator is called _______________
16. In slotted line, power radiates from slot — Yes / No.
17. Matched termination is a device whose $Z_L = $ _______________
18. Matched termination absorbs incident power — Yes / No.
19. Short does not reflect power — Yes / No.
20. VSWR for shorted load is _______________
21. VSWR for matched terminated load is _______________
22. A typical reflex klystron is _______________
23. Spectrum analyzer gives information in _______________
24. Oscilloscope gives information in _______________
25. Absorption type wavemeter exists — Yes / No.
26. Wavemeter may be transmission type — Yes / No.
27. Network analyzer is useful to measure parameters of a network
    — Yes / No.
28. Bolometer is useful to measure _______________
29. Transit time in reflex klystron is _______________
30. Efficiency of reflex klystron is _______________
31. Variable attenuator is useful to control _______________
32. The equivalent circuit of a crystal detector is _______________

33. Micro-ammeter is used as an ___________

34. In double minimum method of VSWR measurement, VSWR is ________

35. VSWR is ____________

36. Attenuation is defined as ____________

37. Waveguide slotted line is always associated with the detector probe
                           – Yes / No.

38. The characteristics of Gunn diode is ___________

39. Magic Tee is a four-port device   – Yes / No.

40. Magic Tee is used as a summer   – Yes / No.

41. Magic Tee can be used as a differencer device   – Yes / No.

42. Circulator is only a two-port device   – Yes / No.

43. Circulator is a multiport device   – Yes / No.

44. An ideal circulator is a matched device   – Yes / No.

45. The S-matrix of a three-port circulator _________

46. Reflex klystron should have a power supply to operate it as an oscillator
                           – Yes / No.

47. The dimensions of standard X-band rectangular waveguide are ______

48. Detector is tunable   – Yes / No.

49. Isolator is connected between source and load   – Yes / No.

50. High power reflex klystrons are used in the laboratories   – Yes / No.

51. A Tee junction has 3 independent ports   – Yes / No.

52. E-plane Tee is called _________________

53. H-plane Tee is called ____________

54. Magic Tee is ___________

55. A microwave junction is interconnection of two or more devices
                           – Yes / No.

56. H, Y, Z parameters are not measured at microwave frequencies
                           – Yes / No.

57. The S-matrix of Magic Tee is _________

58. In Magic Tee $S_{43} = S_{34} =$ _________

59. Hybrid ring may have series junctions   – Yes / No.

60. The S-matrix of an ideal hybrid ring is ____________

61. Directional coupler is a ____________ port device.

62. The S-matrix of a directional coupler is ____________

63. In a directional coupler, all 4 ports are ____________

64. In a directional coupler, there is no coupling between ports __________ and also between ______________ .

65. In a directional coupler, $S_{11} = S_{12} =$ ________________ .

66. $S_{33} = S_{44} =$ ___________ in a directional coupler.

67. The directivity is a measure of how well the forward traveling wave in the primary guide couples only to a specific port of the secondary guide
                           – Yes / No.

68. Coupling factor of a directional coupler is a measure of ratio of power levels in primary and secondary guides.   – Yes / No.
69. Circulator is a _________________ port junction.
70. The S-matrix of a four port circulator is _________________
71. An isolator is a _______________ device.
72. The isolator is called _______________
73. Electronic efficiency of a reflex klystron oscillator is defined as ______
74. The reflex klystron is a _______________ cavity klystron.
75. The efficiency of reflex klystron is _______________
76. Reflex klystron is a low power generator   – Yes / No.
77. The frequency range of reflex klystron is _______________
78. Gunn diode is a _______________
79. Gunn diode is discovered by _______________

Answers for Chapter 9

1. velocity modulator
2. an oscillator
3. re-entrant
4. negative resistance device
5. yes
6. 1.0
7. ∞
8. $\dfrac{(1+|\rho|)}{(1-|\rho|)}$
9. $10 \log_{10}\left(\dfrac{p_i}{p_r}\right)$
10. dc
11. convert ac to dc
12. $\lambda_g = \dfrac{\lambda}{\left[1-\left(\dfrac{\lambda}{\lambda_c}\right)^2\right]^{\frac{1}{2}}}$
13. (2 × broad wall dimension)
14. protecting source from reflected power
15. pad
16. no
17. Z_0
18. yes
19. no
20. ∞
21. 1.0
22. 2K25
23. frequency domain
24. time domain
25. yes
26. yes
27. yes
28. power
29. $\dfrac{\left(n+\dfrac{3}{4}\right)}{f_0}$
30. $p_0/V_a I_c$

power microwave

31. incoming power

32. 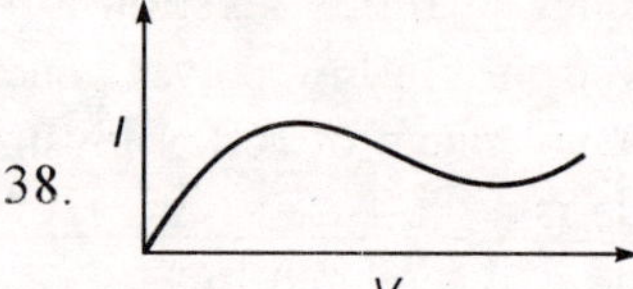

33. indicating meter

34. $\dfrac{\lambda_g}{\pi(d_1 - d_2)}$

35. $V_{\max}/V_{\min}$

36. $10 \log_{10} (p_i/p_0)$

37. yes

38.

39. yes

40. yes

41. yes

42. no

43. yes

44. yes

45. $\begin{bmatrix} 0 & 0 & s_{13} \\ ss_{21} & 0 & 0 \\ 0 & s_{32} & 0 \end{bmatrix}$

46. yes

47. 1.016×2.286 cm

48. yes

49. yes

50. no

51. yes

52. series Tee

53. shunt Tee

54. a hybrid Tee

55. yes

56. yes

57. $\begin{bmatrix} 0 & 0 & s_{13} & s_{14} \\ 0 & 0 & s_{23} & s_{24} \\ s_{31} & s_{32} & 0 & 0 \\ s_{41} & s_{42} & 0 & 0 \end{bmatrix}$

58. 0

59. yes

60. $\begin{bmatrix} 0 & s_{12} & 0 & s_{14} \\ s_{21} & 0 & s_{23} & 0 \\ 0 & s_{32} & 0 & s_{34} \\ s_{41} & 0 & s_{43} & 0 \end{bmatrix}$

61. 4

62. $\begin{bmatrix} 0 & s_{12} & 0 & s_{14} \\ s_{21} & 0 & s_{23} & 0 \\ 0 & s_{32} & 0 & s_{34} \\ s_{41} & 0 & s_{43} & 0 \end{bmatrix}$

63. matched

64. 1 and 3, 2 and 4

65. 0

66. 0

67. yes

68. No

69. multi

70. $\begin{bmatrix} 0 & 0 & 0 & 1 \\ 1 & 0 & 0 & 0 \\ 0 & 1 & 0 & 0 \\ 0 & 0 & 1 & 0 \end{bmatrix}$

71. non-reciprocal

72. uni-line

73. p_{ac}/p_{dc}

74. single

75. 20 to 30%

76. yes

77. 1 to 25 GHz

78. TED device

79. J.B. Gunn

REFERENCES

1. R.E. Collin, "Foundations for Microwave Engineering", McGraw Hill, New Delhi, 1992.
2. K.C. Gupta, "Microwaves", New Age International (P) Ltd., New Delhi, 1995.
3. Samuel Y. Liao, "Microwave Devices and Circuits" Pearson Education, New Delhi, 2003.
4. Joseph Helszajn, "Microwave Engineering : Passive, Active and Non-Reciprocal Circuits", McGraw Hill, New Delhi, 1992.
5. Randy Bancroft, "Microstrip and Printed Antenna Design", PHI, New Delhi, 2006.
6. M.L. Sisodia, G.S. Raghuvanshi, "Basic Microwave Techniques and Laboratory Manual", New Age International (P) Ltd., New Delhi, 2000.
7. E. Hund, "Microwave Communications", McGraw Hill, New Delhi, 1989.
8. Annapurna Das, S.K. Das, "Microwave Engineering", Tata McGraw Hill, New Delhi, 2001.
9. M.J. Howes, D.V. Morgan, "Microwave Devices", John Wiley & Sons, New York, 1976.
10. M. Kulkarni, "Microwave and Radar Engineering", Umesh Publications, New Delhi, 2007.
11. G.P. Srivastava, Vijay Lakshmi Gupta, "Microwave Devices and Circuit Design," PHI, New Delhi, 2006.
12. J.D. Kraus, "Electromagnetics", McGraw Hill, New Delhi, 2005.
13. Rizzi, Peter, "Microwave Engineering : Passive Circuits", PHI, NJ, 1988.
14. Saad. T and R.C. Hansen, "Microwave Engineering", Handbook, Vol. 1, Artech House, Dedham, Mass, 1971.
15. Y.S. Liao, "Engineering Applications of Electromagnetic Theory", West Publishing Company, St. Paul, Monn, 1988.
16. L. Young, "Advances in Microwaves", Academic Press, New York, 1974.
17. I. Bahl, ed., "Microwave Solid-State Circuit Design", John Wiley & Sons, New York, 1988.

18. Samuel Silver, "Microwave Antenna Theory and Design", McGraw Hill, New York, 1949.

19. Robert Shrader, "Electronic Communication", McGraw Hill, New York, 1985.

20. Jordan, E and K.G. Balmain, "Electromagnetic Waves and Radar Systems" PHI, 1968.

21. George Kennedy, "Electronic Communication Systems", McGraw Hill, New York, 1988.

INDEX

A

Absorption type wavemeter, 334
Adjustable shorts, 142
Admittance characteristics, 272
Admittance matrix, 105
Antenna bandwidth, 273
Antenna efficiency, 271
Antenna impedance, 270
Antenna,
    classification of, 274
    effective length of, 270
    parameters of, 269
Array of collinear dipoles feed, 307
Attenuation constant, 19
Attenuation, measurement of, 359, 376
Attenuator, 132, 374
    calibration of, 373
    Insertion loss, 376
    salient features of, 134, 374
    types of, 133

B

Backward wave amplifier (BWA), 62
    principle of operation of, 62
    salient features of, 63
Backward wave oscillator (BWO), 60
    applications of, 61
    principle of operation of, 60
    salient features of, 61
Band designations, 3
Band pass filter, 164
Band stop filter, 164
Bandwidth control,
    methods of, 291

BARITT diode, 88
    principle of operation of, 89
    salient features of, 88
Bends, 135
    salient features of, 136
Bessel's function, 131
BJT, disadvantages of, 252
Bolometers, 380

C

Cassegrain feed, 307
    application of, 308
    advantages of, 308
    disadvantages of, 308
Cavity resonators, 127
    applications of, 129
    salient features of, 128, 221
Characteristic impedance, 180, 184, 185
    salient features of, 180
Cheese antenna, 305
Choke joint, 160, 161
    salient features of, 161
Circuit equation, 59
Circular loop inductor, 255
Circular polarization, 21
Circular spiral inductor, 256
Circular waveguide, 130, 217
    salient features of, 130, 217
Circulator, 137, 270
    applications of, 371
    salient features of, 371
CMOS,
    fabrication of, 252

Coaxial cable, 117
   fields in, 207
   properties of, 206
Coaxial lines, 117, 205
   applications of, 121
   salient features of, 118, 205
Coaxial wavemeters, 334
Continuous loading, 189
Continuously variable attenuators, 133, 374
Copper losses, 192
Corner reflector, 297
   salient features of, 299
Corners, 134
Corrugated horn, 287
   salient features of, 287
Cosecant square beams, 311
Coupling coefficient,
   measurement of, 369
Coupling factor, 146
Crystal detector,
   characteristics of, 342
Crystallization, 193
Cut paraboloid, 304
Cyclotron frequency magnetron, 53

D
Degenerate mode, 214, 220
Deposition, 250
Dielectric constant,
   measurement of, 383, 386
Dielectric lens, 293
   salient features of, 294
Dielectric losses, 193
Dielectric materials, 248
   applications of, 248
   properties of, 248
Different geometries,
   inductance of, 255
Diffusion, 249
Directional characteristics, 269
Directional coupler,
   applications of, 147, 363
   measurement of, 361
   parameters of, 146, 361, 363
   salient features of, 145, 362

   scattering matrix of, 147
Directive gain, 271
Directivity, 147, 271
Discrete circuit, 242
Distortionless lines, 184
   salient features of, 186
Dominant mode, 214, 220

E
Effective area, 272
Electric fields, 4
Electromagnetic field equations, 10
Electron motion,
   formula for, 54
Electron trajectory in magnetron, 54
Electronic circuits,
   types of, 242
Electronic equation, 58
Elevation narrow beam, 312
Elliptical polarization, 21
Epitaxial growth, 249
E-plane metal plate lens, 294
E-plane Tee junction, 152
   salient features of, 153
E-plane Tee,
   scattering matrix of, 154

F
Fabrication,
   steps involved in, 250
Fanned beams, 309
   applications of, 309
Faraday's isolator,
   operation of, 143
Ferrite devices, 158
   applications of, 158
   salient features of, 158
Field effect transistors, 75
   salient features of, 75
Film integrated circuit, 243
Fixed attenuators, 133, 374
Flanges, 162
   salient features of, 162
Flat inductor, 255
Four-port circulator, 137
   applications of, 138

operation of, 137
    salient features of, 138
Free space, 176
    characteristics of, 11, 176
    Maxwell's equations for, 12
Frequency measurements, 354
Frequency meters, 334
Frequency, 16
Front-to-back ratio, 273

G
Guide wavelength,
    measurement of, 346
Gunn diode,
    advantages of, 80
    applications of, 80
    disadvantages of, 80
    principle of operation of, 79
    salient features of, 79
    V-I characteristics of, 351

H
Half-wave dipole feed, 307
Half-wave dipole, 274
    salient features of, 274
High pass filter, 164
Horn antenna,
    types of, 282
Horn feed, 307
    salient features of, 307
Horns,
    applications of, 286
    salient features, 286
H-plane Tee junction, 149
    salient features of, 149
    scattering matrix of, 151
Hybrid integrated circuit, 243, 260
    advantages of, 261
    applications of, 261
    salient features of, 260
Hybrid ring,
    operation of, 160
    salient features of, 160
Hybrid rings (rat race coupler), 159
Hybrid Tee (magic tee), 155
    salient features of, 155

I
Ideal substrates,
    characteristics of, 244
IEEE microwave frequency bands, 3
IMPATT diode, 81
    advantages of, 85
    applications of, 85
    disadvantages of, 85
    efficiency of, 84
    equivalent circuit of, 84
    principle of operation of, 82
    salient features, 84
    V-I characteristics of, 83
Impedance matrix, 104
Insertion loss, 103, 376
    measurements of, 372
Integrated circuit, 242
Interdigitated capacitor, 259
Ion implementation, 249
Isolation,
    measurement of, 369, 372
Circulator, 369
Isolators, 369
    advantages of, 142
    salient features of, 370

K
Klystron, 39
    applications of, 41
    efficiency of, 338

L
Lead capacitance, 35
    effect of, 35
    minimization of, 35
Lead inductance, 34
    effect of, 35
    minimization of, 35
Lens antennas, 292
    functions of, 292
    principle of operation of, 292
Linear polarization, 20
Lithography, 250
Loading,
    types of, 188
Losses lines, 183
    definition of, 183

Lossless transmission lines,
   salient features of, 184
Low pass filter, 164
LSA diode, 80
   salient features, 81
Lumped loading, 189

M
Magic Tee,
   applications of, 156, 367
   salient features of, 366
   scattering matrix of, 156
Magnetic fields, 4
Main lobe,
   beam width of, 274
Matched load,
   salient features of, 140
Matched transmission line, 192
Maxwell's equations, 10, 11
Metal-oxide-metal capacitor, 259
Metals,
   applications of, 246
   characteristics of, 246
   properties of, 246
MIC,
   merits of, 243
Microstrip, 288
Microstripline,
   attenuation in, 229
   advantages of, 229
   disadvantages of, 229
   salient features of, 230
Microstripline,
   geometry of, 228
   salient features of, 227
Microwave antennas, 268
Microwave bipolar transistor, 75
   salient features of, 75
Microwave circuits,
   losses in, 102
Microwave crystals, 380
Microwave devices,
   operation of, 23
Microwave electromagnetic spectrum
   domain, 4
Microwave filters, 163

Microwave integrated circuit, 241
   salient features of, 241
Microwave measurements, 333
   precautions in, 336
Microwave oven, 23
   features of, 24
   functions of, 24
   principle of operation of, 23
   specifications of, 23
   types of, 24
Microwave passive components, 113
Microwave power,
   measurement of, 380
Microwave regions, 3
Microwave semiconductor devices, 74
Microwave sources, 333
Microwave tubes, 34
   limitations of conventional, 34
   salient features of, 37
Microwave,
   advantages of, 2
   applications of, 21
   characteristics of, 1
   definition of, 1
   parameters of, 3
   power flow by, 15
   velocity of, 16
Mismatched transmission line, 192
MMIC fabrication,
   methods of, 249
MMIC materials, 244
MMICs,
   applications of, 244, 253
Monolithic microwave integrated
   circuit, 243
MOSFET,
   advantages of, 252
   fabrication of, 251
Multimode graded-index fibre,
   characteristics of, 232
Multistep-index fibre,
   characteristics of, 231

N
Network analyzer, 336
NMOS fabrication, 253

Non-uniform arrays, 277

O
Offset paraboloid, 306
Open line,
    input impedance for, 188
Open stubs, 197
Open-wire transmission line,
    structure of, 178
Optical fibres, 230
    characteristics of, 230
    types of, 230
Oscilloscopes, 334
Output power gain, 59
Oxidation, 249

P
P mode, 53
Parabolic cylinder, 305
Parabolic reflectors,
    disadvantages of, 304
    feed systems for, 306
    operation of, 301
    salient features of, 303
    types of, 304
Paraboloid, 299
Parallel open-wire transmission lines,
    salient features of, 177
Parallel plates, 209
Passive components,
    fabrication of, 253
Patch antennas, 288, 289
Patch loading, 188
Phase shift constant, 19
Phase shifters,
    salient features of, 159
Phasor form,
    Maxwell's equations in, 12
Photo etching, 250
Pillbox antenna, 305
PIN diode, 90
    applications of, 91
    features of, 91
Planar capacitors, 259
Planar inductor,
    design of, 254

Planar resistors,
    applications of, 254
    design of, 253
    resistance of, 254
Plane reflector, 297
Polarization, 20, 273
    types of, 20
Posts, 161
Power gain, 271
Power meters, 336
Power output, 49
Primary antennas, 274
Primary constants, 178
Propagation constant, 18, 180, 181, 183, 185
    expression for, 19

Q
Quality factor, 220
Quarter-wave monopole,
    salient features of, 275
Quasi-monolithic integrated circuit, 243

R
Radars, 25
Radars,
    types of, 26
Radiation intensity, 270
Radiation losses, 193
Radiation pattern,
    main lobe of, 273
    side lobes of, 273
Radiation resistance, 270
Reaction type wavemeter, 334
Receiving antenna,
    effective length of, 270
Rectangular waveguide resonance
    isolator, 143, 144
    salient features of, 144
Rectangular waveguide, 124
    salient features of, 211
    structure of, 210
Reflection coefficient, 189
Reflection loss, 103
Reflector antennas, 297
Reflex Klystron, 41, 337

analysis of, 47
characteristics of, 339
construction of, 41
efficiency of, 51
mode characteristics of, 341
modes in, 46
principle of operation of, 42
velocity modulation in, 44
voltage characteristics of, 45
Resistive materials, 248
Return loss, 103
RF lines, 181
salient features of, 182
Ridge waveguides,
advantages of, 132
disadvantages of, 132
salient features of, 131

S
S, *Z* and *Y* matrices,
summary of, 106
Sampling oscilloscopes, 334
Satellite, 26
Scattering matrix of series element, 108
Scattering matrix, 97
characteristics of, 99
definition of, 99
of multi-port network, 102
properties of, 97
salient features of, 101
Secondary antennas, 292
Secondary constants, 179
Sector beams, 310
Shaped beam antennas, 309
Shielded pair line, 204
Shorted line,
input impedance for, 188
Shorted stubs, 197
Shunt element, 107
Single mode step-index fibre,
characteristics of, 231
Single stub matching,
design of, 197
Sinusoidal time varying fields, 12
Skin effect, 193
Slot antenna, 278
impedance of, 280

salient features of, 278
Slow wave devices, 55
advantages of, 56
examples of, 56
Slow wave structure, 55
Slow wave, 56
Small and large numbers,
representation of, 22
Smith chart, 195
applications of, 196
calculation of impedance using, 379
description of, 196
Spectrum analyzer, 335
applications of, 336
salient features of, 335
Spill-over, 287
Split-ring of negative resistance
magnetron, 54
applications of, 54
Square spiral inductor, 255
Standard mismatches, 142
Standing wave ratio, 191
reflection coefficient and, 191
salient features of, 192
Static electric field, 4
Static magnetic field, 5
Step attenuators, 133
Step loads, 141
Stripline,
attenuation in, 226
field distribution in, 225
geometry of, 225
Stubs, 196
Substrate materials,
applications of, 245
examples of, 244
properties of, 245

T
Tapered loads, 141
applications of, 141
TE and TM waves,
propagation parameters of, 216
TE waves,
field equations for, 213
Tee junctions, 149

Terminations, 140
Thermo-couplers, 380
3 dB directional coupler,
    scattering matrix of, 148
Three-port circulator,
    scattering matrix of, 139
Time varying electric field, 6
Time varying magnetic field, 10
TM waves,
    field equations of, 214
Torus antenna, 306
Transferred electron devices (TEDs), 78
    applications of, 78
    salient features of, 78
Transit time,
    effect of, 36
    expression for, 36
    minimization of, 36
Transitions, 163
    salient features of, 163
Transmission line parameters, 193, 194
Transmission line,
    applications of, 174
    examples of, 173
    functions of, 173
    input impedance of, 186
    in MICs, 250
    loading of, 188
    losses in, 192
    salient features of, 174
    standing waves on the, 190
Transmission loss, 103
Transmission medium, 176
Transmission type wavemeter, 334
Transmitting antenna,
    effective length of, 270
Transverse electric waves, 212
Transverse electromagnetic waves, 215
    characteristics of, 216
Transverse magnetic waves (TM
    waves), 214
TRAPATT diode, 85
    advantages of, 88
    applications of, 88
    disadvantages of, 88

principle of operation of, 86
    salient features of, 85
Traveling wave magnetrons, 52
    principle of operation of, 52
Traveling wave tube, 56
    principle of operation of, 56
Tuning screws, 161
    salient features of, 161
Tunnel diode amplifier, 77
Tunnel diode oscillator,
    advantages of, 78
Tunnel diode, 76
    application of, 78
    principle of operation of, 76
    salient features of, 76
Twisted pair lines, 204
Twists, 136
Two-cavity Klystron amplifier,
    parts of, 40
    principle of operation of, 40
Two-cavity Klystron, 39
Two-port network,
    scattering matrix of, 99
Two-wire lines, 113
    advantages of, 116
    applications of, 116
    disadvantages of, 116
    salient features of, 114
Two-wire open lines,
    applications of, 203
Two-wire parallel open lines, 177
Two-wire transmission lines,
    definitions of the parameters of, 178
TWT structure, 57
TWT,
    gain parameter of, 59
Typical diploes,
    impedance of, 280

U
Uniform linear arrays, 276
    salient features of, 276
Unitary property, 98
Unknown impedance,
    measurement of, 377

V

Velocity modulation, 38
   method of producing, 38
VSWR measurements, 356

W

Wave equations, 12
Waveguide resonators, 218
   salient features of, 218
Waveguide slots, 280

   definition of, 124
   salient features of, 124
Waveguides,
   salient features of, 346
Wavelength, 17
Wavemeters, 334
Wire inductor, 255

Y

Yagi-uda antenna feed, 307